AF389308

Trends in Plasticity of Metals and Alloys

Trends in Plasticity of Metals and Alloys

Editors

Mikhaïl A. Lebyodkin
Vincent Taupin

MDPI • Basel • Beijing • Wuhan • Barcelona • Belgrade • Manchester • Tokyo • Cluj • Tianjin

Editors
Mikhaïl A. Lebyodkin
CNRS
France

Vincent Taupin
CNRS
France

Editorial Office
MDPI
St. Alban-Anlage 66
4052 Basel, Switzerland

This is a reprint of articles from the Special Issue published online in the open access journal *Metals* (ISSN 2075-4701) (available at: https://www.mdpi.com/journal/metals/special_issues/trends_plasticity).

For citation purposes, cite each article independently as indicated on the article page online and as indicated below:

LastName, A.A.; LastName, B.B.; LastName, C.C. Article Title. *Journal Name* **Year**, *Volume Number*, Page Range.

ISBN 978-3-0365-1561-8 (Hbk)
ISBN 978-3-0365-1562-5 (PDF)

Contents

About the Editors

Mikhail A. Lebyodkin Laboratory of Microstructures and Materials Mechanics (LEM3), CNRS, Université de Lorraine, Arts et Métiers ParisTech, F-57000 Metz, France.

Vincent Taupin Laboratory of Microstructures and Materials Mechanics (LEM3), CNRS, Université de Lorraine, Arts et Métiers ParisTech, F-57000 Metz, France.

Editorial

Trends in Plasticity of Metals and Alloys

Mikhaïl A. Lebyodkin [1,2,*] and Vincent Taupin [1,2]

[1] Laboratoire d'Etude des Microstructures et de Mécanique des Matériaux (LEM3), Université de Lorraine, CNRS, Arts et Métiers ParisTech, 7 rue Félix Savart, 57000 Metz, France; vincent.taupin@univ-lorraine.fr

[2] Center of Excellence "LabEx DAMAS", Université de Lorraine, 7 rue Félix Savart, 57070 Metz, France

* Correspondence: mikhail.lebedkin@univ-lorraine.fr; Tel.: +33-(0)-372-747-771; Fax: +33-(0)-387-315-366

Citation: Lebyodkin, M.A.; Taupin, V. Trends in Plasticity of Metals and Alloys. *Metals* **2021**, *11*, 615. https://doi.org/10.3390/met11040615

Received: 26 March 2021
Accepted: 7 April 2021
Published: 10 April 2021

Publisher's Note: MDPI stays neutral with regard to jurisdictional claims in published maps and institutional affiliations.

Having been at the center of technological progress for thousands of years, metals continue to be a primary material in our lives today. Investigations of the plasticity of crystalline solids, to which conventional metallic materials belong, were marked in the 20th century by a success in the understanding of microscopic mechanisms of plastic flow. This breakthrough was notably due to the development of experimental techniques allowing for the observation of individual defects, dislocations par excellence. The concurrent development of high-performance computers and numerical techniques gave rise to the hope that it would soon be possible to predict the mechanical behavior of bulk material by averaging the motions of individual defects, even if the sample contains myriads of defects.

However, investigations made during the last decades have shown that a fundamental property of plastic flow of solids is the self-organization of crystal defects. As a result, plastic flow is inherently heterogeneous and associated with complex dynamics and the formation of spatial structures in a mesoscopic scale range. Therefore, the understanding of the micro–macro transition inevitably implies investigation of local heterogeneities pertinent to the collective behavior of defects. On the other hand, the same period has seen considerable progress in the development of high-performance metals and alloys that possess enhanced mechanical properties ensured by complex microstructures. Ultrafine-grain materials, metallic glasses, gradient microstructures, etc. are gaining an increasing role in the industry. Therefore, the problem of the self-organization of defects comes to the fore as the characteristic lengths of the microstructure become comparable to the respective scales imposed by the collective deformation processes.

These challenges were a driving force for the development of sophisticated experimental techniques (in situ TEM, electron channeling contrast imaging, digital image correlation (DIC), nano-indentation, etc.), advanced multiscale modeling (molecular dynamics, discrete dislocation dynamics, strain gradient models, etc.), and methods of analysis of the observed and simulated complex spatiotemporal behaviors from the viewpoint of self-organization, with an aim of establishing process–microstructure–property links and filling gaps between the elementary atomic-scale mechanisms and the scale of a laboratory sample.

This Special Issue is composed of ten works presenting several topics within this huge and constantly developing domain. However, they touch upon all its basic pillars: new approaches to long-standing questions and traditional model materials allowing a deeper understanding of the underlying physical mechanisms, novel materials obtained by various processing routes, and advanced experimental and modeling techniques.

Paper [1] revisits a long-standing problem of work hardening processes and the possibility of reaching a steady state in unidirectional plastic deformation. This problem has recently received strong impetus due to the techniques of severe plastic deformation, which have made it possible to extend the attainable Von Mises equivalent strains from a unity value, limiting conventional tensile tests, to extraordinary values of 10^5 in high pressure torsion. The paper provides a comprehensive review of the question and proposes an explanation for a new strain hardening stage revealed in such deformation processes.

Achieving large strains and the elaboration of materials with complex microstructures put forward the need to take into account rotational modes of deformation, associated with disclinations. A review of field dislocation and disclination mechanics—a crystal plasticity model suggested about fifty years ago in a form based on continuous distributions of defects and developed to further include here the dynamics of rotational discontinuities—is given in [2]. Furthermore, the paper considers the application of this concept for modeling grain boundary-mediated plasticity.

Two works concern new approaches to the problem of fatigue behavior of materials [3,4]. In [3], the combination of the digital image correlation and electron backscatter diffraction techniques allows revealing the strong heterogeneities of strain and lattice rotation at the free surface of a tantalum polycrystalline aggregate during low cycle fatigue. The crystal plasticity finite element simulation confirms the strongly heterogeneous local ratcheting. These results allow the proposal of a new criterion for the initiation of local fatigue crack based on the local amount of ratcheting plastic strain. A cyclic plasticity model predicting strain range-dependent behavior is proposed and verified for austenitic steels in [4]. The model considers a virtual back stress, a new internal variable that allows the handling of the material behavior at large strain amplitudes.

Paper [5] revisits the problem of the mechanical characterization of ductile materials via compression tests and the limits of equivalence of tensile and compression test curves. This question being particularly essential in the case of alloys for medical purposes, the proposed compression test method is validated for several biomedical alloys.

Three papers then concern strain localization effects, each work considering this multi-faceted problem under different angles and for distinct scale ranges [6–8]. The "autowave" phenomenon manifested by the occurrence of waves of strain localization during constant-rate deformation is addressed in [6]. Such waves, usually characterized by rather low velocities (<0.1 mm/s), have been observed in a large number of materials. To explain this phenomenon, a general approach based on the wave–particle duality is suggested with the aid of a quasi-particle representation relating waves and quasi-particles through the de Broglie equation. A dual character of self-organized deformation processes manifests itself directly, with no recourse to quantum theory, when a millisecond time resolution is attained. This is the case of [7], devoted to the investigation of the statistics of the local strain-rate field during the Portevin–Le Chatelier effect in an AlMg alloy. In addition to the large deformation bands that are characteristic of plastic instability, the high-frequency one-dimensional DIC technique reveals a wave–intermittency duality manifested by a wavy spatial organization of local plastic strain-rate bursts, which obey power-law statistics tantamount to avalanche-like deformation behavior. Strain localization at a considerably finer spatial scale pertaining to slip bands is considered in [8] for an Al–Cu–Li alloy prone to significant strain localization effects. By virtue of the interferometry technique, which allows the quantification of strain localization by assessing the surface roughness and computer simulations using discrete dislocation dynamics, the authors find evidence that the presence of Mn and Zr dispersoids may drastically homogenize the plastic strain in the investigated material.

Finally, two papers are devoted to nonconventional materials which are attracting strong attention in view of elaborating lighter materials and simultaneously improving mechanical properties [9,10]. A crystalline silver alloy with a complex microstructure associated with nanometric-size porosity obtained by selective chemical etching is studied in [9]. Using electron tomography and a nanoindentation technique, the paper aims at interpreting a highly heterogeneous mechanical behavior characterized by a complex relationship between the indenter penetration depth, the porosity, and the microstructure. The case of non-crystalline metallic materials is considered in [10], where nanoscale plasticity of bulk metallic glasses is studied theoretically for the case of a spherical nano-indenter. The work shows that plastic strain rearrangements, which describe plasticity in the absence of a long-range order, depend on the choice of the disorder. In particular, the conditions for the spatial organization leading to the formation of macroscopic shear bands are discussed.

Acknowledgments: The guest editors would like to thank all authors who contributed to building this Special Issue and the reviewers for their comprehensive comments. The Editors of *Metals* journal are also warmly acknowledged for having proposed this opportunity and for their constant support. The guest editors would like to thank in particular the Editorial Assistant Kinsee Guo for invaluable help all along the preparation of this Special Issue.

Conflicts of Interest: The authors declare no conflict of interest.

References

1. Sevillano, J.G. Dynamic Steady State by Unlimited Unidirectional Plastic Deformation of Crystalline Materials Deforming by Dislocation Glide at Low to Moderate Temperatures. *Metals* **2020**, *10*, 66. [CrossRef]
2. Fressengeas, C.; Taupin, V. Revisiting the Application of Field Dislocation and Disclination Mechanics to Grain Boundary. *Metals* **2020**, *10*, 1517. [CrossRef]
3. Colas, D.; Finot, E.; Flouriot, S.; Forest, S.; Mazière, M.; Paris, T. Experimental and Computational Approach to Fatigue Behavior of Polycrystalline Tantalum. *Metals* **2021**, *11*, 416. [CrossRef]
4. Halama, R.; Fumfera, J.; Gál, P.; Kumar, T.; Markopoulos, A. Modeling the Strain-Range Dependent Cyclic Hardening of SS304 and 08Ch18N10T Stainless Steel with a Memory Surface. *Metals* **2019**, *9*, 832. [CrossRef]
5. Affolter, C.; Thorwarth, G.; Arabi-Hashemi, A.; Müller, U.; Weisse, B. Ductile Compressive Behavior of Biomedical Alloys. *Metals* **2020**, *10*, 60. [CrossRef]
6. Zuev, L.B.; Barannikova, S.A. Quasi-Particle Approach to the Autowave Physics of Metal Plasticity. *Metals* **2020**, *10*, 1446. [CrossRef]
7. Lebyodkin, M.; Bougherira, Y.; Lebedkina, T.; Entemeyer, D. Scaling in the Local Strain-Rate Field during Jerky Flow in an Al-3%Mg Alloy. *Metals* **2020**, *10*, 134. [CrossRef]
8. Carrasco, E.J.; Chevy, J.; Davo, B.; Fivel, M. Effects of Manganese and Zirconium Dispersoids on Starin Localiation in Aluminum Alloys. *Metals* **2021**, *11*, 200. [CrossRef]
9. Champion, Y.; Laurent-Brocq, M.; Lhuissier, P.; Charlot, F.; Jorge, A.M., Jr.; Barsuk, D. Understanding the Interdependence of Penetration Depth and Deformation on Nanoindentaton of Nanoporous Silver. *Metals* **2019**, *9*, 1346. [CrossRef]
10. Tanguy, A.; Chen, P.; Chaise, T.; Nélias, D. Shear Banding in a Contact Problem between Metallic Glasses. *Metals* **2021**, *11*, 257. [CrossRef]

 *metals* 

Review

Dynamic Steady State by Unlimited Unidirectional Plastic Deformation of Crystalline Materials Deforming by Dislocation Glide at Low to Moderate Temperatures

Javier Gil Sevillano

TECNUN and Ceit, Technological Campus, University of Navarra, Paseo M. de Lardizabal 13, 20018 San Sebastián, Spain; jgil@ceit.es

Received: 26 November 2019; Accepted: 23 December 2019; Published: 1 January 2020

Abstract: This paper presents an outline of the quest for the mechanical steady state that an unlimited unidirectional plastic strain applied at low to moderate temperature is presumed to develop in single-phase crystalline materials deforming by dislocation glide, with particular emphasis on its athermal strength limit. Fifty years ago, the study of crystalline plasticity was focused on the strain range covered by tensile tests, i.e., on true strains less than unity; the canonic stress–strain behavior was the succession of stages I, II, and III, the latter supposedly leading to a steady state defining a temperature and strain rate-dependent flow stress limit. The experimentally available strain range was increased up to Von Mises equivalent strains as high as 10 by the extensive use of torsion tests or by combinations of intermittent deformations by wire drawing or rolling with tensile tests during the 1970s. The assumed exhaustion of the strain-hardening rate was not verified; new deformation stages, IV and V, were proposed, and the predicted strength limit for deformed materials was nearly doubled. Since the advent of severe plastic deformation techniques in the 1980s, such a range was still significantly augmented. Strains of the order of several hundreds were routinely reached, but former conclusions relative to the limit of the flow stress were not substantially changed. However, very recently, the plastic strain range has allegedly been expanded to 10^5 true strain units by using torsion under high pressure (HPT), surprisingly for some common metals, without experimental confirmation of having reached any steady state. This overview has been motivated by the scientific and technological interest of such an open-ended story. A tentative explanation for the newly proposed ultra-severe hardening deformation stage is given.

Keywords: plastic deformation; strain hardening; deformation stages; dynamic steady state; crystalline materials; dislocation density limit

---

1. Introduction

This paper presents an overview of the quest for the mechanical steady state that an unlimited unidirectional plastic strain applied at low to moderate temperature is presumed to develop in single-phase crystalline materials which deform by dislocation glide, with particular emphasis on its athermal strength limit. An enormous research work has been undertaken on this subject during the past 50 years.

The plastic deformation of single-phase crystalline materials induces structural changes that bring about important changes of mechanical properties: strain hardening, ductility, toughness, etc. At (approximately) low to moderate homologous temperatures relative to the melting temperature $(T/T_M \lesssim 0.3)$ and at Von Mises equivalent (VME) strain rates outside the creep or the high strain rate regimes $(10^{-5} \text{ s}^{-1} \lesssim \dot{\varepsilon} \lesssim 10^2 \text{ s}^{-1})$, the strain contribution of deformation mechanisms requiring long-range self-diffusion is not relevant. Deformation takes place mainly by dislocation glide on crystallographic slip systems (in some cases complemented by deformation twinning) and the VME flow stress ($\overline{\sigma}$) versus VME plastic strain ($\overline{\epsilon}$) response shows several well-known "deformation stages" (see [1–3] for reviews).

Most often, the strain-hardening rate, SHR, $\overline{\theta} = \partial\overline{\sigma}/\partial\overline{\epsilon}$, steadily decreases from a transient high initial value, particularly when the starting structure is polycrystalline or in general, when the strain is resolved at the meso level by the combined activation of several non-coplanar slip systems (multiple slip). Beyond a short (in terms of strain) elastic–plastic transient, the initial stress–strain curve fits the so-called Voce equation rather well:

$$\frac{\overline{\sigma}_s - \overline{\sigma}}{\overline{\sigma}_s - \overline{\sigma}_0} = exp\left(-\frac{\overline{\theta}_0 \cdot \overline{\epsilon}}{\overline{\sigma}_s - \overline{\sigma}_0}\right). \tag{1}$$

The three parameters of the Voce equation are an ideal elastic limit, $\overline{\sigma}_0$ (the back extrapolation of the flow stress to zero strain), the SHR extrapolated to zero strain, $\overline{\theta}_0$ and an ideal saturation flow stress that would be reached after infinite strain, $\overline{\sigma}_s$. This regime of deformation is identical to the so-called Stage III observed in single crystals, as will be explained below.

In the low temperature/high strain rate range of the $T - \dot{\overline{\epsilon}}$ sub-space here considered, single crystals may show another two deformation stages before Stage III. High-purity single crystals oriented for single slip show an "easy glide" Stage I of very weak SHR relative to the shear modulus G^* ($\overline{\theta}_I/G^* \approx 3 \times 10^{-3}$) (shear modulus appropriate for the calculation of the dislocation line tension with account of the elastic anisotropy of the crystal). After some strain (implying crystal rotation toward multiple slip regions in orientation space), the easy-glide Stage I gives way to a new stage, Stage II, of high and constant SHR. Multicrystals, i.e., polycrystals (usually of big grain size) whose smallest physical dimension L of the deforming specimen is less than about 10 times the grain size, $L\widetilde{<}10D$, show a composite Stage I+II before Stage III fully develops (their crystals deform under relaxed constraints relative to a grain within a true polycrystal). Stage II of single crystals is an athermal hardening stage that is not always present but always detectable as asymptotic behavior. It is followed by another strain stage of the steady decrease of SHR, Stage III. Both the transition from stages II to III and the evolution of the SHR in Stage III are dependent on T and $\dot{\overline{\epsilon}}$ (and, of course, on the material: its chemical composition, crystallographic structure, lattice friction stress level, melting temperature T_M, and stacking-fault energy, SFE, in particular).

Although it is true that such three-stage behavior is much better documented for face-centered cubic (fcc) metals of medium or high SFE than for other materials, it is typical of any crystalline materials deforming exclusively by crystallographic slip (dislocation glide) up to moderate VME strains of the order of 1. Of course, it is masked (it becomes irrelevant) in intrinsically hard materials, i.e., materials with high values of lattice friction stress, because their work hardening weakly contributes to their flow stress.

At the microscopic level, the observed three-stage flow stress curve stems mainly from quasi-athermal interactions of gliding dislocations with other gliding or stored dislocations and with grain boundaries (giving way to statistically stored dislocations, SSD) counteracted in Stage III by thermally activatable dynamic recovery mechanisms of dislocation annihilation. They are the main contributors to the SHR of stages I (from self-interactions in single glide), II of single crystals (chiefly from attractive non-coplanar interactions in multiple glide), and III of single crystals or polycrystals [4]. The interaction with grain boundaries is at the origin of strengthening by the Hall–Petch effect, mostly from the storage of geometrically necessary dislocations, GND, related with the plastic heterogeneity of neighboring grains [2,5]. Therefore, at least if one is interested in the broad picture and not (on this occasion) in the details, for understanding the flow stress–strain evolution, one must confront it with the evolution of two structural variables, the dislocation density and the grain size (setting aside the evolution of crystallographic texture, incorporated in an often weakly changing orientation factor). This will be the scheme followed in this paper; it turns out that revising the plastic behavior at increasing strains amounts to follow the story of this subject in chronological order.

Finally, a mention of the effect of deformation twinning on the stress–strain behavior at large strains should be given. When deformation twinning complements dislocation glide, the original grains are progressively fragmented by the boundaries separating the increasing number of twins from their grain matrix. As a result, a nearly constant SHR due to a dynamic Hall–Petch effect reinforces the dislocation density-related SHR during the strain range where deformation twinning is significant. Such behavior is blatant in the so-called TWIP steels, but it is evident in other easily twinning materials, too (hexagonal close-packed (hcp) and low SFE fcc metals and alloys) [6–8]. The TWIP regime ends when deformation twinning becomes exhausted in the highly deformed refined microstructure; then, the stress–strain returns to Stage III and other possible ensuing stages. At large strains, the effect of deformation twinning is a shift of the SHR versus the flow stress curve to the right. Figure 1 is an example of the described behavior.

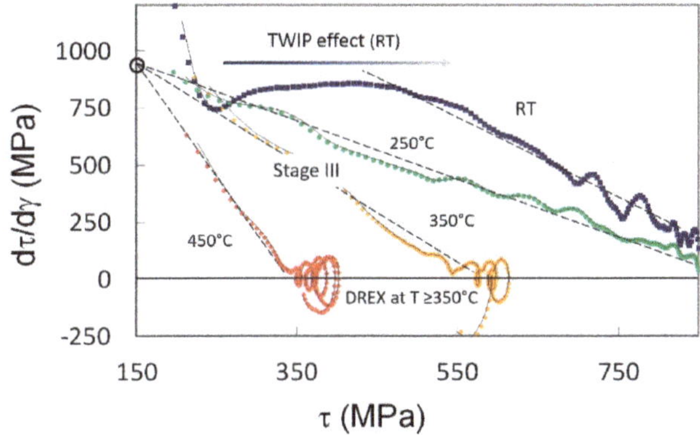

Figure 1. Fe-22%Mn-0.6%C twinning-induced plasticity (TWIP) steel. Strain-hardening rate (SHR) in shear vs. macroscopic shear flow stress from torsion tests at several temperatures and 1.4×10^{-3} s^{-1} shear strain rate. At room temperature (RT), the TWIP activity starts after a short Stage III activity and produces a transient SHR increase. Deformation twinning at this strain rate is irrelevant for T > 150 °C. See ref. [6] for more results on this steel.

2. Small to Moderate Plastic Strains

In the absence of intense deformation twinning or excluding nanograin materials, the strain-induced flow stress change in a single-phase material is largely governed by the strain-induced change of its statistically stored dislocation density, ρ_s, which most often represents the main fraction of the total dislocation density, ρ.

2.1. Statistical Dislocation Storage

Assume a mobile dislocation line that would glide in the absence of any activity of dynamic recovery through the three-dimensional structure of dislocation lines present in a cubic crystal volume $l_x l_y l_z$ with z being the direction normal to the slip plane and x the slip direction. Its crossing of the volume element would lead to

(a) a macroscopic finite strain increment of amount

$$\Delta \bar{\varepsilon} = \Delta u_x / M l_z = b / M l_z \tag{2}$$

where b is the modulus of the Burgers vector of the mobile dislocation (the shear displacement increment Δu_x) and $M \approx 3$ is the texture-dependent orientation factor relating tensile strain with crystallographic slip shear. As deformation texture evolves the value of the orientation factor changes. However, in general, the influence of such a change is not as important as that of the other strain-induced structural changes. Reference [9] treats in detail the evolution of the orientation factor through several strain paths.

(b) an increment of the stored dislocation length per unit volume that, from geometrical reasons, scales with $\sqrt{\rho}$ (geometrical similitude implies that the dislocation length stored per unit area swept by the dislocation, ΔL_S, is proportional to the forest dislocation density piercing the slip plane, $\rho/2$, times a characteristic length of such density, $1/\sqrt{\rho}$):

$$\Delta\rho^+ = l_x l_y \Delta L_S / l_x l_y l_z = \beta \sqrt{\rho}/l_z \tag{3}$$

The scaling factor β is independent of strain as far as the dislocation structure maintains geometrical similitude independently of ρ [2,3].

Now, on account of the so-called "Taylor relationship", between dislocation density and flow stress

$$\bar{\sigma} - \bar{\sigma}_0 = M\alpha G^* b \sqrt{\rho} \tag{4}$$

it is possible to derive the strain-hardening rate, SHR, in such circumstances.

In Equation (4), $\bar{\sigma}_0$ is the flow stress limit of the structure at zero stored dislocation density (it includes the contributions of lattice friction stress and solution strengthening, i.e., it depends on temperature and strain rate), G^* is the shear elastic modulus at the test temperature, and $\alpha \cong 0.3$ is the quasi-athermal and quasi-universal strengthening factor of three-dimensional dislocation structures.

The macroscopic SHR for strain deprived of any dynamic recovery activity would be

$$\bar{\theta}_0 \cong \frac{M^2 \alpha \beta}{2} G^* \approx \beta G^* \tag{5}$$

i.e., a constant SHR as far as the assumptions made are valid, which is in good agreement with experimental findings.

In fact, this is not quite true: the factor α in the Taylor equation slightly depends (through a logarithmic factor) on the dislocation density (it visibly decreases as the dislocation density increases by several orders of magnitude). Similarly, α is mildly sensitive to the spatial distribution of local dislocation density that changes from near-uniform to cellular to subgranular as plastic strain goes on, increasing in that order [2]. Thus, the two effects somewhat compensate each other, and Equation (5) is an acceptable approximation.

The storage parameter β also depends on the spatial distribution of the local dislocation density, as stated above, and plastic deformation does not strictly maintain structural similitude (as strain goes on, the spatial heterogeneity of the dislocation density increases). Nevertheless, the simple dislocation storage model without recovery explains well enough both the experimental observations of the athermal stage II observed in single crystals deforming by multiple slip at low flow stresses and temperatures (insignificant activation of dynamic recovery processes) and the athermal low-temperature asymptotic Stage II of polycrystals.

2.2. Dynamic Recovery

Thermomechanically activatable recovery processes are inherent to plastic strain by dislocation glide. The constant SHR given by Equation (5) is an idealization; together with the process of athermal dislocation density storage given by Equation (3), dynamic processes of dislocation density annihilation are active. Due to the composition of the two effects, the SRH steadily decreases as strain goes on: it is the deformation Stage III mentioned in the introduction.

Empirically, the SHR evolution, which is strongly dependent on temperature and weakly dependent on strain rate, is very well described by the equation:

$$\frac{\overline{\theta}}{\overline{\theta}_0} = 1 - \frac{\overline{\sigma} - \overline{\sigma}_0}{\overline{\sigma}_s^{III}\left(T, \dot{\overline{\epsilon}}\right) - \overline{\sigma}_0} \tag{6}$$

As strain increases, the SHR decreases linearly with the flow stress from an upper athermal limit value $\overline{\theta}_0$ to zero. The flow stress grows monotonically and asymptotically tends to a saturation flow stress $\overline{\sigma}_s^{III} = \overline{\sigma}_s^{III}\left(T, \dot{\overline{\epsilon}}\right)$. The integration of Equation (6) leads to the Voce stress–strain shown in Equation (1).

In terms of the underlying evolution of the dislocation density and on account of Equation (4), Equation (6) corresponds to the Kocks–Mecking equation of evolution of dislocation density in Stage III [3]:

$$d\rho / (d\overline{\epsilon}) = M(\beta \frac{\sqrt{\rho}}{b} - \eta\rho) \tag{7}$$

where the parameter $\eta = \eta\left(T, \dot{\overline{\epsilon}}\right)$ depends on the temperature and strain rate in parallel with the behavior of the saturation flow stress. Kocks and Mecking [3] have discussed in detail such dependence: the process of dynamic recovery occurs via the interaction of mobile dislocations with specific sites of the stored dislocation structure under the action of the local flow stress (the effective stress, superposition of the macroscopic flow stress, and the local internal stress) and thermal activation. The activation energy, constant all along Stage III, shows a logarithmic dependence on effective stress.

The recovery processes intervening in Stage III are most probably more than one. One of them is a partial collapse of stored Orowan loops (remnants of gliding dislocations encircling regions of dislocation density higher than the mean) under the forward stress of gliding dislocations of the same sign; another obvious one is the annihilation of segments of the gliding dislocations with segments of opposite sign from the stored or mobile dislocation density.

Surely cross-slip plays a decisive role in those dynamic recovery processes, hence the observed influence of SFE. A close examination of the two terms of Equation (7) supports this idea: when a dislocation sweeps a surface S when gliding across a crystal volume, the number of storage sites per unit surface determine a stored length per unit volume scaling with the current density of dislocations, i.e., Equation (3). Simultaneously, the gliding dislocation explores a volume of $S{\cdot}h(T, \dot{\overline{\epsilon}})$ above and below the swept surface, where several activatable recovery sites per unit volume reside, each one determining the annihilation of a dislocation length. On account of Equations (2) and (7), the dislocation length eliminated per unit volume should be

$$\Delta\rho^- = -\eta b\rho \sqrt{\rho} \tag{8}$$

which is dimensionally compatible with the product of the explored volume times a number of activatable sites per unit volume related with the dislocation structure, i.e., scaling as $N \sim \rho^{3/2}$, times a fixed annihilated length per site, i.e., a length independent of the scale of the structure.

The Kocks–Mecking model for Stage III is astonishingly simple, effective, and convincing, hence its widespread use in simulations of crystalline plasticity using dislocation densities as internal variables [4].

For a given material, all stress–strain curves collapse in a straight line linking two points, $\overline{\theta}/\overline{\theta}_0 = 1$ on the ordinate with $(\overline{\sigma} - \overline{\sigma}_0)/\left[\overline{\sigma}_s^{III}\left(T, \dot{\overline{\epsilon}}\right) - \overline{\sigma}_0\right] = 1$ on the abscissa in a normalized SHR versus flow stress diagram, as shown in Figure 2. Of the two parameters defining Stage III, the quasi-universal value of the first one [2,3] is:

$$\overline{\theta}_0 / G^* \cong 0.05. \tag{9}$$

The temperature and strain rate dependence of the second term, $\left[\bar{\sigma}_s^{III}\left(T,\ \dot{\bar{\epsilon}}\right)-\bar{\sigma}_0\right]$, needs to be characterized for each material, although its functional form has been given by Kocks and Mecking [3].

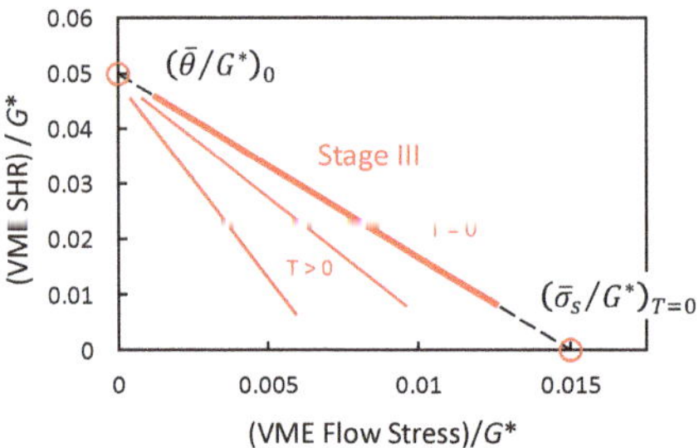

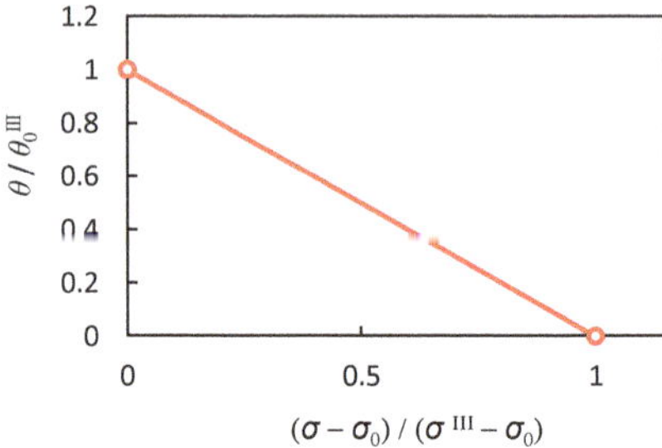

Figure 2. Sketch of Stage III of plastic deformation according to the model of Kocks and Mecking [3].

It is interesting to know, at least approximately, the upper (athermal) limit of the flow stress, governed by the athermal saturation value of Stage III dislocation density. For pure Cu, such a limit flow stress value is:

$$\left[\left(\bar{\sigma}_s^{III}\right)_0 - \bar{\sigma}_0\right]/G^* \cong 0.015. \tag{10}$$

It corresponds to a dislocation density of the order of

$$\left(\rho_s^{III}\right)_0 \cong 10^{16}\cdot\text{m}^{-2}. \tag{11}$$

These two limit values are probably valid for other crystals too, at least as order of magnitude figures, particularly for body-centered cubic (bcc) metals deformed at temperatures above their low-temperature threshold (more in general, when their flow stress is not dominated by the lattice friction stress) and for hexagonal close-packed (hcp) metals.

Fifty years ago, those limits were thought to be the intrinsic limits of strain-induced strengthening of single-phase crystalline materials, at least in many academic backgrounds.

3. "Large" Strains (1 $\widetilde{<} \bar{\bar{\epsilon}} \widetilde{<}$ 10)

Interest in exploring and understanding the microstructural basis of the mechanical effects of large strains in crystalline materials arose in the 1960s. Useful strains in tension and compression tests that were most often used earlier for studying plastic behavior were limited by gross deviations from the near-uniform strain and strain rate: the development of localized deformation (necking in tension) or barreling (from friction) in compression.

Scientific interest in large strains was preceded by the technical need of reliable knowledge of stress–strain behavior for improving large-strain processes such as cold rolling, wire drawing, forging, or sheet-forming (reliable input for numerical simulations and knowledge of the structural state of the product). Torsion tests (no external geometry changes in the deforming specimen) or intermittent tension tests during the incremental accumulation of strain by well-designed rolling or wire drawing passes (quasi-free of strain heterogeneities and redundancies [10]) were already extensively used for such purpose since the 1950s (see [9] for a review). As a result, the domain of large plastic strains became available for study, at least for relatively ductile materials.

It soon became evident that no saturation occurs after Stage III, but that a new deformation stage appears at large plastic strains, "Stage IV", which maintains a weak and decreasing SHR well inside the large strain range, 1$\widetilde{<}\bar{\epsilon}\widetilde{<}$10 [2,3,9,11,12]. Stage IV was inaccessible to studies based on tensile tests because the stage III–IV transition occurs for $\bar{\theta} \cong 0.3\bar{\sigma}$ [13], i.e., for strains larger than the tensile strain determining the onset of mechanical instability in tension, which is given by the Considère criterion, $\bar{\theta} = \bar{\sigma}$.

Figure 3 shows a good set of stress–strain curves of commercial purity copper measured in torsion inside the temperature and strain rate range considered in this paper. The characteristics of the stages III–IV are conspicuous. By using the same scaling employed for Stage III (Figure 2), the stress–strain curves of three fcc metals of varying SFE, measured at different temperatures, collapse rather well in a single master curve encompassing both Stages III and IV (Figure 4). This means that although the athermal mechanisms of dislocation storage in Stage IV differ from those of Stage III (they are much weaker), the temperature dependence of Stage IV is nearly the same as that in Stage III.

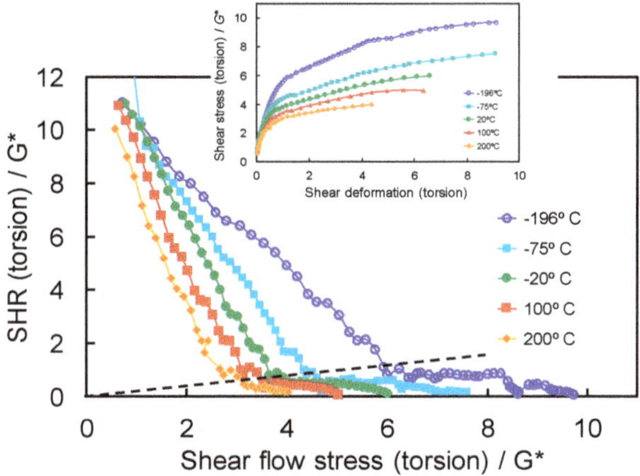

Figure 3. Shear strain-hardening rate vs. shear flow stress in torsion of 99.98% Cu (77, 198, 293, 373, and 473 K) deformed at a Von Mises equivalent (VME) strain rate $\dot{\bar{\varepsilon}} = 8 \times 10^{-3}$ s^{-1}. The inset shows the shear strain vs. shear flow stress curves. Replotted from Alberdi [11].

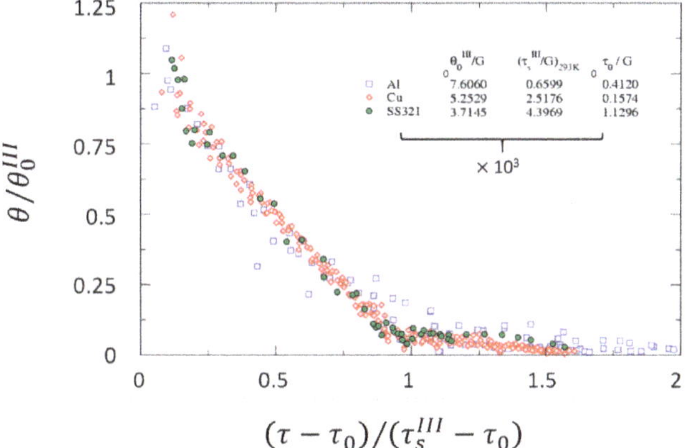

Figure 4. Torsional work-hardening rate vs. flow stress data of 99.98% Cu (77, 198, 293, 373, and 473 K), 1050 Al (77, 198, 293, and 373 K), both deformed at $\dot{\bar{\varepsilon}} = 8 \times 10^{-3}$ s^{-1} and austenitic SS321 stainless steel (20 °C and 200 °C) deformed at $\dot{\bar{\varepsilon}} = 2.3 \times 10^{-3}$ s^{-1}. The numerical data inserted in the figure correspond to crystallographic slip strain vs. CRSS (critical resolved shear stress) behavior; a Taylor orientation factor $M = 3$ should be employed to transform them into VME values (i.e., factors M and M^2 for respectively the flow stress and the SHR). Replotted results from Alberdi [11].

For copper or aluminum, as shown in Figure 5, the asymptotic athermal limit at the apparent end of Stage IV is:

$$\left[\left(\bar{\sigma}_s^{IV}\right)_0 - \bar{\sigma}_0\right]/G^* \cong 0.024. \tag{12}$$

It approximately corresponds to a dislocation density:

$$\left(\rho_s^{IV}\right)_0 \cong 2.6 \times 10^{16} \text{ m}^{-2}. \tag{13}$$

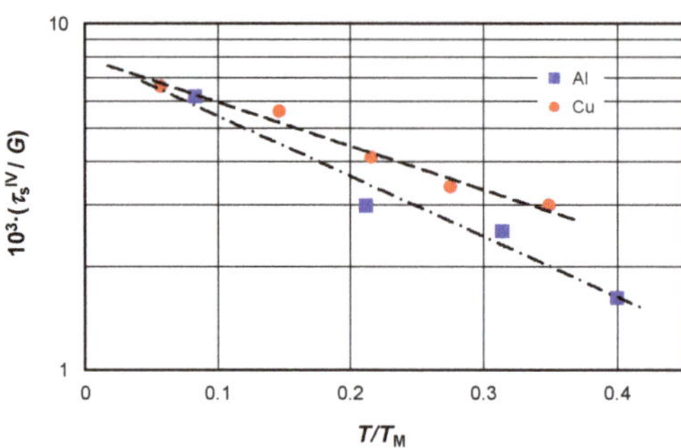

Figure 5. Normalized crystallographic slip shear flow stress (from torsion tests, VME strain rate $\dot{\bar{\varepsilon}} = 8 \times 10^{-3}$ s^{-1}) at the (apparent) saturation of Stage IV of commercial purity copper and aluminum as a function of homologous test temperature. The athermal limit is about 0.008 G^*. Using an orientation factor $M = 3$, the VME athermal saturation flow stress for these two face-centered cubic (fcc) metals would be 0.024 G^*. Data from Alberdi [11].

Either structural damage (precursor of ductile failure) or the emergence of another stage endowed with a powerful dynamic recovery mechanism that soon annihilates the SHR (a possible Stage V, [12]) precludes experimentally approaching those limits in the realm of "large strains". However, a retardation of the apparent reaching of the saturation flow stress by performing the plastic deformation under a hydrostatic pressure (at the basis of the "severe plastic deformation" processes to be described in the next two sections of this paper) was experimentally observed. Thus, it seems that "Stage V" might be a spurious plastic deformation stage, at least for single-phase metallic alloys (however, an unrelated Stage V has been observed in diamond-cubic semiconductors Ge and Si deformed at high temperature, $T \geq 0.65 T_M$ [14]).

The explanation of Stage IV is far from complete. The strain-induced processes of dislocation storage and recovery do not maintain the geometrical similitude-based storage law conducive to Equation (3) or the dynamic annihilation law, as shown in Equation (8). Consequently, a progressive deviation from the Voce stress–strain law as Stage III goes on is expected. Furthermore, the existence of processes additional to the main process responsible for Stage III but weakly contributing to hardening is very plausible; their effect would be only detectable when dynamic recovery has nearly neutralized the main hardening contribution to Stage III. Two of them are as follows.

- Continuous accumulation of debris (e.g., dipolar debris) ancillary to the main process of statistical storage and the dynamic recovery of dislocations (the interaction of mobile dislocations with the current total dislocation density present and with other defects of any dimension) [13],
- Accumulation of a slowly increasing density ρ_g of geometrically necessary dislocations (GND) coupled to the mesoscopic plastic strain gradients inherent to the heterogeneity of the evolving structure [15,16].

However, other authors explain the evolution of hardening along the III–IV–V stages without resorting to any micromechanistic transition affecting the storage or annihilation of dislocation density, i.e., as the deviation from the ideal Stage III mentioned earlier. For instance, Estrin et al. assume that the apparent transition defining the stages is merely the result of a "concerted action" of the deformation of cell interiors and cell walls considered as two phases of a composite, each with their own hardening and softening mechanisms, under the condition of a continuous independent shrinking of the cell size and of the thickness of the cell walls [17]. It may be that Stage IV stems from a combination of factors. The similarity of the thermal activation of the recovery processes working in Stages III and IV deserves further analysis, particularly for explaining the recovery of ρ_g at a microscopic level, as GND density is linked to the presence of (very real) mesoscopic strain gradients. Either the strain gradients tend to disappear as deformation goes on or the GNDs within dislocation cells walls disappear by the transformation of subgrains in grains whose highly misoriented boundaries are not anymore built with lattice dislocations but with GBD, grain boundary dislocations; perhaps, what we observe is a superposition of both phenomena.

4. The Case of Large Strains by Axisymmetric Elongation of BCC and HCP Polycrystals

Stage IV and the limits of strength and dislocation density expressed by Equations (12) and (13) are not the whole story of stress–strain behavior at large strains: the behavior is strain-path dependent. For instance, axisymmetrically elongated (by wire drawing) bcc and hcp metals display a constant SHR beyond the usual Stage III, which is concomitant with the development of a unique structure of elongated cells, subgrains, and grains with ribbon-shaped cross-sections that require a characteristic curling for maintaining the compatibility of the polycrystalline aggregate ("Van Gogh sky" structure [18,19]). Such curling amounts to the axial bending of these ribbons that, combined with their structural refining, amounts to a continuous increase of the mesoscopic gradients, i.e., a continuous increase of GND density contributing to sustaining the SHR up to the largest technologically possible deformations accessible to wire drawing ($\bar{\varepsilon} \approx 10$). Such peculiar behavior is due to the development of a crystallographic texture that promotes the deformation of the crystal units by plane-strain elongation, despite the axially symmetric deformation imposed to the wires. The same metals (e.g., iron [20]) behave differently when deformed through other strain paths. The phenomenon of a combined continuous injection of GND density and enhanced structural refining makes it possible to reach and often surpass the above-stated Stage III athermal limits of flow stress and dislocation density in wire-drawn bcc and hcp metals of high melting point.

The extension of the constant SHR stage of bcc and hcp metals beyond its current wire drawing limit is an interesting question that is not exempt of practical implications, because the sustained SHR would allow for reaching flow stresses and dislocation densities in fine wires well beyond the maximum values available today [21,22].

5. SPD, "Severe Plastic Deformations" ($\tilde{\bar{\varepsilon}} > 10$)

5.1. ECAP

Largely strained bulk specimens are both desirable for the macroscopic testing of properties and attractive for many possible technical applications. The accumulation of a big number of passes leading to $\tilde{\bar{\varepsilon}} < 10$ by wire drawing or rolling only furnish thin filaments or foils less than 100 μm thick; technological constrains preclude the processing of thicker sections of very hard materials. Moreover, often the processing is limited by ductile or brittle fracture nucleated from strain-induced structural damage, because deformation in rolling or wire drawing is imparted under hydrostatic pressure levels below the current VME flow stress value. Similarly, although the torsion of cylindrical bulk specimens does not introduce important shape or size changes of the test specimens, the hydrostatic pressure in free torsion is zero; VME strains of 5 are rarely reached before fracture in free torsion.

Before the 1980s, Segal and co-workers developed in the Soviet Union a process allowing for, ideally, indefinite accumulation of extrusion passes imparting about 1 VME strain per pass to bulk bar-shaped specimens under high pressure without significantly changing their external shape.

The process, which was dubbed ECAE or ECAP, acronyms of equal-channel angular extrusion or pressing, is extrusion through a channel of uniform section with a sharp kink of about 90° [23,24]. When the material goes through the kink of the channel, it endures a simple shear under a hydrostatic compression of the order of the flow stress of the material. Deformation per pass is about unity for a 90° kink, and in practice, the extrusion can be repeated up to near 20 passes, i.e., $\widetilde{\bar{\epsilon}}< 20$ (most materials become so strong after those strains that fail by ductile or brittle fracture further in the process). The process and its potential for modifying the material structure by "severe plastic deformation", SPD, was soon intensively exploited by Valiev and co-workers [25,26]. The possibilities offered by the ECAP process and its variants for exploring the response of materials to very large plastic strains made a big impact, stimulated the invention of other new SPD processes, and triggered the production of a tremendous amount of related bibliography in the last 40 years. Many comprehensive reviews on the subject have been published, e.g., [27–29].

The extension of the stress–strain analysis provided by ECAP does not change the most important conclusions reached from studies in the "large strain" range: Stage IV carries on toward its asymptotic limit without signs of any Stage V. Figure 6 shows results for commercially pure Cu, Ni, Fe, and Ti deformed by ECAP at room temperature (data from refs. [30–46]). Although there is much scatter in the results (inherent to the ECAP process and its analysis: the plastic heterogeneity of the deformed samples, different variants of the strain path used by different authors, testing by tension or by hardness), it seems clear that an apparent saturation of the flow stress occurs near a VME strain of 7. Averaging the available flow stresses beyond $\bar{\varepsilon} > 7$ for each metal, the values compare well with the saturation flow stress of Stage IV, $\left(\bar{\sigma}_s^{IV}\right)_{300\ K}$, obtained by extrapolation to zero SHR, the results from conventional torsion tests, or from rolling at 300 K, e.g., [2,3,9,11,12] and Figures 4 and 5.

The obvious question now was to go beyond the maximum feasible ECAP strains in order to verify whether the saturation of Stage IV is the true dynamic steady state in low/medium temperature plasticity. Surprisingly, an elegant deformation process was available since the early 1930s to settle this question: HPT, torsion under (very) high pressure.

5.2. HPT

Percy Bridgman's main research subject was the physical behavior of materials under high hydrostatic pressure (Nobel prize, 1946). Among many other topics, he was interested in the high-pressure effects on plastic deformation and fracture ([47] is the synthesis of his work on this subject). He developed many devices for high-pressure studies and specially the "Bridgman anvils", where a coin-shaped sample is squeezed in a shallow cavity between two otherwise flat anvils acted by an axial compressive force. The sample diameter matches that of the cavity, and its thickness lightly exceeds the height of the cavity when the two anvils are in contact. Under enough compressive force, material flows plastically and is extruded through the thin spacing between the anvils, which is a situation that creates a hydrostatic stress state in the material that grows exponentially as the spacing decreases, so that hydrostatic pressures of several GPa are easily reached. Applying now a rotation to one of the anvils, you have a torsion test of the disk (dragged by friction) into which in contrast with most other SPD processes, the hydrostatic compression is decoupled from the torque performing the torsion, i.e., it can be controlled by the axial compressive force applied to the anvils independently of the flow stress of the material being tested.

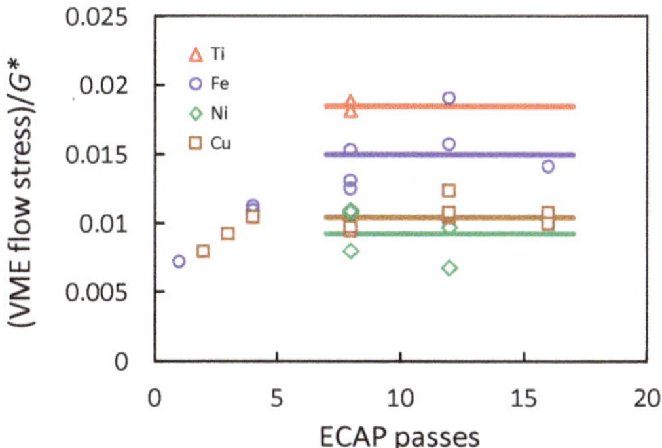

Figure 6. VME flow stress data (ultimate tensile strength, UTS or a third of the Vickers hardness, HV/3) of fcc 99.95% Cu [30–35] and 99.98% Ni [36–39], body-centered cubic (bcc) 99.95% Fe [40–44] and hcp 99.5% (grade 1) Ti [45,46] as a function of the number of equal-channel angular pressing (ECAP) passes performed at 300 K (one pass imparts about one unit of VME strain). Dotted lines are the average of values of flow stress between 8 and 16 passes for each metal; an apparent saturation stress is reached at 300 K beyond 7 ECAP passes for at least Cu, Ni, and Fe.

With enough pressure, the structural damage and crack propagation in the sample can in principle be suppressed, and indefinite simple shear strains can be applied. Moreover, recording the torque versus rotation angle offers the possibility of directly deriving the shear stress–strain curve. On the other hand, shear strain increases linearly from zero at the center of the disk to a maximum at its periphery. By measuring the hardness along the radius of a single deformed sample, many discrete points of the stress–strain curve from zero to the maximum deformation at the outer surface of the sample can easily be obtained.

Bridgman studied the HPT behavior of many materials. However, its device and results were practically overlooked by materials scientists (with the noteworthy exception of the rediscovery of the method made by Erbel, which was virtually overlooked for a long time, too [48,49]), until the publications of Valiev and co-workers, who had started using the HPT device in the 1980s [25,50].

A wealth of studies on the stress–strain, structure, and properties of materials severely strained by HPT has since then inundated the bibliography, VME strains larger than 100 being easily attainable even in materials of poor ductility [29,50–55].

Although ideally indefinite shear strain can be applied by HPT, several experimental problems arise after the application of a few rotations to the mobile anvil (mainly, the maintenance of the geometry of the disk without damage starting from its periphery); the consequence is that most HPT studies are restricted to VME strains smaller than 50. In that strain range and for the materials and test conditions considered in this paper, the stress–strain behavior does not qualitatively differ from that seen with the ECAP process (e.g., see in the reviews cited above [29,49,51–55]). However, quantitatively, the saturation stresses estimated from ECAP are only apparent, the very weak SHR of the terminal Stage IV continues its decline much longer than expected, and the saturation stresses estimated from HPT tests are up to 20% larger (no wonder if one accepts for Stage IV the validity of the Kocks–Mecking model or one of its variants, leading to a Voce-type equation that only asymptotically approaches a steady state).

6. Ultra-SPD, "Ultra-Severe Plastic Deformations" ($\tilde{\epsilon} > 500$)

Quite recently, Edalati [56] has published the microhardness versus strain results of pure fcc (Ni), bcc (Fe), and HCP (Ti) deformed by HPT well beyond 1000 units of shear strain at room temperature ($0.155 \leq T/T_m \leq 0.174$ for these metals), which is a realm that he has dubbed "ultra-SPD strains". Edalati and co-workers have found that around $\tilde{\epsilon} > 500$, the stress–strain curves of the three metals present an inflection point giving access to what looks to be a new deformation stage that significantly strengthens the material over the Stage IV saturation flow stress if the strain is continued far enough (Figures 7 and 8). The strengthening is not due to any pressure-induced phase change (the pressure employed for HPT straining αTi, 2 GPa, is well below that required for its transformation to ωTi). The pattern of behavior is similar for the three metals. The stronger the metal (because of its homologous deformation temperature and SFE), the earlier appears the SHR after the Stage IV saturation plateau.

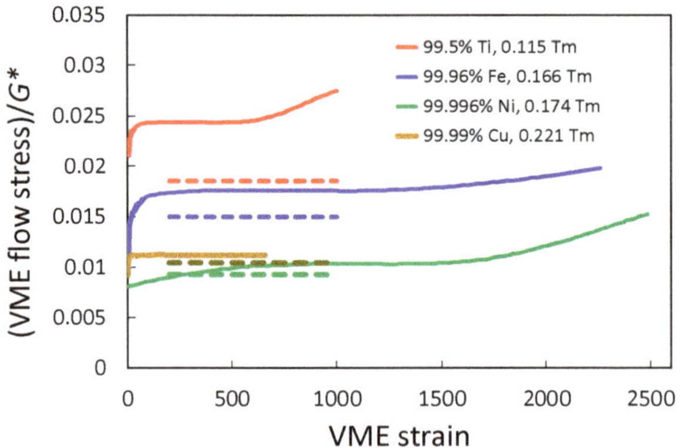

Figure 7. The stress–strain behavior of pure fcc Ni, bcc Fe, and hcp Ti ultra-deformed by torsion under (very) high pressure (HPT) (respectively 2, 6, and 2 GPa applied pressure) at room temperature beyond VME strains of 1000. The VME flow stress has been measured as HV/3, which is one-third of the Vickers microhardness. Replotted from Edalati [56]. The HPT curve of fcc Cu (6 GPa axial pressure) is from Edalati et al. [57]. Dotted lines are the average Stage IV apparent saturation flow stress of metals of similar composition deformed by ECAP in the range $8 \leq \bar{\varepsilon} \leq 16$ (Figure 6). Homologous deformation temperatures are indicated in the figure. The SFE of the four materials at RT increase in the inverse order of their homologous deformation temperature from approx. 45 mJ/m^{-2} of Cu to approx. 180 mJ/m^{-2} of Ti.

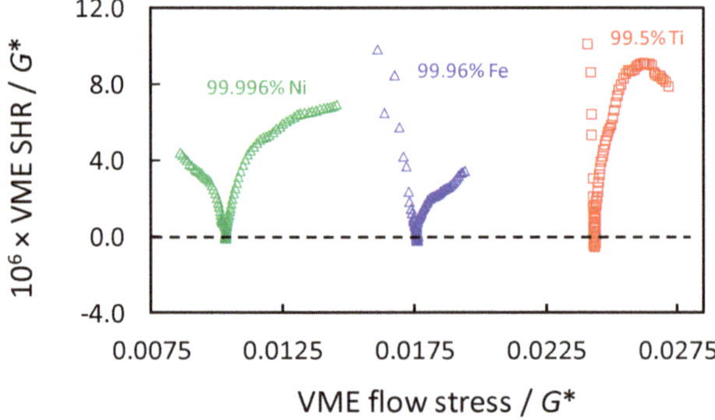

Figure 8. Normalized VME strain hardening rate, SHR/G^* vs. normalized VME flow stress, $\bar{\sigma}/G^*$ for pure Ni, Fe, and Ti ultra-SPD deformed by HPT at room temperature. Derived from the stress–strain curves measured by Edalati [56], as shown in Figure 7. Saturation of Stage IV and emergence of a new hardening stage.

Confirmation that we are dealing with a genuine deformation stage demands further research. However, let us speculate about possible microstructural sources that could explain the observed behavior.

We should first point out that the ultra-SPD SHR is awfully weak: less than 10^{-5} G^* for Ni, Fe or Ti at 300 K, i.e., two orders of magnitude smaller than the weak SHR observed in Stage I of easy-glide single crystals.

Secondly, as this new stage appears only after the saturation of Stage IV, the origin of it cannot anymore be related with the new statistical storage of gliding dislocation lengths, which were fully annihilated by concurrent Stage III dynamic recovery.

Possible origins of such weak SHR are:

- An artefact of the HPT process: as told before, the simple shear of disks by HPT develops a macroscopic gradient of shear strain. The accommodation of such a gradient requires a continuous storage of a density of redundant GND. The gradient is enormous after $\bar{\varepsilon} > 500$. The gradient can be avoided by using rings of appropriate geometry (as used by Erbel [48]).
- Hardening from accumulation of an increasing density of point defects or their clusters arising by the interaction of gliding dislocations with other dislocation lines or from the processes of dynamic recovery of dislocation lengths. Such point-like defects, i.e., vacancy clusters, represent weak obstacles for gliding dislocations.
- A sudden transition from a laminar shear plastic flow to a turbulent one at the mesoscopic microstructural level; it would trigger the accumulation of a density of GND until reaching some new steady state.
- A transition from the control of the flow stress by the average dislocation density to its control by the current grain size (refined by fragmentation and by the imposed deformation, counteracted by specific recovery mechanisms). The transition would take place at some critical size of the grains.

The first of these four tentative contributions to ultra-SPD behavior cannot be discarded. After a macroscopic shear strain γ at the periphery of the disk, the shear strain gradient in the HPT sample amounts to

$$\chi = \gamma / R \tag{14}$$

where R is the sample radius. The accommodation of a plastic gradient by redundant dislocations requires the storage of a GND density of about

$$\rho_g = M_g \cdot (\chi/b) \tag{15}$$

with $M_g \geq 1$ is an orientation-dependent factor accounting for resolving a simple shear by a combination of dislocations from different slip systems. In the HPT case,

$$\left(\rho_g\right)_{HPT} = M_g \cdot (\chi/b) \ = \ M_g (\gamma/bR) \cong M_g (M_{shear}\bar{\varepsilon}/bR). \tag{16}$$

The orientation factor for macroscopic shear strain is $M_{shear} \cong 1.5$. We can roughly estimate the VME flow stress for ultra-SPD at room temperature for a material such as Fe (i.e., $\left(\bar{\sigma}_s^{IV}\right)_{RT}/G^* = 0.017$), ignoring any possible dynamic recovery of the GND density as the effect of the addition of the two dislocation densities:

$$\left(\bar{\sigma}_{uSPD}\right)_{RT}/G^* \ = \ \sqrt{0.017^2 + \left[M_{tension}\alpha b \sqrt{\left(\rho_g\right)_{HPT}}\right]^2}. \tag{17}$$

Assuming $M_g \cong 1.5$ (the orientation factor for simple shear), $M_{tension} \cong 3$, $\alpha \cong 0.3$, $R = 5$ mm, and $b = 0.25 \times 10^{-9}$ m, the result is shown in Figure 9 together with the experimental Fe curve from Edalati [56]. The hardening in excess of the Stage IV saturation flow stress is of the right order of magnitude, but the pattern of behavior does not match the experimental one; the gradient contribution to the flow stress follows a continuous power law right from the approach of the flow stress to its saturation value at strains smaller than 100.

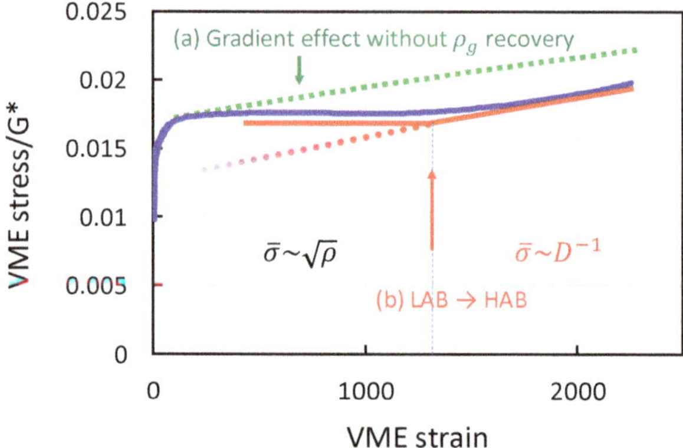

Figure 9. Expected behavior at ultra-severe plastic deformation (SPD) by assuming flow stress control from either (**a**) saturated statistically stored dislocation density plus geometrically necessary dislocations (GND) density from the HPT shear gradient, Equation (17) or (**b**) from the 3D dislocation density (cells or subgrains) during Stage IV and from the grain size beyond Stage IV (ultra-SPD range).

The ultra-SPD burst of hardening is only manifest after a long stress plateau indicative of a transition of the process controlling the strain hardening. This points to the third and fourth origins of the ultra-SPD hardening as plausible explanations, because the second one also represents, as the first, a continuous increase of hardening obstacles as far as dislocation interactions occur.

A hypothetical transition from a laminar to a turbulent shear flow needs experimental confirmation; besides microstructural evidences (in the form, perhaps, of shear bands), probably it would be accompanied by a transition of crystallographic texture, too.

Finally, the fourth explanation looks most interesting: one of the main scientific and technical fascinations of SPD processes is their capability of nanostructuring initially single-crystalline or coarse-polycrystalline materials. It is well known since the seminal work of Langford and Cohen [58], which has been many times confirmed thereafter, that the structure of dislocation cells developed in Stage III shrinks in size and develops an increasing misorientation between neighboring cells; the cells progressively sharpen its walls until becoming low-angle boundaries (subgrain walls) and finally, with the progression of Stage IV, the majority of them become high-angle boundaries of new grains of near-nanometric size that SPD continues refining [9,29,51]. The grain size evolution shows an exponential decline until an apparent temperature-dependent saturation grain size sets in after shear strains of the order of 50 [59–62]. Such an alleged terminal grain size is in the nanometric range ($\approx$100 nm). Only by ultra-SPD processing has the absence of flow stress saturation been detected; it may be that the grain size is still decreasing at a very slow pace and becomes only conspicuous when the colossal ultra-SPD strain range is explored. Quantitative microstructural studies are needed in order to confirm a decrease of grain size concomitant with the ultra-SPD hardening stage. Anyway, if we assume that it is the case, we can advance an explanation.

The classical static Hall–Petch relationship ($\bar{\sigma} \sim D^{-1/2}$), which is mainly justifiable by the strengthening from a grain-size-dependent intragranular dislocation density level, deviates toward a $\bar{\sigma} \sim D^{-1}$ relationship upon further grain size refining. Something akin to it could be at the base of the ultra-SPD transition.

As the strain progresses in Stage IV, most microstructural boundaries have converted into true grain boundaries (misorientation larger than $\approx 12°$), and grain size has become sub-micrometric. The flow stress is controlled by the stress for long-range dislocation glide through the intragranular dislocation forest, as shown in Equation (4), which is equivalent to a critical dislocation line curvature. If the grain size is big enough relative to the effective inter-dislocation distance, the mobile dislocation line ends by reaching the grain boundary where it is either integrated as an extrinsic defect there or totally or partially penetrates the neighboring grains. The maximum curvature required for the intragranular travel of mobile dislocation lines is imposed by the Stage IV dislocation density at saturation. If the grain size was being continuously refined upon straining through a monotonic strain path, a critical grain size would be reached for which the fitting of the curvature of its cross-section would be more flow stress demanding for the dislocation than its gliding through the intragranular dislocation density. From that point on, the flow stress would be controlled by the dynamically evolving grain size through an approximate $\bar{\sigma} \sim D^{-1}$ relationship [22,63,64]. This possible behavior is sketched in Figure 9, too.

Once the Stage IV saturation is approached, the grain fragmentation processes related with new net dislocation storage are not anymore operative. The further evolution of grain size merely comes from its size and shape change resulting from the strain path-dependent mesoscopic strain (very effective in a wire drawing of bcc or hcp metals but very inefficient in the simple shear deformation), which is counteracted by processes of dynamic annihilation of the GB surface. For instance, the process of shear-coupled boundary migration, SCBM, both significantly contributes to the plastic deformation and to the GB surface annihilation at low temperature when grain size is in the nanometric range, particularly in macroscopic shear [59]. It happens likewise with grain boundary sliding, GBS, which is particularly favored in simple shear because of the progressive orientation of the GB toward the macroscopic shear plane (contributes to strain and weakens the shape change relative to the macroscopic one), but very ineffective in wire drawing, which orients the microstructure in the axial direction. For our explanation to be valid, these dynamic GB recovery processes should not be able to completely suppress grain refining coming from the effect of the macroscopic strain.

Figure 10 compares the stress–strain evolution of "pure" iron and "pure" titanium either deformed by wire drawing [58,65] or by HPT [56,60,66]. The trend of curves qualitatively agrees with the above given description [22,58,63,64,67]. In large-strain simple shear (HPT) of bcc iron, the minimum average dimension of deforming grains would evolve, in the absence of dynamic recovery processes, as

$$\frac{D}{D_c} \cong \frac{1}{\sqrt{3}(\epsilon - \epsilon_c)}. \tag{18}$$

In wire drawing, instead,

$$e^{-(\epsilon-\epsilon_c)/2} \geq \frac{D}{D_c} \geq e^{-(\epsilon-\epsilon_c)}. \tag{19}$$

The suffix c indicates the transition of control of the flow stress by glide of the mobile dislocations through the dislocation forest to dislocation bowing-controlled glide. In fact, both Equations (15) and (16) must incorporate a lag factor that is much smaller than 1 for shear and about 1 for axisymmetric drawing, on account of the GB-related processes contributing to strain and dynamic recovery. It is easy to see that the shear lag factor must be very small in order to match the SHR observed in ultra-SPD ($\bar{\theta}_{uSPD} < 10^{-5} G^*$). It is interesting to remark that Hosokawa et al. [68] who, as noted by Edalati [56], first observed SHR after a large plateau of the HPT stress–strain curve of Fe, also related such SHR revival with the hardening of iron by large-strain wire drawing and its relationship with the grain size evolution enforced by the macroscopic shape change.

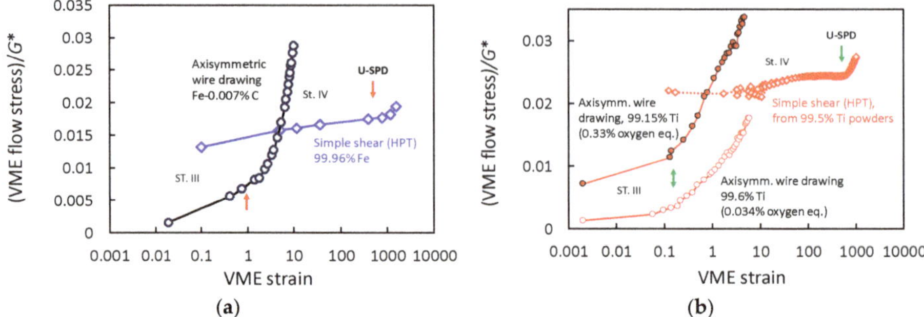

Figure 10. Stress–strain of (**a**) Fe and (**b**) Ti deformed by axisymmetric drawing (Fe-0.007%C, Langford and Cohen [58], 99.15% and 99.6% Ti, Biswas [65], UTS from tensile tests) and by simple shear (HPT, 99.96% Fe and 99.5% Ti, Edalati [56,66], HV/3). Notice the logarithmic scale of the abscissa. The difference in the flow stresses at small/moderate strains may be due to differences in the friction-like flow stress fraction (chemical composition, interstitial content in particular) and initial grain size of the materials, the different testing method for flow stress determination, and the unaccounted compressive deformation of the HPT disk at the beginning of the HPT shear test.

The case of fcc nickel is remarkable. No conclusive information can be obtained from available wire drawing stress–strain curves (a maximum VME strain of about 2 [69]). However, two independent flow stress versus rolling strain curves (plane strain elongation) do show an awakening of strain hardening after a Stage IV plateau reminiscent of the transition observed in HPT deformation at very much larger strains [70,71]. Figure 11 displays one of these rolling deformation stress–strain results together with a wire-drawing curve and the HPT results previously shown. The similarity with the behavior exhibited by iron or titanium (Figure 10) is evident. In rolling, in the absence of dynamic recovery, the evolution of the minimum average dimension of deforming grains would be:

$$\frac{D}{D_c} = e^{-(\varepsilon - \varepsilon_c)}. \tag{20}$$

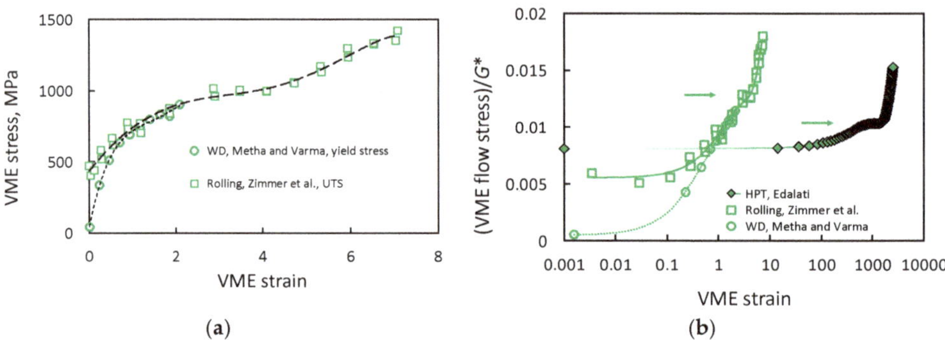

Figure 11. Stress–strain of commercial purity nickel (Ni 200, 99.4% Ni) deformed by (**a**) wire drawing (replotted from [69] and rolling (replotted from) [71]) compared with (**b**) the behavior of 99.996% Ni deformed by HPT (ultra-severe SPD) (replotted from [56]). In figure (**b**) arrows point to the seeming ultra-SPD transition.

According to our tentative explanation, the critical grain size for the ultra-SPD transition would be obtained from equating the critical bowing stress in the grain [22] with the saturation flow stress due to the terminal dislocation density of Stage IV given by Equation (12):

$$\frac{M\tau_c}{G^*} = \frac{\varphi M}{2\pi(1-v)}\frac{b}{D_c}\left(ln\frac{D_c}{\varphi b} + 0.7\right) = 0.024. \tag{21}$$

A shape-factor $\varphi \widetilde{>} 1$ relating the required local critical bowing dislocation diameter with the average grain size has been introduced in Equation (21). With a Poisson ratio of $v = 0.3$, $b = 0.25$ nm, and $M = 3$, the predicted average critical grain size would be $D_c = 40\ \varphi$ nm.

7. Beyond Current Ultra-SPD?

One may now wonder what would happen with the strength and structure of materials if strains by wire drawing or by HPT were continued still much further, to 100,000 or more in the latter case (Edalati [56] has imparted HPT shear strains up to 10^5 to several materials). Difficulties for guaranteeing experimental results free from artifacts after such enormous strain levels surely are considerable. However, theoretical and numerical simulations can shed some light on this question [21,22,72]:

(a) About the strengthening by statistically stored dislocations in the absence of macroscopic (externally imposed) plastic strain gradients, we have the athermal limits of dislocation density and flow stress given by Equations (12) and (13). Those values are extrapolations of experimental results. Experimental values of dislocation densities obtained after plastic deformation in conditions close to athermal are in the same range: 10^{16} m^{-2} after shock experiments or 10^{17} m^{-2} after deformation at 4.2 K. Similarly, atomistic simulations of shock deformation of pure Cu or Ni show peak non-equilibrium dislocation densities up to 10^{18} m^{-2} that decrease to 10^{16} m^{-2} after relaxation (for references, see [21]). Thus, the figures given by Equations (12) and (13) are reasonably confirmed.

(b) Non-redundant geometrically necessary dislocation density is immune to most of the recovery mechanisms to which statistically stored dislocation density is vulnerable (mechanisms involving the annihilation of dislocation segments of opposite sign). However, the GND density can also be absorbed and digested in high-angle grain boundaries to the price of a change in their defect structure and associated intergranular misorientation. Disregarding such possible recovery, an absolute "natural" limit to dislocation density of any kind, redundant or not, is inherent to the discrete, atomistic constitution of the crystals. Surprisingly, a variety of different criteria for establishing such limits (geometric, mechanical, thermodynamic) yield a very narrow interval for its value [21]:

$$0.2 \leq \rho_{max}\cdot b^2 \leq 0.3 \tag{22}$$

i.e., about 4×10^{18} m^{-2} for the metals considered in this paper. From Equation (3) with the α value adequate to the high figure of the dislocation density, the maximum athermal flow stress associated to such dislocation density would be

$$[(\overline{\sigma} - \overline{\sigma}_0)_{max}/G^*]_{T=0} \cong 0.3. \tag{23}$$

However, remember that as explained above, such a limit cannot be reached by the plastic deformation of single-phase crystalline materials developing Stage IV because of the emergence of a nanograin structure after large strains and its control of the flow stress (Figure 9).

What should we expect to happen in an alloy where the saturation of Stage IV does not occur? The obvious answer is amorphization. It does not seem casual that the absolute limit value given above for the flow stress coincides with the ideal strength for the irreversible deformation of metallic glasses [21].

(c)   However, increasing the dislocation density does not seem to be the actual microstructural path for amorphization by deformation. Amorphization by deformation is experimentally observed in many alloys with a high glass-forming ability as a transition from a nanograined to a disordered glassy structure.

Increasing the strains by ultra-SPD HPT or by wire drawing, etc., beyond their current top highs (Figures 7 and 10) is certainly leading to a strength limit, but which one?

Many materials seem to reach a steady state too before amorphization—for instance, pure metals or its dilute alloys. For instance, iron powders deformed by high-energy ball milling at room temperature (high strain rate, non-monotonic strain path) saturate at about 3.5 GPa (0.05 G^*), well above the apparent saturation value for Stage IV. Its structure is constituted of equiaxed nanograins.

The structure of iron wires drawn to $\bar{\varepsilon}$ = 10 after the accumulation of a big number of passes is constituted of axisymmetrically oriented elongated nanograins of an equiaxed section randomly rotated about the ⟨110⟩ crystallographic direction. If such a structure were geometrically perfect (ideal crystallographic domains of infinite axial length and separated by perfect, flat intergranular axially oriented walls), it would be unable to develop elongation-coupled grain boundary migration or grain rotation, i.e., it would be resistant to dynamic recovery in the ultra-SPD stage; continued elongation would lead to further strengthening until reaching some critical structural refinement for attaining the disordered state. However, the real structure is far from ideal, and given its extraordinarily fine grain size, it offers opportunity for recovery mechanisms to significantly contribute to arrive at a steady state of a limited GB surface per unit volume. Numerical MD simulations intended to emulate the observed imperfect fibrous structure of drawn Fe wires up to grain sizes in their cross-section as fine as 3 nm (Figure 12) suggest that the VME flow stress of pure iron wires would saturate at about 5 GPa (0.08 G^*) [22]. This is an estimation of the athermal VME flow stress limit of iron following this unique strain path (the MD result is 6.75 GPa for the saturation stress, of which 1.75 GPa corresponds to a dynamic friction stress because of the very high strain rate employed in the MD simulations (10^8 s^{-1}), as stated in the caption of Figure 12).

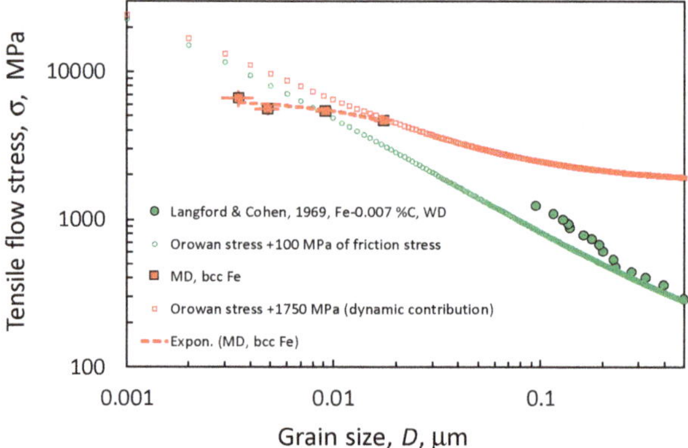

Figure 12. Experimental VME flow stress vs. grain size in the cross-section of <110> textured axisymmetrically drawn iron wires (Fe-0.007%C, from Langford and Cohen [58]) and MD results (molecular dynamics simulations) of similar (but finer) structures of pure iron from ref. [22]. Superposed: the critical stress for bowing a dislocation to the curvature of the grains (the Orowan equation) plus a 100 MPa additive term representing the friction stress from the lattice (appropriate for Fe-0.007%C) or a 1.75 GPa dynamic friction stress (appropriate for the MD simulations because of the very high strain rate employed in the calculations, 10^8 m^{-1} [22]).

8. Conclusions

- SPD processes have extended the experimental study of deformation of materials to levels of VME strains unimaginable half a century ago: from 1 to 10^5 VME strain.

- The effect of plastic deformation on the flow stress and internal structure of materials is path-dependent. One must be aware that both ECAP and HPT, the most performant SPD processes for attaining very large strains in bulk samples, impart simple shear deformation (similar to conventional torsion tests). It is convenient to compare the ECAP or HPT results with results from other more traditional processes imparting large strains, such as rolling (plane-strain elongation, equivalent to pure shear) and wire drawing (axisymmetrical elongation).

- For bcc or hcp materials, wire drawing is much more effective at strengthening or structural refining than SPD processes, because of the unique internal mesoscopic strain pattern associated to a texture that strongly favors internal plane strain elongation, requiring grain interfolding. Ultra-SPD strained Fe or Ti by HPT to a VME of 4000 do not achieve the strength levels obtained by drawing to a VME strain of 10.

- The results of SPD processes applied to metals refute the existence of a true deformation Stage V, prematurely putting an end to the slow progression of Stage IV toward its alleged steady state. Under the high hydrostatic pressures of SPD processes, such progression continues its path much further than expected from the behavior observed in conventional tension or torsion tests.

- Recently published results of ultra-SPD stress–strain behavior of several pure metals show that a hardening transition occurs after a very large plateau of constant flow stress. The SHR of the new deformation stage is very weak, but after enormous shear strains of the order of 10^3, the flow stress significantly increases above the plateau stress of Stage IV. If the authenticity and generality of the new stage is confirmed, it will offer the possibility of strengthening bulk materials by HPT to unforeseen levels.

- One plausible explanation (among others) for the ultra-SPD transition has been proposed in this paper: In Stage IV, the previous cellular and subgrain structure evolves to a near-exclusively grain structure of shrinking average size as deformation goes on. Inside the grains, the dislocation density is limited to the maximum value allowed in Stage IV as far as its crossing requires a curvature of the gliding dislocations larger than the average curvature of the encircling grain cross-section. Beyond that point, fully crossing the whole grain section requires a flow stress that exceeds the critical stress for cutting through the intragranular dislocation density unable to increase its value, i.e., the grain size takes control of the flow stress. In wire drawing, the reduction of the grain cross-section is very rapid and the SHR is very high; in HPT, the decline is very weak, and the SHR is very small.

Table 1 summarizes the approximate maximum (athermal) values of the saturation VME flow stress of single-phase polycrystals according to this paper.

Table 1. Approximate athermal values of the saturation VME flow stress and their associated determinant microstructural features of plastically deformed single-phase polycrystals.

Stage	$[(\sigma_s - \sigma_0)/G^*]_{T=0}$	ρ_s, m^{-2}	Comments
III	0.015	10^{16}	Limited by max. disloc. density
IV	0.024	2.6×10^{16}	
ultra-SPD	Probably ≤ 0.08	2.6×10^{16}	Limited by sat. grain size, $D_s \approx 3\,\mathrm{nm}$
bcc and hcp wire-drawing	0.08	2.6×10^{16}	
absolute ideal limit	0.3	4×10^{18}	Limited by amorphization

The room-temperature shear moduli used for normalizing the flow stresses in this figure and throughout this paper are 42.1, 78.9, 64, and 43.6 GPa for, respectively, Cu, Ni, Fe, and αTi [73].

9. Suggestions for Further Research Work

Despite the big amount of research published on the topic of this paper, many open questions remain. Some very obvious among others are:

(a) The detailed mechanistic explanation of the recovery processes of Stage III responsible for the validity of the Voce stress–strain equation;

(b) The systematic study and characterization of the structural basis of Stage IV and its dependence of on temperature, strain rate, and SFE;

(c) Confirmation of the authenticity of the recently proposed ultra-SPD stage and study of its micromechanistic origin is, of course, crucial.

(d) If the authenticity of the new stage is confirmed, its full characterization will be an excellent research opportunity. The influence of chemical composition on this hardening transition should be indeed addressed; the influence of both purity and alloying also call for study. Similarly, the influence of the homologous deformation temperature, strain rate, and SFE on the emergence as well as the stability of the strength after the strains imparted would merit a the research effort.

Funding: This research received no external funding.

Acknowledgments: The author acknowledges Mikhail Lebyodkin for his kind invitation to write this paper. He also gratefully thanks George Langford for having generously shared his works with him since beginning the research. This paper is dedicated to him on occasion of the 50th anniversary of the publication, with Morris Cohen, of his famous and inspiring paper on iron wire drawn to large strains (ref. [58] of this paper).

Conflicts of Interest: The author declares no conflict of interest.

References

1. Basinski, S.J.; Basinski, Z.S. Plastic Deformation and Work Hardening. In *Dislocations in Solids*; Nabarro, F.R.N., Ed.; Amsterdam-Oxford: North Holland, The Netherlands, 1979; Volume 4, pp. 261–362.

2. Gil Sevillano, J. Flow Stress and Work Hardening. In *Materials Science and Technology. A Comprehensive Treatment*; Cahn, R.V., Haasen, P., Kramer, R.J., Mughrabi, H., Eds.; Plastic Deformation and Fracture of Materials; VCH: Weinheim, Germany, 1993; Volume 4.

3. Kocks, U.F.; Mecking, H. Physics and phenomenology of strain hardening: The FCC case. *Prog. Mater. Sci.* **2003**, *48*, 171–273. [CrossRef]

4. Kubin, L.P. *Dislocations, Mesoscopic Simulations and Plastic Flow*; Oxford Series on Materials Modelling; Oxford University Press: Oxford, UK, 2019.

5. Cordero, Z.C.; Knight, B.E.; Schuh, C.A. Six Decades of the Hall–Petch Effect—A Survey of Grain-Size Strengthening Studies on Pure Metals. *Int. Mater. Rev.* **2016**, *61*, 495–512. [CrossRef]

6. De las Cuevas, F.; Gil Sevillano, J. Effects of temperature and strain rate in strain hardening in torsion of a twinning-induced plasticity steel. *Mater. Sci. Technol.* **2019**, *35*, 669–679. [CrossRef]

7. Rohatgi, A.; Vecchio, K.S.; Gray, G.T., III. The influence of stacking fault energy on the mechanical behaviour of Cu and Cu-Al alloys: Deformation twinning, work hardening and dynamic recovery. *Metall. Mater. Trans. A* **2001**, *32*, 135–145. [CrossRef]

8. Salem, A.A.; Kalidindi, S.R.; Doherty, R.D.; Semiatin, S.L. Strain hardening due to deformation twinning in alpha-titanium: Mechanisms. *Metall. Mater. Trans. A* **2006**, *37*, 259–268. [CrossRef]

9. Gil Sevillano, J.; Van Houtte, P.; Aernoudt, E. Large strain work hardening and textures. *Prog. Mater. Sci.* **1980**, *25*, 69–412. [CrossRef]

10. Backofen, W.A. *Deformation Processing*; Addison Wesley: Boston, MA, USA, 1972.

11. Alberdi Garitaonandia, J.M. Grandes Deformaciones Plásticas en Frío en Policristales de Cobre y Aluminio (torsión). Ph.D. Thesis, Faculty of Sciences, University of Navarra, San Sebastián, Spain, 1984.

12. Zehetbauer, M.; Seumer, V. Cold work hardening in stages IV and V in F.C.C. metals—I. Experiments and interpretation. *Acta Metall. Mater.* **1993**, *41*, 577–588. [CrossRef]

13. Rollett, A.D.; Kocks, U.F. A review of the stages of work hardening. In *Dislocations 93*; Rabier, J., George, A., Bréchet, Y., Kubin, L.P., Eds.; Solid State Phenomena; Trans Tech Publications Ltd.: Baech, Switzerland, 1993; Volume 35–36, pp. 1–18.

14. Siethoff, H.; Schroeter, W.; Metallkunde, Z. New phenomena in the plasticity of semiconductors and FCC metals at high temperatures. I.-Theoretical Models. *Z. Metallkd.* **1984**, *75*, 475–481.

15. Aldazabal, J.; Alberdi, J.M.; Gil Sevillano, J. Stage IV: Microscopic or mesoscopic effect? In *Nanomaterials by Severe Plastic Deformation: Proceedings of the Conference Nanomaterials by Severe Plastic Deformation-nanoSPD2, Vienna*; Zehetbauer, M., Valiev, R.Z., Eds.; Wiley-VCH: Weinhein, Germany, 2004; pp. 65–71.

16. Kok, S.; Beaudoin, A.J.; Tortorelli, D.A. On the development of stage IV hardening using a model based on the mechanical threshold. *Acta Mater.* **2002**, *50*, 1653–1667. [CrossRef]

17. Estrin, Y.; Toth, L.S.; Molinari, A.; Bréchet, Y. A dislocation-based model for all hardening stages in large strain deformation. *Acta Mater.* **1998**, *46*, 5509–5522. [CrossRef]

18. Gil Sevillano, J.; Matey Muñoz, L.; Flaquer Fuster, J. Ciels de Van Gogh et propriétés mécaniques. *J. Phys. IV* **1998**, *8*, Pr4-Pr155–Pr4-Pr165. [CrossRef]

19. Gil Sevillano, J.; González, D.; Martínez-Esnaola, J.M. Heterogeneous deformation and internal stresses developed in BCC wires by axisymmetric elongation. *Mater. Sci. Forum* **2007**, *550*, 75–84. [CrossRef]

20. Young, C.M.; Anderson, L.J.; Sherby, O.D. On the steady state flow stress of iron at low temperature and large strains. *Metall. Trans.* **1974**, *5*, 519–520. [CrossRef]

21. Gil Sevillano, J. On the limits of strain hardening by plastic deformation. In *Proceeding International Symposium on Plastic Deformation and Texture Analysis Alcoy (Spain) 2012*; Amigó Borrás, V., Ed.; Editorial Universitat Politècnica de València: Valencia, Spain, 2012; pp. 1–15.

22. Gil Sevillano, J.; Aldazabal, J.; Aldazabal, I. Elasto-plastic behaviour of a columnar structure of nanocrystalline iron with sharp ‹011› texture. *Materialia* **2018**, *2*, 218–230.

23. Segal, V.M. Invention Certificate of the USSR, No. 575,892 (1977) and Methods of investigation of strain-deformed state in processes of plastic change of shape. Ph.D. Thesis, Physico-Technical Institute of Acad. Sci. of Belorus, SSR, Minsk, Russia, 1974.

24. Segal, V.M.; Reznikov, V.I.; Drobyshevkij, A.E.; Kopylov, V.I. Plastic treatment of metals by plastic shear. *Russ. Metall.* **1981**, *1*, 115–123.

25. Valiev, R.Z.; Kaibyshev, O.A.; Kuznetsov, R.I.; Musalimov, R.S.; Tsenev, N.K. Low-temperature superplasticity of metallic materials. *Dokl. Akad. Nauk. SSSR* **1988**, *301*, 864–866.

26. Valiev, R.Z.; Krasilnikov, N.A.; Tsenev, N.K. Plastic deformation of alloys with submicron-grained structure. *Mater. Sci. Eng. A* **1991**, *137*, 35. [CrossRef]

27. Valiev, R.Z.; Islamgaliev, R.K.; Alexandrov, I.V. Bulk nanostructured materials from severe plastic deformation. *Prog. Mater. Sci.* **2000**, *45*, 103–189. [CrossRef]

28. Valiev, R.Z.; Langdon, T.G. Principles of equal-channel angular pressing as a processing tool for grain refinement. *Prog. Mater. Sci.* **2006**, *51*, 881–981. [CrossRef]

29. Estrin, Y.; Vinogradov, A. Extreme grain refinement by severe plastic deformation: A wealth of challenging results. *Acta Mater.* **2013**, *61*, 782–817. [CrossRef]

30. Dalla Torre, F.H.; Pereloma, E.V.; Davies, C.H.J. Strain hardening behaviour and deformation kinetics of Cu deformed by equal channel angular extrusion from 1 to 16 passes. *Acta Mater.* **2006**, *54*, 1135–1146. [CrossRef]

31. Figueiredo, R.B.; Langdon, T.G. The development of superplastic ductilities and microstructural homogeneity in a magnesium ZK60 alloy processed by ECAP. *Mater. Sci. Eng. A* **2006**, *430*, 151–156. [CrossRef]

32. Vinogradov, A.; Suzuki, T.; Hashimoto, S.; Kitagawa, K.; Kuznetsov, A.; Dobatkin, S. Structure and mechanical properties of submicrocrystalline copper produced by ECAP to very high strains. *Mater. Sci. Forum* **2006**, *503–504*, 971–976. [CrossRef]

33. Wang, Y.L.; Lapovok, R.; Wang, J.T.; Qi, Y.S.; Estrin, Y. Thermal behaviour of copper processed by ECAP with and without back pressure. *Mater. Sci. Eng. A* **2015**, *628*, 21–29. [CrossRef]

34. Valiev, R.Z.; Alexandrov, I.V.; Zhu, Y.T.; Lowe, T.C. Paradox of strength and ductility in metals processed by severe plastic deformation. *J. Mater. Res.* **2002**, *17*, 5–8. [CrossRef]

35. Mishra, A.; Kad, B.K.; Gregori, F.; Meyers, M.A. Microstructural evolution in copper subjected to severe plastic deformation: Experiments and analysis. *Acta Mater.* **2007**, *55*, 13–28. [CrossRef]

36. Kulczyk, M.; Pachla, W.; Swiderska-Sroda, A.; Krasilnikov, N.; Diduszko, R.; Mazur, S.A.; Lojkowski, W.; Kurzydlowski, K.J. Combination of ECAP and Hydrostatic Extrusion for UFG Microstructure Generation in Nickel. *Solid State Phenom.* **2006**, *114*, 51–56. [CrossRef]

37. Zhilyaev, A.P.; Nurismalova, G.V.; Baró, M.; Valiev, R.Z.; Langdon, T.G. Thermal stability and microstructural evolution in ultra-fined nickel after equal-channel angular pressing (ECAP). *Metall. Mater. Trans. A* **2002**, *33*, 1865–1868. [CrossRef]

38. Raju, K.M.; Krishna, M.G.; Padmanabhan, K.A.; Muraleedharan, K.; Gurao, N.P.; Wilde, G. Grain size and grain boundary character disribution in ultra-fine grained (ECAP) nickel. *Mater. Sci. Eng. A* **2008**, *491*, 1–7. [CrossRef]

39. Liu, F.; Yuan, H.; Goel, S.; Liu, Y.; Wang, J.T. Bulk nanolaminated nickel: Preparation, microstructure, mechanical property, and thermal stability. *Metall. Mater. Trans. A* **2017**, *49*, 576–594. [CrossRef]

40. Muñoz Bolaños, J.A.; Higuera Cobos, O.F.; Cabrera Marrero, J.M. Strain hardening behavior of ARMCO iron processed by ECAP. *IOP Conf. Ser. Mater. Sci. Eng.* **2014**, *63*, 1–8.

41. Muñoz, J.A.; Higuera, O.F.; Cabrera, J.M. Microstructural and mechanical study in the plastic zone of ARMCO Iron processed by ECAP. *Mater. Sci. Eng.* **2017**, *697*, 24–36. [CrossRef]

42. Han, B.Q.; Lavernia, E.J.; Mohamed, F.A. Dislocation structure and deformation in iron processed by equal-channel-angular pressing. *Metall. Mater. Trans. A* **2004**, *35*, 1343–1350. [CrossRef]

43. Segal, V.M.; Reznikov, V.I.; Kopylov, V.I.; Pavlik, D.A.; Malyshev, V.F. *Protsessy Plasticheskogo Strukturoobrazovaniya Metallov*; Nauka i Tekhnika: Minsk, Russia, 1994; p. 232.

44. Sus-Ryszkowska, M.; Wejrzanowski, T.; Pakieła, Z.; Kurzydłowski, K.J. Microstructure of ECAP severely deformed iron and its mechanical properties. *Mater. Sci. Eng.* **2004**, *369*, 151–156. [CrossRef]

45. Vinogradov, A.Y.; Stolyarov, V.V.; Hashimoto, S.; Valiev, R.Z. Cyclic behavior of ultrafine-grain titanium produced by severe plastic deformation. *Mater. Sci. Eng. A.* **2001**, *318*, 163–173. [CrossRef]

46. Zhao, X.; Fu, W.; Yang, X.; Langdon, T.G. Microstructure and properties of pure titanium processed by equal-channel angular pressing at room temperature. *Scr. Mater.* **2008**, *59*, 542–545. [CrossRef]

47. Bridgman, P.W. *Studies in Large Plastic Flow and Fracture. With Special Emphasis on the Effects of Hydrostatic Pressure*; Harvard Univ. Press: Cambridge, MA, USA, 1952.

48. Edalati, K.; Horita, Z. A Review on High-Pressure Torsion (HPT) from 1935 to 1988. *Mater. Sci. Eng. A* **2016**, *652*, 325–352. [CrossRef]

49. Erbel, S. *Mechanizm Zmian Własności Metali Poddanych Wielkim Odkształceniom*; Wydawnictwa PW: Warszawa, Poland, 1976.

50. Bryla, K.; Edalati, K. Historical studies by Polish scientist on ultrafine-grained materials by severe plastic deformation. *Mater. Trans.* **2019**, *60*, 1553–1560. [CrossRef]

51. Pippan, R.; Wetscher, F.; Hafok, M.; Vorhauer, A.; Zhyliaev, I.S. The limits of refinement by severe plastic deformation, the limits of refinement by plastic deformation. *Adv. Eng. Mater.* **2006**, *8*, 1046–1056. [CrossRef]

52. Zhyliaev, A.P.; Langdon, T.G. Using high-pressure torsion for metal processing: Fundamentals and applications. *Prog. Mater. Sci.* **2008**, *53*, 893–979. [CrossRef]

53. Starink, M.J.; Cheng, X.; Yang, S. Hardening of pure metals by high-pressure torsion: A physically based model employing volume-averaged defect evolutions. *Acta Mater.* **2013**, *61*, 183–192. [CrossRef]

54. Kawasaki, M. Different models of hardness evolution in ultrafine-grained materials processed by high-pressure torsion. *J. Mater. Sci.* **2013**, *49*, 18–34. [CrossRef]

55. Balasubramanian, N.; Langdon, T.G. The strength–grain size relationship in ultrafine-grained metals. *Metall. Mater. Trans A* **2017**, *47*, 5827–5838. [CrossRef]

56. Edalati, K. Metallurgical Alchemy by ultra-severe plastic deformation via high-pressure torsion process. *Mater. Trans.* **2019**, *60*, 1221–1229. [CrossRef]

57. Edalati, K.; Cubero-Sesin, J.M.; Alhamidi, A.; Mohamed, I.F.; Horita, Z. Influence of severe plastic deformation at cryogenic temperature on grain refinement and softening of pure metals: Investigation using high-pressure torsion. *Mater. Sci. Eng. A* **2014**, *613*, 103–110. [CrossRef]

58. Langford, G.; Cohen, M. Strain hardening of iron by severe plastic deformation. *Trans. ASM* **1969**, *62*, 623–638.

59. Pippan, R.; Scheriau, S.; Taylor, A.; Hafok, M.; Hohenwarter, A.; Bachmaier, A. Saturation of fragmentation during severe plastic deformation. *Annu. Rev. Mater. Res.* **2010**, *40*, 319–343. [CrossRef]

60. Edalati, K.; Horita, Z. High-pressure torsion of pure metals: Influence of atomic bond parameters and stacking fault energy on grain size and correlation with hardness. *Acta Mater.* **2011**, *59*, 6831–6836. [CrossRef]

61. Edalati, K.; Horita, Z. Significance of homologous temperature in softening behavior and grain size of pure metals processed by high-pressure torsion, Mater. *Sci. Eng.* **2011**, *528*, 7514–7523. [CrossRef]

62. Edalati, K.; Akama, D.; Nishio, A.; Lee, S.; Yonenaga, Y.; Cubero-Sesin, J.M.; Horita, Z. Influence of dislocation-solute atom interactions and stacking fault energy on grain size of single-phase alloys after severe plastic deformation using high-pressure torsion. *Acta Mater.* **2014**, *69*, 68–77. [CrossRef]
63. Gil Sevillano, J.; Aldazabal, I.; Luque, A.; Aldazabal, J. Atomistic simulation of the elongation response of a ⟨011⟩ oriented columnar nano-grain bcc Fe polycrystalline sample. *Meccanica* **2016**, *51*, 401–413. [CrossRef]
64. Langford, G.; Cohen, M. Calculation of cell-size strengthening of wire-drawn iron. *Metall. Trans.* **1970**, *1*, 1478–1480. [CrossRef]
65. Biswas, C.P. Strain Hardening of Titanium by Severe Plastic Deformation. Ph.D. Thesis, Massachusetts Institute of Tecnology, Cambridge, MA, USA, 1973.
66. Edalati, K.; Horita, Z.; Fujiwara, H.; Ameyama, K. Cold consolidation of ball-milled titanium powders using high-pressure torsion. *Metall. Mater. Trans. A* **2010**, *41*, 3308–3317. [CrossRef]
67. Rack, H.J.; Cohen, M. Strain hardening of iron–titanium alloys at very large strains. *Mater. Sci. Eng.* **1970**, *6*, 320–326. [CrossRef]
68. Hosokawa, A.; Ii, S.; Tsuchiya, K. Work hardening and microstructural development during high-pressure torsion in pure iron. *Mater. Trans.* **2014**, *55*, 1097–1103. [CrossRef]
69. Mehta, S.; Varma, S.K. Structure-sensitive properties during room-temperature wire drawing at various speeds in nickel 200. *J. Mater. Sci.* **1992**, *27*, 3570–3574. [CrossRef]
70. Nuttall, J.; Nutting, J. Structure and properties of heavily cold-worked fcc metals and alloys. *Metal Sci.* **1978**, *12*, 430–438. [CrossRef]
71. Zimmer, W.H.; Hecker, S.S.; Rohr, D.L.; Murr, L.E. Large strain plastic deformation of commercially pure nickel. *Met. Sci.* **1983**, *17*, 198–208. [CrossRef]
72. Zepeda-Ruiz, L.A.; Stukowski, A.; Oppelstrup, T.; Bulatov, V.V. Probing the limits of metal plasticity with molecular dynamics simulations. *Nature* **2017**, *550*, 492–495. [CrossRef]
73. Frost, H.J.; Ashby, M.F. *Deformation Mechanism Maps: The Plasticity and Creep of Metals and Ceramics*; Pergamon Press: Oxford, UK, 1982.

Article

Revisiting the Application of Field Dislocation and Disclination Mechanics to Grain Boundaries

Claude Fressengeas and Vincent Taupin *

Université de Lorraine, CNRS, LEM3, 7 rue Félix Savart, 57070 Metz, France; claude.fressengeas@univ-lorraine.fr
* Correspondence: vincent.taupin@univ-lorraine.fr

Received: 28 September 2020; Accepted: 9 November 2020; Published: 16 November 2020

Abstract: We review the mechanical theory of dislocation and disclination density fields and its application to grain boundary modeling. The theory accounts for the incompatibility of the elastic strain and curvature tensors due to the presence of dislocations and disclinations. The free energy density is assumed to be quadratic in elastic strain and curvature and has nonlocal character. The balance of loads in the body is described by higher-order equations using the work-conjugates of the strain and curvature tensors, i.e., the stress and couple-stress tensors. Conservation statements for the translational and rotational discontinuities provide a dynamic framework for dislocation and disclination motion in terms of transport relationships. Plasticity of the body is therefore viewed as being mediated by both dislocation and disclination motion. The driving forces for these motions are identified from the mechanical dissipation, which provides guidelines for the admissible constitutive relations. On this basis, the theory is expressed as a set of partial differential equations where the unknowns are the material displacement and the dislocation and disclination density fields. The theory is applied in cases where rotational defects matter in the structure and deformation of the body, such as grain boundaries in polycrystals and grain boundary-mediated plasticity. Characteristic examples are provided for the grain boundary structure in terms of periodic arrays of disclination dipoles and for grain boundary migration under applied shear.

Keywords: plasticity; disclination; dislocation; grain boundary

---

1. Introduction

Disclinations and dislocations were simultaneously introduced by Volterra, more than a century ago [1]. Dislocations are crystal defects arising from translational lattice incompatibility, as measured by the Burgers vector. Disclinations are defects originating in the rotational incompatibility of the lattice characterized by the Frank vector [2]. Disclinations have long been neglected in the field theory of crystal defects, due to their diverging elastic energy, which precludes their occurrence as single defects [3]. However self-screened configurations, such as disclination dipoles, involve bounded elastic energy and may appear frequently [4,5]. Indeed, they have been proposed for describing grain boundaries in various polycrystals, by deriving the disclination density field from crystal orientation maps, as obtained by electron backscattered diffraction [6,7]. Disclination dipoles were also exhibited in planar configurations formed by C_{60} molecular layers on $WO_2/W(110)$ substrates [8]. Generally, disclinations are liable for complementing dislocations in the description of the lattice structure when single-valued elastic rotation fields do not exist, as in polycrystals. Disclination dipoles were shown to provide a good description of grain boundaries [9–12].

In addition to obstacles to plastic deformation by dislocation glide and twinning, grain boundaries can also contribute to plastic deformation. In ultra-fine-grained metals or in materials lacking dislocation slip systems, they can become plasticity mediators, not only by supplying dislocation sources, but also via migration mechanisms. For example, when a shear stress is applied to a bicrystal, one crystal can grow at the expense of the other by the normal motion of the boundary, inducing plastic shear in the crystal traversed by the boundary. Such shear/migration coupling is characterized by the ratio β of the displacements parallel and normal to the boundary. Surface-dislocation-based models [13,14] captured well the variations of this coupling factor as a function of the misorientation from the Burgers vector's content of the interface. Molecular dynamics methods [14–16] provided elementary mechanisms for boundary migration under stress. As considered here, field models involving dislocation and disclination densities are also candidates for the modeling of grain boundaries and their plasticity. The size of the investigated microstructures can be of the order of (but can be larger than) the dimensions typically considered in atomistic simulations. The field models allow analyzing the fine structure of the grain boundaries and the distribution of elastic energy at an atomic resolution length scale. It is also possible to model boundary migration by the motion of arrays of disclination dipoles, and to model dissipative and diffusive phenomena using disclination and dislocation transport equations [17].

In the present paper, we provide a review of the mechanical theory of dislocation and disclination fields [18] and show its interest in dealing with crystalline bodies subject to grain boundary-mediated plasticity. We limit the presentation to infinitesimal transformations. A more complete presentation of the model may be found in the framework of generalized disclinations [19]. By a field theory, we mean a theory of continuously distributed crystal defect densities using the tools of the mathematical theory of partial differential equations and boundary problem solving. As such, we capitalize on the earlier elasto-static theories of dislocation fields [20,21] and of dislocation and disclination fields [2]. Both theories are strongly connected: the latter reduces to the former when the disclination density field vanishes. We also benefit from the mechanical theory of dislocation fields [22], which additionally deals with dislocation-mediated plasticity. In the latter, the material displacement and dislocation density fields are derived from a set of partial differential equations complemented with initial and boundary conditions, provided constitutive information for the elastic behavior and dislocation motion is supplied [23]. This theory was shown to capture well complex features of elasto-plastic deformation such as size effects [24,25], directional hardening [24,26], plastic strain localization [27], dislocation cores [28,29] and patterning of dislocation ensembles [30,31].

When the dislocation velocity vanishes, the mechanical theory of dislocation fields [22] reduces to the elasto-static theory of dislocations [20,21]. The present framework of disclination and dislocation fields reduces to the mechanics of dislocation fields when the disclination density is ignored. It also reduces to the elasto-static crystal theory of dislocations and disclinations [2] when the velocities of dislocations and disclinations are ignored. A previous theory of plasticity accounting for both the translational and rotational aspects of lattice incompatibility was proposed by Kossecka and deWit [32,33]. Driving forces for disclination motion were discussed in [34]. The theory offered kinematic guidelines, but it was not used to solve realistic boundary value problems. In the present work, we try to use the field disclination and dislocation mechanics model in situations where the dynamic interplay between dislocations and disclinations is essential to the understanding of the plastic deformation of the material.

The outline of the paper is the following. Mathematical notations are provided in Section 2. A review of the incompatible elastic dislocation and disclination model [2] is given in Section 3. Static coupling between dislocations and disclinations occurs at this level, due to the continuity condition for dislocation densities. The equations of balance of momentum and moment of momentum are given in Section 4, which further includes the presentation of nonlinear nonlocal elastic constitutive laws that can be of interest for core problems. Section 5 is devoted to the transport properties of dislocations and disclinations. Dynamic coupling occurs at this stage, because disclination mobility yields a source/sink term for dislocations. In Section 6, guidance of the Clausius–Duhem inequality allows defining the driving forces for dislocation and disclination motion, and helps providing appropriate constitutive relations between the driving forces and defects velocities. Section 7 shows possible algorithms used for the solution of the model equations. A plane "edge-wedge" model is detailed in Section 8, and used in Sections 9 and 10 to investigate the structure and mobility of grain boundaries seen as periodic arrays of disclination dipoles.

2. Mathematical Notations

A bold symbol is used for a tensor, as in: **A**. To avoid possible ambiguities, an upper arrow can be used for a vector: $\vec{V}$. The transpose of a tensor **A** is $\mathbf{A}^t$. Tensor subscript indices are expressed with respect to a basis $(\mathbf{e}_i, i = 1, 2, 3)$ of a rectangular Cartesian coordinate system. Arrays of one/two dots denote contraction of the respective number of "adjacent" indices on two immediately neighboring tensors. For instance, the tensor **A.B** with components $A_{ik}B_{kj}$ comes from the dot product of tensors **A** and **B**, and **A** : **B** = $A_{ij}B_{ij}$ is the inner product. The cross product of a second-order tensor **A** and a vector **V**, the **div** and **curl** operators acting on second-order tensors are defined by:

$$(\mathbf{A} \times \mathbf{V})_{ij} = e_{jkl} A_{ik} V_l \tag{1}$$

$$(\mathbf{div}\,\mathbf{A})_i = A_{ij,j} \tag{2}$$

$$(\mathbf{curl}\,\mathbf{A})_{ij} = e_{jkl} A_{il,k}. \tag{3}$$

where $e_{jkl} = \mathbf{e}_j \cdot (\mathbf{e}_k \times \mathbf{e}_l)$ is a component of the Levi–Civita tensor **X**. It is 1 if the jkl permutation is even, -1 if it is odd, 0 otherwise. The comma followed by a component index denotes a spatial derivative with respect to the Cartesian coordinate. A vector $\vec{A}$ is associated with tensor **A** through the inner product of **A** with the tensor **X**:

$$(\vec{A})_k = -\frac{1}{2}(\mathbf{X} : \mathbf{A})_k = -\frac{1}{2}e_{kij} A_{ij} \tag{4}$$

$$(\mathbf{A})_{ij} = -(\mathbf{X} \cdot \vec{A})_{ij} = -e_{ijk}(\vec{A})_k. \tag{5}$$

The trace of the second-order tensor **A** is $tr(\mathbf{A}) = A_{ii}$. The symmetric/skew-symmetric parts are $\mathbf{A}^{sym}$ and $\mathbf{A}^{skew}$. The hydrostatic part is $\mathbf{A}^{hyd} = \frac{1}{3}tr(\mathbf{A})\mathbf{I}$, where **I** is the identity tensor, and its deviatoric part: $\mathbf{A}^{dev} = \mathbf{A} - \mathbf{A}^{hyd}$. An upper dot represents a material time derivative.

3. Review of the Incompatible Elasto-Static Defect Theory

3.1. Incompatibility in the Dislocation Model

In the proposed framework, continuity of the displacement vector field **u**, along with continuity of its derivatives, is assumed at any point, including between atoms, in a simply connected body that is elasto-plastically deformed, except perhaps along interfaces where continuity of the derivatives may not hold. The rotation vector

$$\vec{\omega} = \frac{1}{2}\mathbf{curl}\,\mathbf{u} \tag{6}$$

is also defined at any point. These statements assume in a dual way that the material is capable of transmitting stresses and couple stresses, possibly below inter-atomic length scales. Such continuity was very early proposed [35,36] when defining pressure and stresses in quantum systems. It was later used by [37] when defining the stress tensor in quantum mechanics. By definition, couple stresses (moments of stress) at a point come from mechanical stresses existing at distant material points. As such, couple stresses introduce nonlocality in the mechanical description of the material. The distortion tensor is the gradient of the displacement **U** = **grad u**. It is curl-free:

$$\mathbf{curl}\,\mathbf{U} = 0. \tag{7}$$

If dislocations are present, the elastic $\mathbf{U}_e$ and plastic $\mathbf{U}_p$ components of **U** contain incompatible parts, which are not curl-free, but are div-free. Indeed, if dislocations thread a patch S in the body, a constant discontinuity **b** in the elastic displacement exists across S, which manifests itself as a closure defect along the circuit C surrounding S:

$$\mathbf{b} = [\![\mathbf{u}_e]\!] = \int_C \mathbf{U}_e \cdot \mathbf{dl}. \tag{8}$$

b is referred to as the resulting Burgers vector of this dislocation ensemble. Applying Stoke's theorem yields:

$$\mathbf{b} = \int_C \mathbf{U}_e \cdot \mathbf{dl} = \int_S \mathbf{curl}\, \mathbf{U}_e \cdot \mathbf{n} dS, \tag{9}$$

where **n** is the unit normal to S. The discontinuity **b** is also characterized in a pointwise continuous manner by introducing Nye's dislocation density tensor α:

$$\mathbf{curl}\, \mathbf{U}_e = \alpha \tag{10}$$

such that

$$\mathbf{b} = \int_S \alpha \cdot \mathbf{n} dS. \tag{11}$$

Thus, dislocation lines threading S and having a non-zero resulting Burgers vector induce the existence of an incompatible part $\mathbf{U}_e^{\perp}$ of the elastic distortion tensor. Applying the Stokes-Helmholtz decomposition of square-integrable tensor fields with square-integrable first order derivatives [38], we can uniquely define the square-integrable tensor and vector χ and **z** such that the elastic distortion $\mathbf{U}_e$ reads:

$$\mathbf{U}_e = \mathbf{curl}\, \chi + \mathbf{grad}\, \mathbf{z}. \tag{12}$$

The curl of $\mathbf{U}_e$ extracts **curl** χ and removes **grad z** while, conversely, taking the divergence extracts **grad z** and removes **curl** χ. Thus, Equation (10) involves only **curl** χ, and

$$\mathbf{curl}\, \mathbf{U}_e^{\perp} = \mathbf{curl}\, \mathbf{curl}\, \chi = \alpha. \tag{13}$$

The term **grad z** is the compatible (curl-free) part $\mathbf{U}_e^{\|}$ of the elastic distortion $\mathbf{U}_e$, and **z** is the compatible elastic displacement $\mathbf{u}_e^{\|}$, up to a constant. However, Equation (13) is insufficient to identify $\mathbf{U}_e^{\perp}$ from a given dislocation field α. To ensure completeness of the identification, $\mathbf{U}_e^{\perp}$ must vanish if $\alpha = 0$. Following [38,39], we thus augment Equation (13) with the supplementary condition $\mathbf{div}\, \mathbf{U}_e^{\perp} = 0$ together with $\mathbf{U}_e^{\perp} \cdot \mathbf{n} = 0$ on the external boundary with unit normal **n**. This allows ensuring that the solution does not contain any gradient part. Applying the curl operator to Equation (13), we have:

$$\mathbf{curl}\, \mathbf{curl}\, \mathbf{U}_e^{\perp} = \mathbf{grad}\, \mathbf{div}\, \mathbf{U}_e^{\perp} - \mathbf{div}\, \mathbf{grad}\, \mathbf{U}_e^{\perp} = \mathbf{curl}\, \alpha. \tag{14}$$

Then, using the side condition $\mathbf{div}\, \mathbf{U}_e^{\perp} = 0$, we get:

$$\mathbf{div}\, \mathbf{grad}\, \mathbf{U}_e^{\perp} = -\mathbf{curl}\, \alpha. \tag{15}$$

Equation (15) is a Poisson-type equation for the unique determination of the unknown tensor $\mathbf{U}_e^{\perp}$. We could have demonstrated in a similar way the existence of an incompatible part of the plastic distortion, $\mathbf{U}_p^{\perp}$, which is opposite to the incompatible elastic distortion $\mathbf{U}_e^{\perp}$ to ensure the compatibility of the distortion **U** and the continuity of the displacement **u**. An additional compatible component $\mathbf{U}_p^{\|}$ may also exist, together with a compatible plastic displacement $\mathbf{u}_p^{\|}$. We can then write the following equations:

$$\mathbf{U} = \mathbf{U}_e + \mathbf{U}_p \tag{16}$$

$$\mathbf{U}_e = \mathbf{U}_e^{\perp} + \mathbf{U}_e^{\parallel} \tag{17}$$

$$\mathbf{U}_p = \mathbf{U}_p^{\perp} + \mathbf{U}_p^{\parallel} \tag{18}$$

$$0 = \mathbf{U}_e^{\perp} + \mathbf{U}_p^{\perp}. \tag{19}$$

From the above, we see that:

$$\boldsymbol{\alpha} = -\mathbf{curl}\,\mathbf{U}_p^{\perp} \tag{20}$$

similarly to Equation (13). The continuity condition:

$$\mathbf{div}\,\boldsymbol{\alpha} = 0 \tag{21}$$

derives naturally from Equations (10) and (13). The divergence-free character of $\boldsymbol{\alpha}$ reflects the fact that dislocation lines cannot end within the material. They should either exit the material, or be closed lines. As will be shown later in Section 3.2, dislocation lines may possibly end on disclination lines. Now, in view of defining disclinations in the next section, we introduce curvatures and examine how they appear in the present dislocation model. The strain tensor $\boldsymbol{\epsilon}$ is the symmetric part of $\mathbf{U}$, the rotation tensor $\boldsymbol{\omega}$ its skew-symmetric part, corresponding to the rotation vector $\vec{\omega}$ through the relation:

$$\boldsymbol{\omega} = -\mathbf{X} \cdot \vec{\omega}, \; \vec{\omega} = -\frac{1}{2}\boldsymbol{\omega} : \mathbf{X}. \tag{22}$$

With this, Equation (7) yields:

$$\mathbf{curl}\,\boldsymbol{\epsilon} + div(\vec{\omega})\mathbf{I} - \mathbf{grad}^t\vec{\omega} = 0 \tag{23}$$

where $\mathbf{grad}\vec{\omega}$ is the curvature tensor $\boldsymbol{\kappa}$:

$$\boldsymbol{\kappa} = \mathbf{grad}\,\vec{\omega}. \tag{24}$$

The trace of Equation (23) yields:

$$div(\vec{\omega}) = 0, \tag{25}$$

and implies that $tr(\boldsymbol{\kappa}) = 0$, meaning that the curvature tensor is deviatoric. Please note that transposing Equation (23) and then taking the curl yields:

$$\mathbf{curl}\,\mathbf{curl}^t\,\boldsymbol{\epsilon} = 0. \tag{26}$$

This is the Saint-Venant compatibility condition. Applying the above procedure to Equation (10), we find:

$$\mathbf{curl}\,\boldsymbol{\epsilon}_e + div(\vec{\omega}_e)\mathbf{I} - \mathbf{grad}^t\vec{\omega}_e = \boldsymbol{\alpha} \tag{27}$$

and:

$$\boldsymbol{\eta} = \mathbf{curl}\,\mathbf{curl}^t\boldsymbol{\epsilon}_e = \mathbf{curl}\,(\boldsymbol{\alpha}^t - \frac{1}{2}tr(\boldsymbol{\alpha})\mathbf{I}). \tag{28}$$

The incompatibility tensor $\boldsymbol{\eta}$ is a possible measure of the incompatibility due to the presence of dislocations. To terminate this section, we introduce the elastic curvature tensor $\boldsymbol{\kappa}_e$ as the gradient of the elastic rotation $\vec{\omega}_e$:

$$\boldsymbol{\kappa}_e = \mathbf{grad}\,\vec{\omega}_e, \tag{29}$$

which reflects both the continuity of the rotation vector and the curl-free character of the tensor κ_e:

$$\text{curl } \kappa_e = 0. \tag{30}$$

Using κ_e, Equation (27) can be expressed as:

$$\text{curl } \epsilon_e + tr(\kappa_e)\mathbf{I} - \kappa_e^t = \alpha. \tag{31}$$

It is important to note for what follows that the above equation holds even if κ_e is not a gradient. Finally, relations similar to Equations (29) and (31) can be obtained by using Equation (19) for the plastic strain ϵ_p, rotation $\vec{\omega}_p$ and curvature κ_p:

$$\kappa_p = \text{grad } \vec{\omega}_p \tag{32}$$
$$-\text{curl } \epsilon_p - tr(\kappa_p)\mathbf{I} + \kappa_p^t = \alpha. \tag{33}$$

3.2. Incompatibility in the Dislocation and Disclination Model

We now assume that elastic rotation discontinuities $[\![\vec{\omega}_e]\!]$ exist in addition to the displacement discontinuities $[\![\mathbf{u}_e]\!] = \mathbf{b}$ reflected by the dislocation densities. Thus, the elastic curvature tensor κ_e is not a gradient anymore. This is the situation met along grain boundaries in polycrystals. Integrating κ_e along closed circuits in the manner of Equation (9) leads to the angular closure defect ϕ known as the Frank vector:

$$\phi = [\![\vec{\omega}_e]\!] = \oint_C \kappa_e \cdot \mathbf{dr} = \int_S \text{curl } \kappa_e \cdot \mathbf{n}dS. \tag{34}$$

We introduce a tensor field θ such that:

$$\text{curl } \kappa_e = \theta \tag{35}$$

and

$$\phi = \int_S \theta \cdot \mathbf{n}dS. \tag{36}$$

θ is the disclination density tensor [2]. In the presence of disclinations, the elastic displacement discontinuity reads:

$$[\![\mathbf{u}_e]\!] = \oint_C [\epsilon_e - (\kappa_e^t \times \mathbf{r})^t] \cdot \mathbf{dr} + \phi \times \mathbf{r}_0 = \mathbf{b} + \phi \times \mathbf{r}_0. \tag{37}$$

Thus, the discontinuity $[\![\mathbf{u}_e]\!]$ derives from the motion of a rigid body with translation $\mathbf{b}$ and rotation ϕ, in agreement with Weingarten's theorem [40]. The Burgers vector can also be expressed as a function of dislocation and disclination density tensors:

$$\mathbf{b} = \int_S (\alpha - (\theta^t \times \mathbf{r})^t) \cdot \mathbf{n}dS. \tag{38}$$

When both dislocations and disclinations exist, the incompatibility tensor η is:

$$\eta = \text{curl curl}^t \, \epsilon_e = \text{curl } (\alpha^t - \frac{1}{2}tr(\alpha)\mathbf{I}) + \theta. \tag{39}$$

As similarly detailed above for $\mathbf{U}_e^{\perp}$, we must ensure that the incompatible part $\kappa_e^{\perp}$ of the curvature tensor is null if $\theta = 0$. In this aim, Equation (35) must be replaced with:

$$\text{curl } \kappa_e^{\perp} = \theta, \tag{40}$$

together with the side condition $\text{div } \kappa_e^{\perp} = 0$ and boundary condition $\kappa_e^{\perp} \cdot \mathbf{n} = 0$ on external boundaries. These conditions provide a unique solution $\kappa_e^{\perp}$ of the Poisson equation:

$$\operatorname{div} \operatorname{grad} \kappa_e^{\perp} = -\operatorname{curl} \theta. \tag{41}$$

A compatible part $\kappa_e^{\parallel}$ can also exist, but it is not involved in Equations (40) and (41). Therefore, one has the following equations:

$$\kappa_e = \kappa_e^{\perp} + \kappa_e^{\parallel} \tag{42}$$

$$\kappa_p = \kappa_p^{\perp} + \kappa_p^{\parallel} \tag{43}$$

$$\kappa = \kappa_e^{\parallel} + \kappa_p^{\parallel} \tag{44}$$

$$0 = \kappa_e^{\perp} + \kappa_p^{\perp}. \tag{45}$$

The incompatible plastic curvature is opposed to the incompatible elastic curvature to ensure that the total curvature is compatible and the material continuous. Then:

$$\operatorname{curl} \kappa_p^{\perp} = -\theta \tag{46}$$

$$\operatorname{div} \operatorname{grad} \kappa_p^{\perp} = \operatorname{curl} \theta. \tag{47}$$

The continuity equation for disclinations:

$$\operatorname{div} \theta = 0 \tag{48}$$

derives from Equations (35) and (40). It mathematically expresses that disclination lines should not end within the body and form closed loops. The continuity equation for dislocations Equation (21) becomes:

$$\operatorname{div} \alpha = \theta : \mathbf{X}. \tag{49}$$

It means that dislocations may now terminate on twist-disclination lines. Please note that if the disclination density is null, Equations (31), (35), (38) and (49) reduce to Equations (11), (21), (27) and (30), and the elasto-static model of dislocations and disclinations reduces to the above dislocation model.

4. Elasticity in Incompatible Media Containing Dislocations and Disclinations

4.1. Field Equations

We consider describing the mechanical equilibrium of a simply connected body of volume V and external surface S containing dislocations and disclinations, and submitted to external loading. Since the disclination densities are associated with the incompatibility of the elastic curvatures, we are interested in a framework involving, in addition to stresses, the work-conjugates of the curvatures, i.e., the couple stresses. We therefore follow the footsteps of [41] to derive the mechanical balance equations. Let $\mathbf{n}$ be the outward unit normal to S, $\mathbf{t}$ be the traction vector (force per unit area) and $\mathbf{m}$ the moment vector (couple per unit area) acting on S. As they are non-essential for the present purposes, body forces and couples are neglected for the sake of simplicity, as well as inertia forces. The conservation of momentum and moment of momentum of the body therefore imposes the following conditions:

$$\int_S \mathbf{t}\, dS = 0 \tag{50}$$

$$\int_S (\mathbf{m} + \mathbf{r} \times \mathbf{t}) = 0. \tag{51}$$

Assuming that the tractions and moments at boundaries can be written as $\mathbf{t} = \mathbf{T} \cdot \mathbf{n}$ and $\mathbf{m} = \mathbf{M} \cdot \mathbf{n}$, where $\mathbf{T}$ and $\mathbf{M}$ are respectively the stress and couple-stress tensors, we find by substituting in Equations (50) and (51) and applying the divergence theorem:

$$\int_V \mathbf{div\,T}\,dv \;=\; 0 \tag{52}$$

$$\int_V (\mathbf{div\,M} - \mathbf{X}:\mathbf{T})\,dv \;=\; 0. \tag{53}$$

Assuming sufficient continuity, the pointwise local equations are:

$$\mathbf{div\,T} \;=\; 0 \tag{54}$$

$$\mathbf{div\,M} - \mathbf{X}:\mathbf{T} \;=\; 0. \tag{55}$$

Equations (54) and (55) are balance of momentum and moment of momentum equations where the stress tensor $\mathbf{T}$ is not necessarily symmetric. These two equations can be combined to form a higher-order balance equation coupling the symmetrical part of the stress tensor $\mathbf{T}^{sym}$ and the deviatoric part of the couple-stress tensor $\mathbf{M}^{dev}$ [41]:

$$\mathbf{div\,T}^{sym} + \frac{1}{2}\mathbf{curl\,div\,M}^{dev} = 0, \tag{56}$$

but involving neither the skew-symmetric part $\mathbf{T}^{skew}$ of the stress tensor nor the hydrostatic part $\mathbf{M}^{hyd}$ of the couple-stress tensor. $\mathbf{T}^{skew}$ may be obtained from Equation (55), but $\mathbf{M}^{hyd}$ is altogether absent from the mechanical balance equations and is therefore indeterminate in such a couple-stress theory [41–43]. A more general presentation of the theory involving not only curvatures and couple stresses but also strain gradients and their work-conjugates ("hyper-stresses" or "double-stresses") allows avoiding such indeterminacy. It features crystal defects more general than disclinations, and referred to as generalized disclinations. The differences between a couple-stress disclination theory involving only curvatures and a hyper-stress generalized disclination theory involving both curvatures and strain gradients are discussed in [44].

The mechanical dissipation in the material is the difference between the power of the applied forces and the rate of change of the stored energy:

$$D = \int_S (\mathbf{v}\cdot\mathbf{T} + \dot{\boldsymbol{\omega}}\cdot\mathbf{M})\cdot\mathbf{n}\,dS - \int_V \dot{\psi}\,dv = \int_S (v_i T_{ij} + \omega_i M_{ij})n_j\,dS - \int_V \dot{\psi}\,dv. \tag{57}$$

Here, $\mathbf{v}$ is the material velocity and ψ the specific free energy density. Using the balance Equations (54) and (55) together with the divergence theorem, we can write:

$$D = \int_D [(v_i T_{ij} + \omega_i M_{ij})_{,j} - v_i T_{ij,j} - \omega_i(M_{ij,j} - e_{ikl}T_{kl}) - \dot{\psi}]\,dv. \tag{58}$$

Then we use $\omega_i = -e_{imn}\omega_{mn}/2$ to substitute the rotation vector for the rotation tensor through Equation (22), and progressively derive:

$$D \;=\; \int_D (T_{ij}v_{i,j} + M_{ij}\dot{\omega}_{i,j} - \frac{1}{2}e_{ikl}e_{imn}\dot{\omega}_{mn}T_{kl} - \dot{\psi})\,dv \tag{59}$$

$$D \;=\; \int_D (T_{ij}v_{i,j} + M_{ij}\dot{\omega}_{i,j} - \frac{1}{2}(\delta_{km}\delta_{ln} - \delta_{kn}\delta_{lm})\dot{\omega}_{mn}T_{kl} - \dot{\psi})\,dv \tag{60}$$

$$D \;=\; \int_D (T_{ij}v_{i,j} + M_{ij}\dot{\omega}_{i,j} - T_{ij}\dot{\omega}_{ij} - \dot{\psi})\,dv \tag{61}$$

$$D \;=\; \int_D (T_{ij}\dot{e}_{ij} + M_{ij}\dot{\omega}_{i,j} - \dot{\psi})\,dv \tag{62}$$

$$D \;=\; \int_D (\mathbf{T}^{sym}:\dot{e} + \mathbf{M}^{dev}:\dot{\kappa} - \dot{\psi})\,dv. \tag{63}$$

The Clausius–Duhem inequality imposes that the dissipation be positive, or zero for reversible processes: $D \geq 0$. In Equation (63) the inner product between symmetric and skew-symmetric tensors is null. Hence the strain rate tensor $\dot{\epsilon}$ extracts the symmetric part of the stress tensor $\mathbf{T}$, and the deviatoric curvature rate tensor $\dot{\kappa}$ extracts the deviatoric part of the couple-stress tensor. A free energy density function ψ is taken as:

$$\psi = \psi(\epsilon_e, \kappa_e), \tag{64}$$

with contributions from the elastic strain ϵ_e and curvature κ_e. A more detailed description of this free energy will be given below in Section 4.3. For the time being, it is sufficient to note that, differentiating Equation (64) on the one hand, and considering elastic reversible processes such that $D = 0$ in Equation (63) on the other hand, leads to identifying the symmetric part of the stress tensor $\mathbf{T}^{sym}$ and deviatoric part of the couple-stress tensor $\mathbf{M}^{dev}$ with the partial derivatives of the free energy density with respect to the elastic strain and curvature:

$$\dot{\psi} \;=\; \frac{\partial \psi}{\partial \epsilon_e} : \dot{\epsilon}_e + \frac{\partial \psi}{\partial \kappa_e} : \dot{\kappa}_e \tag{65}$$

$$\dot{\psi} \;=\; \mathbf{T}^{sym} : \dot{\epsilon}_e + \mathbf{M}^{dev} : \dot{\kappa}_e \tag{66}$$

$$\mathbf{T}^{sym} \;=\; \frac{\partial \psi}{\partial \epsilon_e} \tag{67}$$

$$\mathbf{M}^{dev} \;=\; \frac{\partial \psi}{\partial \kappa_e}. \tag{68}$$

Equations (67) and (68) will be the constitutive relationships describing the elastic behavior of the material.

4.2. Solving the Incompatible Elasticity Problem

Consider again a simply connected body of volume V containing dislocations and disclinations, submitted to tractions and moments on its external surface S with outward unit normal $\mathbf{n}$. Tractions are imposed on a part S_t of S and rotations/displacements on a part S_u:

$$\mathbf{T} \cdot \mathbf{n} \;=\; \mathbf{t}_d \text{ on } S_t \tag{69}$$

$$\mathbf{u}_e \;=\; \mathbf{u}_d \text{ on } S_u \tag{70}$$

$$\omega_e \;=\; \omega_d \text{ on } S_u. \tag{71}$$

We are interested in describing the state of stress and couple-stress in the body in the presence of dislocation and disclination density fields (α, θ). The incompatible elastic strain and curvature $(\epsilon_e^{\perp}, \kappa_e^{\perp})$ are assumed to be known from the solution of the Poisson Equations (15) and (41). They usually do not allow satisfying the balance Equation (56). Recalling the decomposition of the elastic strain and curvature in the sum of incompatible, $(\epsilon_e^{\perp}, \kappa_e^{\perp})$, and compatible, $(\epsilon_e^{\parallel}, \kappa_e^{\parallel})$, components, Equation (56) can be written as:

$$0 \;=\; \operatorname{div} \mathbf{T}^{sym}(\epsilon_e^{\parallel}, \kappa_e^{\parallel}) + \frac{1}{2} \operatorname{curl} \operatorname{div} \mathbf{M}^{dev}(\epsilon_e^{\parallel}, \kappa_e^{\parallel}) + \mathbf{f}^{\perp}, \tag{72}$$

$$\mathbf{f}^{\perp} \;=\; \operatorname{div} \mathbf{T}^{sym}(\epsilon_e^{\perp}, \kappa_e^{\perp}) + \frac{1}{2} \operatorname{curl} \operatorname{div} \mathbf{M}^{dev}(\epsilon_e^{\perp}, \kappa_e^{\perp}). \tag{73}$$

$\mathbf{f}^{\perp}$ is to be seen as a volumetric density of body forces associated with dislocations and disclinations. In Equations (72) and (73), the symmetric part of the stress tensor and the deviatoric part of the couple-stress tensor are obtained from the constitutive relations Equations (67) and (68). The solution of Equation (72) is the compatible elastic displacement field.

4.3. Elasticity in Dislocation/Disclination Cores

Elastic constitutive laws usually relate the elastic strain tensor to the stress tensor, both taken at the same material point. As such, they are referred to as "local" constitutive laws. They are valid in a statistical sense if the characteristic size of the volume element actually envisioned at the material point is large enough to include a large number of atoms in a perfectly ordered crystal. Local laws may fail at describing correctly elasticity if this characteristic size reduces to the order of a few atomic lattice spacings, particularly if the volume element involves crystal defects such as dislocations and disclinations. Since the mechanical behavior of a particular atom depends on its interactions with all neighboring atoms, whose location is strongly disturbed by the presence of these defects, defining the stress tensor in such a local manner may then become impossible. Nonlocal elastic laws are then needed. Nonlocal elastic laws are relationships using convolution integrals, where the stresses at a given material point depend on the deformation state in some neighborhood, whose extent is controlled by a nonlinear kernel function [45–49]. Such nonlocal elastic laws have been successful in retrieving phonon dispersion curves and regularizing elastic fields in the core region of crystal defects. Convolution integrals can be coupled with higher-order elastic laws to derive nonlocal elastic constitutive laws that are of interest in the core regions of dislocations and disclinations, while local laws may be sufficient in the non-defected regions. The occurrence of different elastic behavior in non-defected and defected regions was already assumed in grain and phase boundaries [50–53]. Following [54] and relying on Eringen-type convolution integrals applied to strains and curvatures, we propose below nonlocal laws describing the elastic behavior in the core region of dislocations and disclinations, and reducing to local laws in the perfectly ordered parts of the crystal.

Nonlocal Convolution Integrals

We introduce a free energy density functional at a point $\mathbf{r}$ in a body V in the form:

$$
\begin{aligned}
\psi(\mathbf{r}) \;=\;\; & \frac{1}{2}C_{ijkl}(\mathbf{r})\epsilon^e_{ij}(\mathbf{r})\epsilon^e_{kl}(\mathbf{r}) + \frac{1}{2}A_{ijkl}(\mathbf{r})\kappa^e_{ij}(\mathbf{r})\kappa^e_{kl}(\mathbf{r}) + \ldots \\[4pt]
\ldots \;+\;\; & \frac{1}{2}\epsilon^e_{ij}(\mathbf{r})\int_V \mathcal{K}(\mathbf{R}')D_{ijkl}(\mathbf{r},\mathbf{r}')\epsilon^e_{kl}(\mathbf{r}')dV' + \ldots \\[4pt]
\ldots \;+\;\; & \frac{1}{2}\kappa^e_{ij}(\mathbf{r})\int_V \mathcal{K}(\mathbf{R}')B_{ijkl}(\mathbf{r},\mathbf{r}')\kappa^e_{kl}(\mathbf{r}')dV'.
\end{aligned}
\tag{74}
$$

The first two pointwise terms on the r.h.s. include symmetric positive-definite quadratic forms with $\mathbf{A}$ and $\mathbf{C}$ given at point $\mathbf{r}$. This leads to the linear elastic laws used and discussed in [2,18,54] in the framework of models for dislocations and disclinations. Nonlocal contributions are added in the form of Eringen-type convolutions, which are the last two terms of the r.h.s. [55]. In such terms, the location vectors $\mathbf{r}'$ are points distinct from point $\mathbf{r}$. The difference vectors $\mathbf{R}' = (\mathbf{r}' - \mathbf{r})$ and their norms $R' := (\mathbf{R}'.\mathbf{R}')^{1/2}$ are also defined. The spatial range of nonlocality is controlled by the nonlocal kernel $\mathcal{K}(\mathbf{R}')$, while the elasticity tensors $\mathbf{B}$ and $\mathbf{D}$ depend on $(\mathbf{r},\mathbf{r}')$ and the elastic curvatures/strains are taken at points $\mathbf{r}'$. The proposed nonlocality is motivated by the strong coupling existing in the core region of defects between displacement/strain and rotation/curvature at distant locations $\mathbf{r}$ and $\mathbf{r}'$ (see Figure 1). One can express the strain induced at point $\mathbf{r}'$ by the curvature at point $\mathbf{r}$:

$$
\epsilon^e_{kl}(\mathbf{r}') = -e_{kmn}\kappa^e_{ml}(\mathbf{r})R'_n, \quad \epsilon_e(\mathbf{r}') = -(\kappa^t_e(\mathbf{r}) \times \mathbf{R}')^t.
\tag{75}
$$

Note the presence of the above nonlocal term in the definition of the Burgers vector in Equation (37) above. Similarly, we can express the curvature at point $\mathbf{r}'$ due to the strain at point $\mathbf{r}$:

$$
\kappa^e_{kl}(\mathbf{r}') = -e_{kmn}\epsilon^e_{ml}(\mathbf{r})\frac{R'_n}{R'^2}, \quad \kappa_e(\mathbf{r}') = -\frac{1}{R'^2}(\epsilon_e(\mathbf{r}) \times \mathbf{R}')^t.
\tag{76}
$$

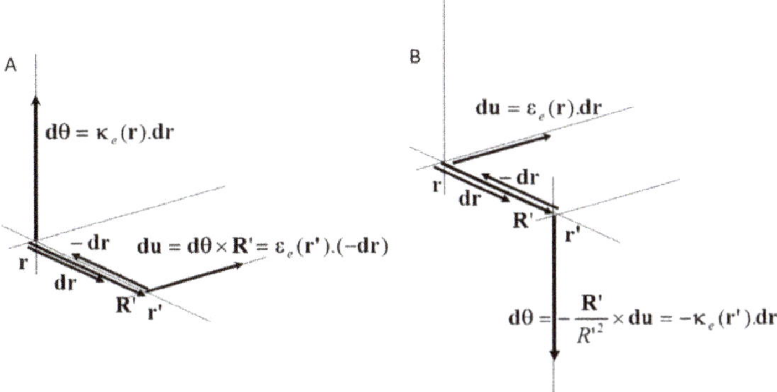

Figure 1. Nonlocal strains and curvatures. (**A**) Nonlocal shear strain at location $\mathbf{r'}$ induced by the rotation $\mathbf{d\omega}_e$ arising from a curvature tensor at location $\mathbf{r}$; (**B**) Nonlocal rotation and curvature at location $\mathbf{r'}$ from the displacement $\mathbf{du}$ induced by a shear strain tensor at location $\mathbf{r}$.

The above nonlocal relations between strains and curvatures can be used to substitute curvatures and strains at points $\mathbf{r}$ for strains and curvatures at points $\mathbf{r'}$, respectively. This allows greatly simplifying the convolutions in the free energy functional Equation (74), in such a way that nonlocality applies only to elasticity tensors while all strains and curvatures are defined at point $\mathbf{r}$ (the reader is referred to ref [55] for more details):

$$\psi = \frac{1}{2}C_{ijkl}\epsilon^e_{ij}\epsilon^e_{kl} + \frac{1}{2}A_{ijkl}\kappa^e_{ij}\kappa^e_{kl} + \frac{1}{2}\overline{D}_{ijkl}\epsilon^e_{ij}\kappa^e_{kl} + \frac{1}{2}\overline{B}_{ijkl}\kappa^e_{ij}\epsilon^e_{kl}, \tag{77}$$

$$\overline{\mathbf{D}}(\mathbf{r}) = \int_V \mathcal{K}(\mathbf{R'})(\mathbf{D}^t \times \mathbf{R'})^t dV', \tag{78}$$

$$\overline{\mathbf{B}}(\mathbf{r}) = \int_V \mathcal{K}(\mathbf{R'})(\mathbf{B} \times \frac{\mathbf{R'}}{R'^2})^t dV'. \tag{79}$$

By differentiating Equation (77) with respect to the elastic strain and curvature tensors, we find the stress and couple-stress tensors:

$$T^{sym}_{ij} = \frac{\partial \psi}{\partial \epsilon^e_{ij}} = C_{ijkl}\epsilon^e_{kl} + \frac{1}{2}(\overline{D}_{ijkl} + \overline{B}_{klij})\kappa^e_{kl} \tag{80}$$

$$M^{dev}_{ij} = \frac{\partial \psi}{\partial \kappa^e_{ij}} = A_{ijkl}\kappa^e_{kl} + \frac{1}{2}(\overline{B}_{ijkl} + \overline{D}_{klij})\epsilon^e_{kl}. \tag{81}$$

Using finally $\overline{E}_{ijkl} = \frac{1}{2}(\overline{D}_{ijkl} + \overline{B}_{klij})$, we can write:

$$T^{sym}_{ij} = \frac{\partial \psi}{\partial \epsilon^e_{ij}} = C_{ijkl}\epsilon^e_{kl} + \overline{E}_{ijkl}\kappa^e_{kl} \tag{82}$$

$$M^{dev}_{ij} = \frac{\partial \psi}{\partial \kappa^e_{ij}} = A_{ijkl}\kappa^e_{kl} + \overline{E}_{klij}\epsilon^e_{kl}. \tag{83}$$

We end up with relationships similar to those proposed in [54], except that the present elastic tensors $(\overline{\mathbf{B}}, \overline{\mathbf{D}}, \overline{\mathbf{E}})$ involve nonlocal integrals. As proposed by Eringen [47,48], we assume $\mathbf{D} = \mathbf{C}$. We also impose $\overline{B}_{ijkl} = \overline{D}_{klij}$, which allows ensuring that the free energy functional is symmetric positive-definite and that the elastic solution is stable. Interestingly, stability of the elastic strain and curvature fields is expected at the convoluted level from the nonlocal character of the formulation, whereas instability would occur in the limit of a purely local formulation [55].

5. Transport of Dislocations and Disclinations

In this Section, we are interested in the kinematics of dislocations and disclinations in the context of the continuous setting set forth in Section 3 for their representation by smooth density fields. In this aim, we take advantage of the transport framework for the spatio-temporal evolution of the defects density fields that derives from the conservation of the Burgers and Frank vectors during their motion. Being unquestionable from a kinematic point of view, this framework, referred to as transport of dislocation and disclination densities, provides a natural basis for the dynamic description of plasticity through dislocation and disclination motion. Since disclinations are mainly distributed along grain boundaries in polycrystals, it provides in particular a basis for describing the contribution to plasticity of grain boundary motion.

Consider a material surface S delimited by a closed circuit C. Let $\mathbf{f}$ be the disclination flux used to measure the rate of inflow into S of disclination lines, carrying along with them their corresponding Frank vectors $\boldsymbol{\phi}$, through a line element $\mathbf{dx}$ of curve C. Let $\mathbf{V}_\theta$ be the velocity of the disclinations with respect to the material, and $\mathbf{l}$ the unit vector along the disclination lines. Then the disclination tensor is: $\boldsymbol{\theta} = \boldsymbol{\phi} \otimes \mathbf{l}$. In the absence of a disclination source term, the conservation of the Frank's vector content demands that the rate of change of the Frank's vector of all disclination lines threading S be equal to the total disclination flux across C:

$$\frac{d}{dt} \int_S \boldsymbol{\theta} \cdot \mathbf{n}\, dS = \int_C \mathbf{f} \cdot \mathbf{dx}. \tag{84}$$

For small transformations, the pointwise statement corresponding to Equation (84) is:

$$\dot{\boldsymbol{\theta}} = \mathbf{curl}\, \mathbf{f} \tag{85}$$

where $\dot{\boldsymbol{\theta}}$ represents the time derivative of the disclination density tensor. The rate of inflow of Frank vectors across the surface $\mathbf{l} \times \mathbf{dx}$ is:

$$\mathbf{f}.\mathbf{dx} = \boldsymbol{\phi}(\mathbf{V}_\theta.(\mathbf{l} \times \mathbf{dx})). \tag{86}$$

Indeed, suppose first that the disclination velocity $\mathbf{V}_\theta$ is in the plane $(\mathbf{l}, \mathbf{dx})$. Then, the disclination does not enter (or exit) surface S, and it is clear from Equation (86) that the flux $\mathbf{f}$ is zero as it should be. Conversely, $\mathbf{f}$ is always non-zero when $\mathbf{V}_\theta$ has a non-zero component normal to the plane $(\mathbf{l}, \mathbf{dx})$, i.e., when the disclination enters or exits surface S, and it changes sign according to whether the disclination enters or exits S. We can then obtain from Equation (86):

$$\mathbf{f} = -\boldsymbol{\phi} \otimes \mathbf{l} \times \mathbf{V}_\theta = -\boldsymbol{\theta} \times \mathbf{V}_\theta. \tag{87}$$

Consequently, the local statement of balance Equation (85) becomes:

$$\dot{\boldsymbol{\theta}} + \mathbf{curl}\, (\boldsymbol{\theta} \times \mathbf{V}_\theta) = 0. \tag{88}$$

Equation (88) is a transport law for the disclination density tensor $\boldsymbol{\theta}$. It can be understood as an evolution equation for $\boldsymbol{\theta}$, provided the disclination velocity $\mathbf{V}_\theta$ is known as a function of the stress/couple-stress state from constitutive statements. Its meaning is that through the curl term, the incompatible part of the disclination flux incrementally feeds the disclination density. Comparing Equations (46) and (88) it follows, after time differentiation of Equation (46) that the cross product $\boldsymbol{\theta} \times \mathbf{V}_\theta$ can be identified with the plastic curvature rate tensor $\dot{\boldsymbol{\kappa}}_p$, up to a gradient:

$$\dot{\boldsymbol{\kappa}}_p = \boldsymbol{\theta} \times \mathbf{V}_\theta + \mathbf{grad}\, \dot{\omega}_p^*. \tag{89}$$

At mesoscale, the term $\dot{\boldsymbol{\kappa}}_p^* = \mathbf{grad}\, \dot{\omega}_p^*$ can be given the significance of a statistical plastic curvature rate, meaning that a non-vanishing plastic curvature rate may be obtained even when the disclination density $\boldsymbol{\theta}$ vanishes at this scale [56]. Indeed, using space-time running averages of the disclination density tensor, disclination velocity $\mathbf{V}_\theta$ and plastic curvature rate tensor $\dot{\kappa}_p$ over a domain of mesoscopic size, allows writing the mesoscopic plastic curvature rate $\overline{\dot{\kappa}_p}$ as:

$$\overline{\dot{\kappa}_p} = \overline{\boldsymbol{\theta} \times \mathbf{V}_\theta} \;=\; \overline{\boldsymbol{\theta}} \times \overline{\mathbf{V}_\theta} + \dot{\kappa}_p^*, \tag{90}$$

where overbars indicate averaged variables. It is seen that $\overline{\dot{\kappa}_p}$ may be non-zero when the net disclination density vanishes at mesoscale ($\overline{\boldsymbol{\theta}} = 0$), in which case it becomes:

$$\overline{\dot{\kappa}_p} = \dot{\kappa}_p^* \,. \tag{91}$$

Dropping the overbars for convenience, the plastic curvature rate may be generally written as:

$$\dot{\kappa}_p = \boldsymbol{\theta} \times \mathbf{V}_\theta + \dot{\kappa}_p^*. \tag{92}$$

As we shall discuss below, the compatible part $\dot{\kappa}_p^{\parallel}$ of $\dot{\kappa}_p$ increments the history-dependent compatible plastic curvature produced by the motion of dislocations and disclinations. Equation (92) accounts for plasticity processes in relation to the mobility of grain boundaries that cannot be described by a pure dislocation-based model.

In the theory of dislocations, the dislocation density tensor $\boldsymbol{\alpha}$ satisfies a transport equation similar to Equation (88), in the absence of a source term [22,57]:

$$\dot{\boldsymbol{\alpha}} + \mathbf{curl}\,(\boldsymbol{\alpha} \times \mathbf{V}_\alpha) = 0. \tag{93}$$

Here, $\mathbf{V}_\alpha$ denotes the dislocation velocity, and the cross product $\boldsymbol{\alpha} \times \mathbf{V}_\alpha$ can be identified with the plastic distortion rate $\dot{\mathbf{U}}_p$, up to a gradient. However, $\dot{\mathbf{U}}_p$ is undefined in a theory of disclinations, because its skew-symmetric part $\dot{\omega}_p$ has discontinuities. Hence, the dislocation transport Equation (93) does not account for curvature incompatibility and has to be modified in a disclination theory. For small transformations, a version correctly accounting for curvature incompatibility is readily obtained from the time derivative of Equation (33):

$$\dot{\boldsymbol{\alpha}} = -\mathbf{curl}\,\dot{\epsilon}_p + \dot{\kappa}_p^t - tr(\dot{\kappa}_p)\mathbf{I} = -\mathbf{curl}\,\dot{\epsilon}_p + \mathbf{s}_\theta. \tag{94}$$

Here the plastic strain rate $\dot{\epsilon}_p$ is

$$\dot{\epsilon}_p = (\boldsymbol{\alpha} \times \mathbf{V}_\alpha)^{sym} + \dot{\epsilon}_p^*, \tag{95}$$

where $\dot{\epsilon}_p^*$ is a gradient tensor acquiring significance at mesoscale, as discussed above for the mesoscopic plastic curvature rate tensor $\dot{\kappa}_p^*$. The compatible part of the plastic strain rate, $\dot{\epsilon}_p^{\parallel}$, is not involved in Equation (94) but it nevertheless increments the compatible plastic strain produced by the motion of dislocations through the lattice. Note importantly in Equation (94) the presence of a dislocation source/sink term, $\mathbf{s}_\theta \;=\; \dot{\kappa}_p^t - tr(\dot{\kappa}_p)\mathbf{I}$, due to the mobility of disclinations. Such a term offers supplementary relaxation mechanisms [4,58], and will be shown to be useful later on in the prediction of shear-coupled boundary migration. When the dislocation/disclination velocities and mesoscopic contributions to plasticity ($\dot{\kappa}_p^*, \dot{\epsilon}_p^*$) are known from the stress/couple-stress state through constitutive relations, Equations (88), (92), (94) and (95) can be used as evolution equations for the disclination and dislocation densities. Please note that Equations (88) and (94) are first order hyperbolic equations, which implies that their solutions have propagative character, as experimentally verified in [31]. Furthermore, hyperbolicity of these equations as a strong impact on the algorithms devoted to their numerical solution from finite elements or spectral approximations [59–61], as detailed below in Section 7.3.

6. Constitutive Relations for Dislocation and Disclination Mobility

In this Section, we account for thermodynamic requirements on positiveness of the mechanical dissipation to formulate constitutive relations for the mobility of dislocations and disclinations by following a procedure proposed in [62]. Substituting Equations (65) and (66) in Equation (63), and using the additivity of elastic and plastic strain rates and curvature rates tensors, we find the dissipation to be:

$$D = \int_D (\mathbf{T}^{sym} : \dot{\boldsymbol{\epsilon}}_p + \mathbf{M}^{dev} : \dot{\boldsymbol{\kappa}}_p) dv \geq 0. \tag{96}$$

To identify driving forces associated with dislocation and disclination velocities, we write down the plastic strain and curvature rate tensors $(\dot{\boldsymbol{\epsilon}}_p, \dot{\boldsymbol{\kappa}}_p)$ in terms of dislocation and disclination fluxes as in Equations (92) and (95) and substitute in Equation (96):

$$D = \int_D (\mathbf{T}^{sym} : (\boldsymbol{\alpha} \times \mathbf{V}_\alpha)^{sym} + \mathbf{T}^{sym} : \dot{\boldsymbol{\epsilon}}_p^* + \mathbf{M}^{dev} : (\boldsymbol{\theta} \times \mathbf{V}_\theta) + \mathbf{M}^{dev} : \dot{\boldsymbol{\kappa}}_p^*) dv, \tag{97}$$

which reads in component form:

$$D = \int_D (\frac{1}{2} T_{ij}^{sym} (e_{jkl}\alpha_{ik}V_l^\alpha + e_{ikl}\alpha_{jk}V_l^\alpha) + T_{ij}^{sym}\dot{\epsilon}_{ij}^{p,*} + M_{ij}^{dev} e_{jkl}\theta_{ik}V_l^\theta + M_{ij}^{dev}\dot{\kappa}_{ij}^{p,*}) dv, \tag{98}$$

or also:

$$D \quad = \quad \int_D (e_{jkl}T_{ij}^{sym}\alpha_{ik}V_l^\alpha + T_{ij}^{sym}\dot{\epsilon}_{ij}^{p,*} + e_{jkl}M_{ij}^{dev}\theta_{ik}V_l^\theta + M_{ij}^{dev}\dot{\kappa}_{ij}^{p,*}) dv. \tag{99}$$

Equivalently, Equation (99) is in intrinsic form:

$$D \quad = \quad \int_D (\mathbf{F}_\alpha \cdot \mathbf{V}_\alpha + \mathbf{T}^{sym} : \dot{\boldsymbol{\epsilon}}_p^* + \mathbf{F}_\theta \cdot \mathbf{V}_\theta + \mathbf{M}^{dev} : \dot{\boldsymbol{\kappa}}_p^*) dv, \tag{100}$$

where we wrote:

$$\mathbf{F}_\alpha \quad = \quad \mathbf{T}^{sym} \cdot \boldsymbol{\alpha} : \mathbf{X}; \quad F_l^\alpha = e_{jkl} T_{ij}^{sym}\alpha_{ik} \tag{101}$$

$$\mathbf{F}_\theta \quad = \quad \mathbf{M}^{dev,t} \cdot \boldsymbol{\theta} : \mathbf{X}; \quad F_l^\theta = e_{jkl} M_{ij}^{dev}\theta_{ik}. \tag{102}$$

$\mathbf{F}_\alpha$ and $\mathbf{F}_\theta$ are the driving forces for dislocation and disclination motion. Using the dyadic notation $\boldsymbol{\alpha} = \mathbf{b} \otimes \mathbf{t}$ for the dislocation density ($\mathbf{b}$ is the Burgers vector, $\mathbf{t}$ the line vector), $\mathbf{F}_\alpha$ can be expressed as: $\mathbf{F}_\alpha = \mathbf{T}^{sym} \cdot \mathbf{b} \times \mathbf{t}$, a form similar to the Peach-Köhler force for discrete dislocations. Similarly for the disclination density tensor, $\boldsymbol{\theta} = \boldsymbol{\phi} \otimes \mathbf{l}$ ($\boldsymbol{\phi}$ is Frank's vector, $\mathbf{l}$ the line vector), and $\mathbf{F}_\theta = \mathbf{M}^{dev,t} \cdot \boldsymbol{\phi} \times \mathbf{l}$. This is a Peach-Köhler-type relationship for disclinations. Thus, $\mathbf{F}_\theta$ drives grain boundary motion in the present disclination theory. It is produced by the couple stresses, the stresses being not directly involved, and is normal to the disclination line. Positiveness of the mechanical dissipation rate Equation (99) can be satisfied by choosing the simple constitutive relations:

$$\mathbf{F}_\alpha \quad = \quad B_\alpha \mathbf{V}_\alpha; \quad B_\alpha \geq 0 \tag{103}$$

$$\mathbf{F}_\theta \quad = \quad B_\theta \mathbf{V}_\theta; \quad B_\theta \geq 0 \tag{104}$$

$$\dot{\boldsymbol{\epsilon}}_p^* \quad = \quad \lambda_\epsilon \mathbf{T}^{sym}; \quad \lambda_\epsilon \geq 0 \tag{105}$$

$$\dot{\boldsymbol{\kappa}}_p^* \quad = \quad \lambda_\kappa \mathbf{M}^{dev}; \quad \lambda_\kappa \geq 0, \tag{106}$$

where $(B_\alpha, B_\theta, \lambda_\epsilon, \lambda_\kappa)$ are positive material parameters. Equations (103)–(106) have the form of Newtonian viscous drag relationships, but $(B_\alpha, B_\theta, \lambda_\epsilon, \lambda_\kappa)$ are generally not simple constants. They may involve non-linearity, or/and hardening laws reflecting for example thermally assisted obstacle overcoming through adequate constitutive statements. Please note that both the glide and climb components of the Peach-Köhler force are included in Equation (103). λ_ϵ describes plasticity through both dislocation glide and climb at mesoscale. Similarly, λ_κ describes plasticity induced by grain boundary motion at mesoscale and driven by $\mathbf{M}^{dev}$ in Equation (106). Possible expressions for λ_κ were proposed in [56] by averaging Equation (104).

7. Solution Algorithms

In the present Section, we propose two algorithms for the numerical solution of the model. The first version solves for the incompatible plastic strains and curvatures at each time step. The second version is a reduced, rate form of the latter, numerically faster, but where incompatible fields are not determined at each step.

7.1. Complete Algorithm

The complete model involves the following equations, gathered from the above sections:

$$\mathbf{curl}\,\kappa_p^{\perp} = -\boldsymbol{\theta} \tag{107}$$

$$\mathbf{curl}\,\epsilon_p^{\perp} = -\boldsymbol{\alpha} + \kappa_p^{t} - tr(\kappa_p)\mathbf{I} \tag{108}$$

$$\mathbf{T} = \mathbf{C}:(\epsilon - \epsilon_p) + \mathbf{D}:(\kappa - \kappa_p) \tag{109}$$

$$\mathbf{M}^{dev} = \mathbf{A}:(\kappa - \kappa_p) + \mathbf{B}:(\epsilon - \epsilon_p) \tag{110}$$

$$0 = \mathbf{div}\,\mathbf{T}^{sym} + \frac{1}{2}\mathbf{curl}\,\mathbf{div}\,\mathbf{M}^{dev} \tag{111}$$

$$\dot{\kappa}_p = \boldsymbol{\theta} \times \mathbf{V}_\theta + \dot{\kappa}_p^{*} \tag{112}$$

$$\dot{\epsilon}_p = (\boldsymbol{\alpha} \times \mathbf{V}_\alpha)^{sym} + \dot{\epsilon}_p^{*} \tag{113}$$

$$\dot{\boldsymbol{\theta}} = -\mathbf{curl}\,\dot{\kappa}_p \tag{114}$$

$$\dot{\boldsymbol{\alpha}} = -\mathbf{curl}\,\dot{\epsilon}_p + \dot{\kappa}_p^{t} - tr(\dot{\kappa}_p)\mathbf{I} \tag{115}$$

$$\mathbf{V}_\alpha = \frac{1}{B_\alpha}\mathbf{T}^{sym} \cdot \boldsymbol{\alpha} : \mathbf{X} \tag{116}$$

$$\mathbf{V}_\theta = \frac{1}{B_\theta}\mathbf{M}^{dev,t} \cdot \boldsymbol{\theta} : \mathbf{X} \tag{117}$$

$$\dot{\epsilon}_p^{*} = \lambda_\epsilon \mathbf{T}^{sym} \tag{118}$$

$$\dot{\kappa}_p^{*} = \lambda_\kappa \mathbf{M}^{dev}, \tag{119}$$

with the boundary conditions:

$$\mathbf{T} \cdot \mathbf{n} = \mathbf{t}_d \text{ on } S_t \tag{120}$$

$$\mathbf{u} = \mathbf{u}_d \text{ on } S_u \tag{121}$$

$$\omega = \omega_d \text{ on } S_u. \tag{122}$$

The unknowns are the displacement field $\mathbf{u}$ and the dislocation/disclination density fields $(\boldsymbol{\alpha}, \boldsymbol{\theta})$. They are known at the initial time. The incompatible component $\kappa_p^{\perp}$ of κ_p is solution of Equation (107), with the side conditions mentioned in Equation (40). $\epsilon_p^{\perp}$ can be determined from Equation (108), with the side conditions $\mathbf{div}\,\epsilon_p^{\perp} = 0, \epsilon_p^{\perp} \cdot \mathbf{n} = 0$ on S. Being history-dependent, $\kappa_p^{\parallel}$ and $\epsilon_p^{\parallel}$ are initially set to zero without loss of generality. Furthermore, using the constitutive laws Equations (109) and (110), the balance of momentum and moment of momentum problem Equation (111) can be solved for the compatible displacement, as shown in Section 4.2. Then, the plastic strain rate and curvature rate $(\dot{\epsilon}_p, \dot{\kappa}_p)$ are found from Equations (112) and (113), where Equations (116)–(119) have been used to evaluate the dislocation/disclination velocities and the mesoscopic terms. Only the compatible parts $(\dot{\epsilon}_p^{\parallel}, \dot{\kappa}_p^{\parallel})$ are employed for updating the plastic strain and curvature tensors. Finally, the dislocation/disclination densities are updated using Equations (114) and (115) and the procedure is iterated at the next time step.

7.2. Reduced Algorithm

Using the rate form of the model equations leads to a simpler and faster incremental algorithm. The main reason and difference is that the decomposition of plastic strains and curvatures into incompatible and compatible parts is possibly accounted for at the initial time, but not later. By taking the time derivative of Equations (109)–(111) and ignoring Equations (107) and (108), the algorithm uses the following equations:

$$0 = \mathbf{div}\,\dot{\mathbf{T}}^{sym} + \frac{1}{2}\mathbf{curl}\,\mathbf{div}\,\dot{\mathbf{M}}^{dev} \tag{123}$$

$$\dot{\mathbf{T}} = \mathbf{C} : (\dot{\boldsymbol{\epsilon}} - \dot{\boldsymbol{\epsilon}}_p) + \mathbf{D} : (\dot{\boldsymbol{\kappa}} - \dot{\boldsymbol{\kappa}}_p) \tag{124}$$

$$\dot{\mathbf{M}}^{dev} = \mathbf{A} : (\dot{\boldsymbol{\kappa}} - \dot{\boldsymbol{\kappa}}_p) + \mathbf{B} : (\dot{\boldsymbol{\epsilon}} - \dot{\boldsymbol{\epsilon}}_p) \tag{125}$$

$$\dot{\boldsymbol{\kappa}}_p = \boldsymbol{\theta} \times \mathbf{V}_\theta + \dot{\boldsymbol{\kappa}}_p^* \tag{126}$$

$$\dot{\boldsymbol{\epsilon}}_p = (\boldsymbol{\alpha} \times \mathbf{V}_\alpha)^{sym} + \dot{\boldsymbol{\epsilon}}_p^* \tag{127}$$

$$\dot{\boldsymbol{\theta}} = -\mathbf{curl}\,\dot{\boldsymbol{\kappa}}_p \tag{128}$$

$$\dot{\boldsymbol{\alpha}} = -\mathbf{curl}\,\dot{\boldsymbol{\epsilon}}_p + \dot{\boldsymbol{\kappa}}_p^t - tr(\dot{\boldsymbol{\kappa}}_p)\mathbf{I} \tag{129}$$

$$\mathbf{V}_\alpha = \frac{1}{B_\alpha}\mathbf{T}^{sym} \cdot \boldsymbol{\alpha} : \mathbf{X} \tag{130}$$

$$\mathbf{V}_\theta = \frac{1}{B_\theta}\mathbf{M}^{dev,t} \cdot \boldsymbol{\theta} : \mathbf{X} \tag{131}$$

$$\dot{\boldsymbol{\epsilon}}_p^* = \lambda_\epsilon \mathbf{T}^{sym} \tag{132}$$

$$\dot{\boldsymbol{\kappa}}_p^* = \lambda_\kappa \mathbf{M}^{dev}, \tag{133}$$

with the boundary conditions:

$$\dot{\mathbf{T}} \cdot \mathbf{n} = \dot{\mathbf{t}}_d \text{ on } S_t \tag{134}$$

$$\dot{\mathbf{u}} = \dot{\mathbf{u}}_d \text{ on } S_u \tag{135}$$

$$\dot{\boldsymbol{\omega}} = \dot{\boldsymbol{\omega}}_d \text{ on } S_u. \tag{136}$$

All fields $(\mathbf{u}, \boldsymbol{\alpha}, \boldsymbol{\theta})$ are given at the initial time, as well as the plastic strain and curvature $(\boldsymbol{\epsilon}_p, \boldsymbol{\kappa}_p)$ arbitrarily set to zero without inconvenience. All fields can then be updated using the above rate equations. Clearly, the solutions obtained from these two algorithms may differ substantially because their continuity and differentiability properties are different.

7.3. Numerical Implementation

The finite element implementation couples a conventional Galerkin scheme for the solution of the equilibrium problem with a mixed Galerkin-Least Squares (GLS) scheme for the numerical approximation of the dislocation/disclination transport equations. Applying a standard Galerkin method to the convective transport equations leads to high frequency, propagating, numerical instabilities arising at sharp gradients. The Least-Squares (LS) finite element method, based on minimizing the norm of the residual, has the potential for eliminating such drawbacks, because it damps high frequency oscillations. Explicit LS methods have been considered so far [60,61]. However, damping is also a drawback of LS methods, because it leads to undue solution relaxation in the long-term. Therefore, an explicit Galerkin-Least-Squares (GLS) method involving convenient linear combinations of the Galerkin and LS algorithms has been proposed [60,61].

A constant drawback of finite element methods is that three-dimensional simulations are exceedingly demanding on computational resources. Spectral methods based on Fast Fourier

Transform (FFT) algorithms were shown to be efficient at solving three-dimensional problems in periodic media and therefore appear as attractive alternatives to finite elements. However, the Gibbs oscillations they inherently produce in regions of strong spatial gradients reinforce the numerical instability of the solutions to the transport equations with respect to high frequency perturbations. Spectral low-pass filters in the Fourier space were shown to be efficient at damping these perturbations, and consequently at providing accurate and stable solutions to the transport problem [59]. FFT algorithms for the solution of the equilibrium problems are also available [55,63].

8. A Plane Edge-Wedge Model

As an illustration of the model and a basis for the next two Sections, we consider a bi-dimensional model with a distribution of wedge disclinations first discussed in [18]. In the orthonormal frame $(\mathbf{e_1}, \mathbf{e_2}, \mathbf{e_3})$, we take the disclination tensor to be: $\boldsymbol{\theta} = \theta_{33}\mathbf{e_3} \otimes \mathbf{e_3}$. The continuity Equation (48) then implies that the disclination density θ_{33} only depends on the coordinates (x_1, x_2). Thus, the incompatibility Equation (46) reduces to:

$$\theta_{33} = \kappa^{p,\perp}_{31,2} - \kappa^{p,\perp}_{32,1}, \tag{137}$$

and the Poisson Equation (47) is:

$$\kappa^{p,\perp}_{31,11} + \kappa^{p,\perp}_{31,22} = +\theta_{33,2} \tag{138}$$

$$\kappa^{p,\perp}_{32,11} + \kappa^{p,\perp}_{32,22} = -\theta_{33,1}. \tag{139}$$

The involved plastic curvatures are only $(\kappa^p_{31}, \kappa^p_{32})$ and the disclination transport Equation (88) reads:

$$\dot{\theta}_{33} = \dot{\kappa}^p_{31,2} - \dot{\kappa}^p_{32,1}. \tag{140}$$

We ignore the mesoscopic plastic curvature rate $\dot{\kappa}^*_p$ in Equation (92) as we want to deal with nanoscale problems. Therefore, the plastic curvature rates are:

$$\dot{\kappa}^p_{31} = -\theta_{33}V^\theta_2 \tag{141}$$

$$\dot{\kappa}^p_{32} = +\theta_{33}V^\theta_1. \tag{142}$$

The constitutive relation Equation (104) for the disclination velocities then yields:

$$V^\theta_1 = +\frac{1}{B_\theta}M^{dev}_{32}\theta_{33} \tag{143}$$

$$V^\theta_2 = -\frac{1}{B_\theta}M^{dev}_{31}\theta_{33}, \tag{144}$$

and the plastic curvature rates read:

$$\dot{\kappa}^p_{31} = \frac{1}{B_\theta}M^{dev}_{31}\theta^2_{33} \tag{145}$$

$$\dot{\kappa}^p_{32} = \frac{1}{B_\theta}M^{dev}_{32}\theta^2_{33}, \tag{146}$$

while the transport equation becomes:

$$\dot{\theta}_{33} = \frac{1}{B_\theta}(M^{dev}_{31}\theta^2_{33})_{,2} - \frac{1}{B_\theta}(M^{dev}_{32}\theta^2_{33})_{,1}. \tag{147}$$

In the dislocation transport Equation (94), the source/sink term s_θ generates only edge dislocations $(\alpha_{13}, \alpha_{23})$. Ignoring also the mesoscopic plastic strain rate $\dot{\epsilon}_p^*$ in Equation (95), the motion of these edge dislocations produces the plastic strain rates:

$$\dot{\epsilon}_{11}^p = -\alpha_{13} V_2^\alpha \tag{148}$$

$$\dot{\epsilon}_{12}^p = \dot{\epsilon}_{21}^p = \frac{1}{2}(\alpha_{13} V_1^\alpha - \alpha_{23} V_2^\alpha) \tag{149}$$

$$\dot{\epsilon}_{22}^p = +\alpha_{23} V_1^\alpha, \tag{150}$$

and the plastic rotation rate:

$$\dot{\omega}_3^p = -\frac{1}{2}(\alpha_{13} V_1^\alpha + \alpha_{23} V_2^\alpha). \tag{151}$$

The above relations Equations (148) and (150) show that climb of edge dislocations $(\alpha_{13}, \alpha_{23})$ produces the extension rates $(\dot{\epsilon}_{11}^p, \dot{\epsilon}_{22}^p)$, while their glide produces the shear $\dot{\epsilon}_{12}^p$ in Equation (149). The dislocation transport Equation (94) therefore yields:

$$\dot{\alpha}_{13} = \dot{\epsilon}_{11,2}^p - \dot{\epsilon}_{12,1}^p + \dot{\kappa}_{31}^p \tag{152}$$

$$\dot{\alpha}_{23} = \dot{\epsilon}_{21,2}^p - \dot{\epsilon}_{22,1}^p + \dot{\kappa}_{32}^p. \tag{153}$$

Using Equations (141), (142), (148), (149) and (150), these edge dislocation generation rates read:

$$\dot{\alpha}_{13} = -(\alpha_{13} V_2^\alpha)_{,2} - \frac{1}{2}(\alpha_{13} V_1^\alpha - \alpha_{23} V_2^\alpha)_{,1} - \theta_{33} V_2^\theta \tag{154}$$

$$\dot{\alpha}_{23} = -(\alpha_{23} V_1^\alpha)_{,1} + \frac{1}{2}(\alpha_{13} V_1^\alpha - \alpha_{23} V_2^\alpha)_{,2} + \theta_{33} V_1^\theta. \tag{155}$$

If all other dislocation densities are initially absent, the model therefore involves only α_{13} and α_{23} densities. The continuity condition Equation (49) then implies a planar distribution of these dislocations. The relation Equation (103) provides the dislocation velocities for climb:

$$V_1^\alpha = +\frac{1}{B_\alpha} T_{22}^{sym} \alpha_{23} \tag{156}$$

$$V_2^\alpha = -\frac{1}{B_\alpha} T_{11}^{sym} \alpha_{13}, \tag{157}$$

and glide:

$$V_1^\alpha = +\frac{1}{B_\alpha} T_{12}^{sym} \alpha_{13} \tag{158}$$

$$V_2^\alpha = -\frac{1}{B_\alpha} T_{12}^{sym} \alpha_{23}. \tag{159}$$

The corresponding plastic strain rates are:

$$\dot{\epsilon}_{11}^p = \frac{1}{B_\alpha} T_{11}^{sym} \alpha_{13}^2 \tag{160}$$

$$\dot{\epsilon}_{12}^p = \dot{\epsilon}_{21}^p = \frac{1}{B_\alpha} T_{12}^{sym} (\alpha_{13}^2 + \alpha_{23}^2) \tag{161}$$

$$\dot{\epsilon}_{22}^p = \frac{1}{B_\alpha} T_{22}^{sym} \alpha_{23}^2, \tag{162}$$

while the transport equation can be reformulated as:

$$\dot{\alpha}_{13} \quad = \quad +\frac{1}{B_\alpha}(T_{11}^{sym}\alpha_{13}^2)_{,2} - \frac{1}{2B_\alpha}(T_{12}^{sym}(\alpha_{13}^2 + \alpha_{23}^2))_{,1} + \frac{1}{B_\theta}M_{31}^{dev}\theta_{33}^2 \tag{163}$$

$$\dot{\alpha}_{23} \quad = \quad -\frac{1}{B_\alpha}(T_{22}^{sym}\alpha_{23}^2)_{,1} + \frac{1}{2B_\alpha}(T_{12}^{sym}(\alpha_{13}^2 + \alpha_{23}^2))_{,2} + \frac{1}{B_\theta}M_{32}^{dev}\theta_{33}^2. \tag{164}$$

The relevant stress and deviatoric couple-stress components in the model are $(T_{11}, T_{12}, T_{21}, T_{22})$ and $(M_{31}^{dev}, M_{32}^{dev})$. The balance of momentum Equation (56) read here:

$$T_{11,1}^{sym} + T_{12,2}^{sym} + \frac{1}{2}(M_{31,12}^{dev} + M_{32,22}^{dev}) \quad = \quad 0 \tag{165}$$

$$T_{21,1}^{sym} + T_{22,2}^{sym} - \frac{1}{2}(M_{31,11}^{dev} + M_{32,21}^{dev}) \quad = \quad 0. \tag{166}$$

Thus, the loading applied at the boundaries of the domain may include shear/tension/ compression in the directions $(\mathbf{e}_1, \mathbf{e}_2)$. Couple stresses and rotations along $\mathbf{e}_3$ can also be prescribed. The elastic constitutive laws Equations (82) and (83) yield:

$$T_{ij}^{sym} \quad = \quad \frac{\partial \psi}{\partial \epsilon_{ij}^e} = C_{ijkl}\epsilon_{kl}^e + D_{ijkl}\kappa_{kl}^e \tag{167}$$

$$M_{ij}^{dev} \quad = \quad \frac{\partial \psi}{\partial \kappa_{ij}^e} = A_{ijkl}\kappa_{kl}^e + B_{ijkl}\epsilon_{kl}^e. \tag{168}$$

Substituting $(\epsilon_{ij} - \epsilon_{ij}^p, \omega_{i,j} - \kappa_{ij}^p)$ for $(\epsilon_{ij}^e, \kappa_{ij}^e)$ in the constitutive Equations (167) and (168), and introducing the resultant stresses and couple stresses in the balance Equations (165) and (166), one obtains two fourth-order partial differential equations for the material displacements (u_1, u_2). As already mentioned, the boundary conditions comprise the prescription of tractions and moments, or/and displacements and rotations on the surface of the body. Please note that no boundary condition is specified on the disclination and dislocation densities.

9. Structure and Elastic Energy of Symmetric Tilt Boundaries

Copper bi-crystals with $\sum 37(610)$ and $\sum 5(310)$ symmetrical tilt boundaries of misorientation $18.9°$ and $36.9°$ were simulated in [64]. The bi-crystals are free of loads and periodic boundary conditions are imposed on the vertical boundaries in Figure 2. The characteristic size of the finite element mesh is 0.5Å. The wedge disclination density fields are shown in Figure 2, where they are superimposed on the relaxed atomic structure as predicted by atomistic calculations [14]. They consist of periodic arrays of wedge-disclination dipoles located at the edges of the structural units. Dilatations/contractions regions are associated with negative/positive disclinations. The local occurrence of large elastic strains suggests that a finite strain setting could be in order. Nevertheless, the predicted elastic fields are in good agreement with atomistic simulations and experiment. The elastic energy is essentially located in the dilatation/contraction areas in the core region of defects. In agreement with atomistic simulations, the energy is contained in the first atomic rows forming the grain boundary, essentially at structural units. Comparing the elastic fields of the $\sum 37(610)$ boundary and of the $\sum 5(310)$ boundary in Figure 2 shows that energy is more localized near the interface when the structural units are close to each other, due to elastic screening effects. Furthermore, the continuous description predicts features such as the shear deformation and curvature across/along the grain boundaries. As shown in Figure 3, there is a good agreement between quantitative estimates of the excess energy of grain boundaries using our model, and measurements from atomistic simulations and experiment [65–67]. Satisfactory agreement is obtained for all misorientations. Notably, the energy cusps predicted for the $\sum 5(310)$ and $\sum 5(210)$ grain boundaries originate in the self-screening of the elastic fields when the separation distance between structural units becomes small.

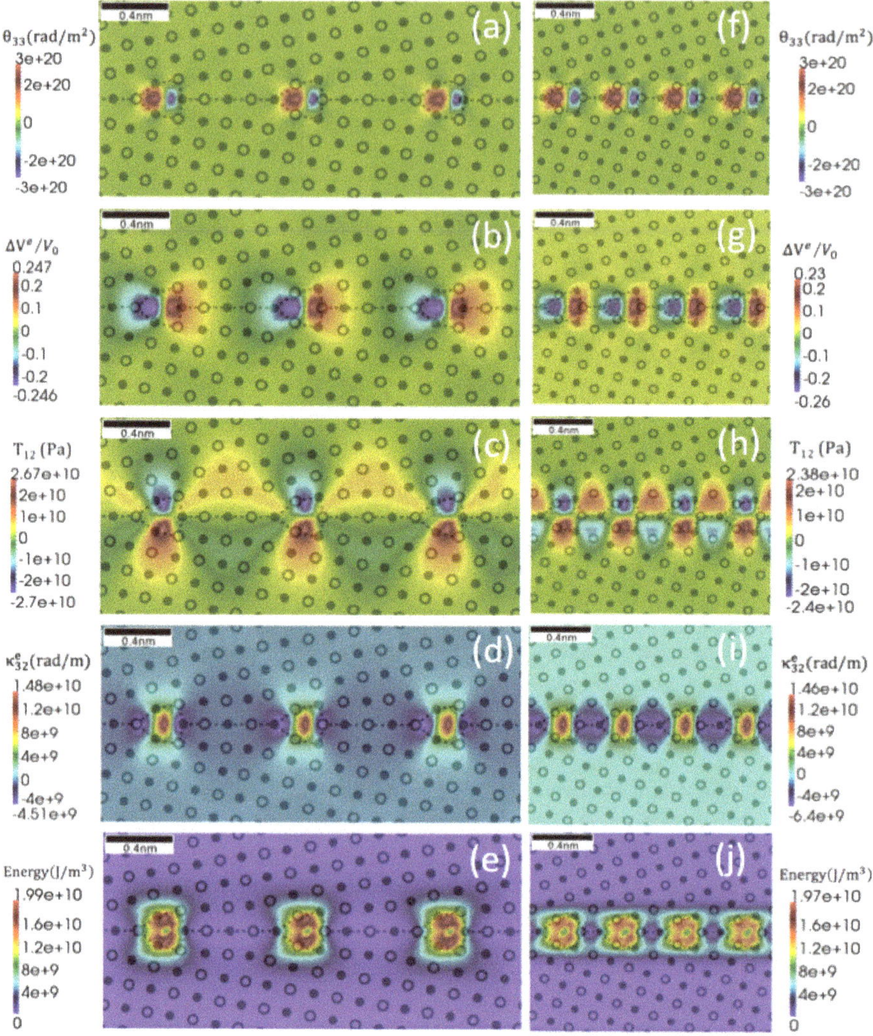

Figure 2. Disclination model for a $\sum 37(610)$ symmetrical $<001>$ tilt boundary of misorientation 18.9° (**left column**) and of a $\sum 5(310)$ symmetrical $<001>$ tilt boundary of misorientation 36.9° (**right column**). The wedge-disclination density θ_{33} (rad/m^2) is shown on top of the relaxed atomic structure (**a,f**); Elastic dilatations/contractions (**b,g**); Shear stress field T_{12} (Pa) (**c,h**); Elastic curvature κ_{32} (rad/m) (**d,i**); Elastic energy density (J/m^3) (**e,j**). Atomic structures are taken from reference [14].

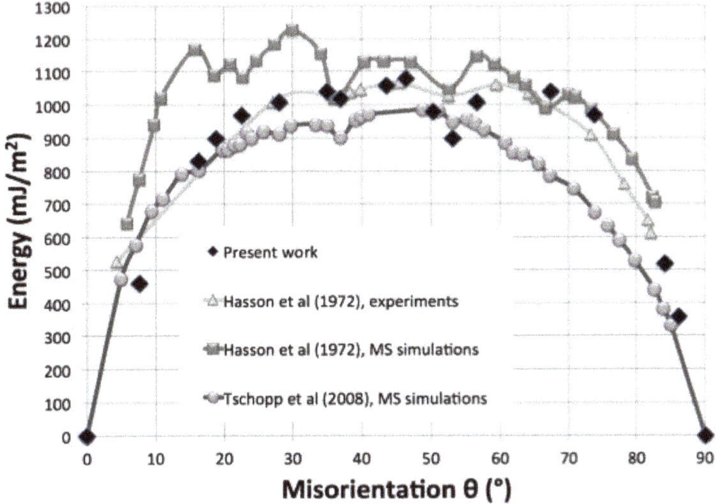

Figure 3. Excess energy of <001> tilt grain boundaries in copper bi-crystals. Black diamonds: present work; Gray triangles/squares: experimental measurements/molecular statics predictions [65]; Gray circles: molecular statics predictions [66].

10. Shear-Coupled Migration of Symmetric Tilt Boundaries

As already suggested above, shear-induced normal grain boundary migration is a fascinating example of grain boundary-mediated plasticity. It was experimentally observed at moderate temperature in Al [68,69]. Following [17], we illustrate this mechanism by submitting the copper bicrystal featuring the $\Sigma 5(310)$ symmetric <001> tilt boundary studied above to shear parallel to the boundary. In Figure 4, the wedge-disclination dipoles array is observed to move downward. This motion generates edge dislocation dipoles α_{13} (line direction normal to the figure, Burgers vector along the boundary). Due to the shear stress T_{12} these edge dipoles glide along the boundary, generate the plastic shear strain field ϵ_{12}^p seen in Figure 4c and annihilate at mid-distance, as Figure 4a,b show. Meanwhile, the downward migration of the disclination dipoles produces a plastic tilt rotation ω_3^p shown in Figure 4d. Striking similitude is observed between the heterogeneity of the plastic shear/rotation fields presented in Figure 4 and those obtained by post-processing atomistic simulations [16,70,71]. Because the boundary moves downward while positive shear is produced, the shear-coupling factor β, which corresponds to the ratio between the parallel translation of the two crystals and the normal migration of the boundary, is negative. The value $\beta \approx -0.94$ measured in Figure 4b is in reasonable agreement with atomistic simulations and experiments [14,68]. The predicted variations of the shear-coupling factor in the full range of misorientations are presented in Figure 3e. The model retrieves the two <001> and <110> branches found in experiments [68] and obtained by molecular dynamics simulations and surface-dislocation-based models [13,14].

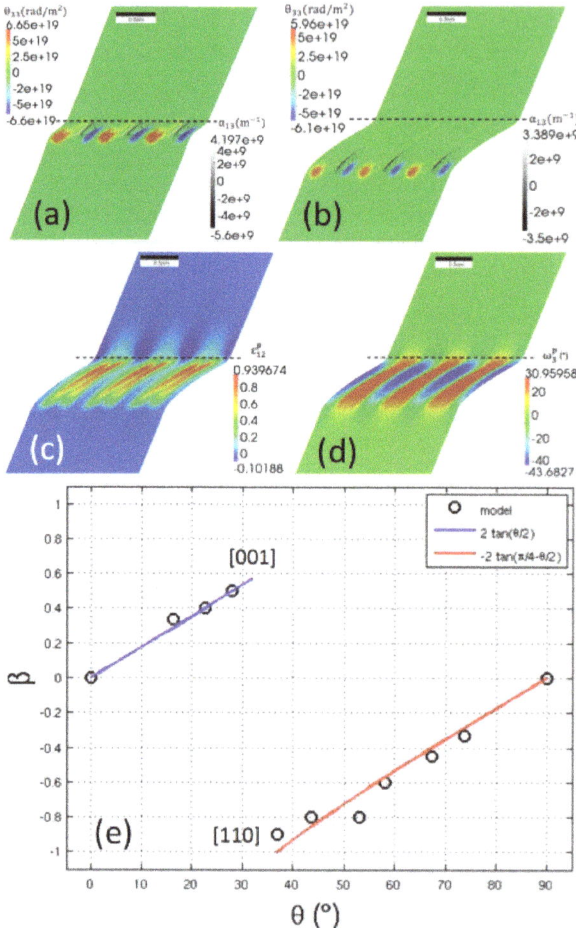

Figure 4. Disclination model for shear-coupled boundary migration. (**a,b**): Wedge-disclination density θ_{33} (rad/m^2) (color coded) and edge dislocation density α_{13} (m^{-1}) (black/white contours), during migration of a $\Sigma5(310)$ symmetric <001> tilt grain boundary of misorientation 36.87°. Two snapshots are shown at 1 ms (**a**) and 10 ms (**b**); (**c,d**): Plastic shear strain ϵ_{12}^{p} (**c**) and plastic rotation $\omega_{3}^{p}(°)$ (**d**) generated during the migration. In (**a–d**), the black dashed-line shows the initial position of the grain boundary. (**e**): Shear-coupling factor as a function of grain boundary misorientation. The circles show values obtained with the disclination model while the solid lines show the <001> and <110> shear deformation modes proposed in [14].

To further analyze the role of disclinations in plasticity, the model was used for simulating shear-coupled migration in forsterite. More specifically, we have used a periodic wedge-disclination array in the particular case of a (011)/<100> tilt boundary with misorientation 60°. This boundary was previously modeled at the atomic scale [72]. Figure 5 presents the disclination density field used to model the grain boundary structure [73]. Wedge-disclination densities are placed on the edges of structural units identified in [72]. The disclination arrangements result in a self-screening quadrupolar configuration, such that most of the energy is contained within the quadrupoles (see Figure 5b). The predicted excess energy of 1.3 J/m^2 is in good quantitative agreement with the value 1.28 J/m^2 measured from atomistic calculations [72]. When a shear stress is applied, the disclination array moves normal to the boundary, as shown in Figure 5. As detailed above for copper bi-crystals, the migration generates dislocation dipoles whose glide parallel to the boundary generates plastic shear within the region swept by the boundary. A more comprehensive study of the disclination structure, energy and mobility of grain boundaries as a function of their misorientation is clearly needed to eventually reach an accurate description of the rheology of olivine aggregates. EBSD-based disclination field imaging, as introduced in [6,7], applied to naturally deformed olivine aggregates should also be helpful in firmly establishing such models.

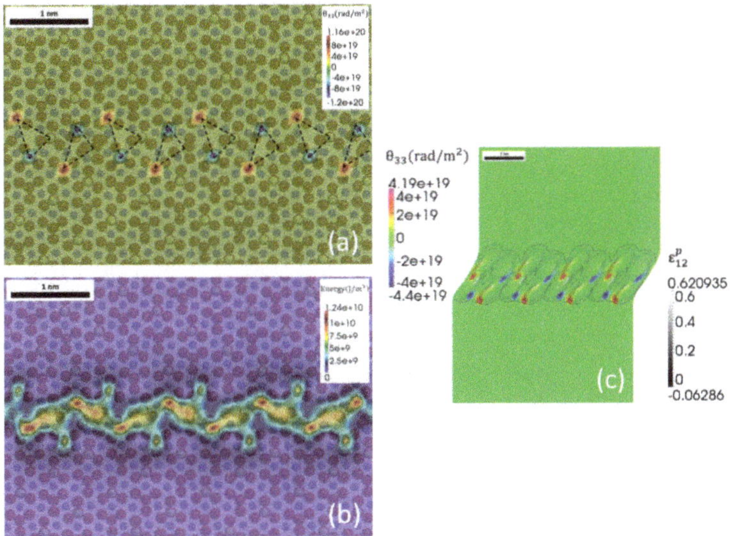

Figure 5. Disclination model for a (011)/<100>, 60° tilt boundary in forsterite. (a) Initial wedge-disclination density (rad·m^{-2}) superimposed on top of the atomic structure. Triangles show the structural units; (b) Elastic energy (J/m^3); (c) Migration of the (011)/<100> tilt boundary. The wedge-disclination density θ_{33} (rad·m^{-2}) is shown in color. The plastic shear generated by migration is shown with gray contours.

11. Concluding Remarks

The elasto-plastic model of dislocations and disclinations essentially consists of a set of partial differential equations, where the unknowns are the material displacement and dislocation/disclination density fields resolved at an atomic resolution length scale, and evolving at the time scale of the elementary dissipative mechanisms [18]. It is nonlocal in space and time in the conventional variables of compatible continuum mechanics and it provides a nonlocal incompatible generalization of the latter leading to well-posed problems in dislocation and disclination dynamics. It extends the elasto-plastic model of dislocations [22] by dealing, in addition to translational discontinuities and dislocations, with rotational discontinuities and disclinations. which allows grain boundary modeling in polycrystals. It also extends the elasto-static model of dislocations and disclinations [2], by dealing with plasticity mediated by transport of dislocations/disclinations within a consistent thermomechanical framework.

It has been sometimes believed that the assumption of continuity of the field variables is no longer valid at nanoscale and below, when the discrete atomistic nature of matter becomes apparent, and that continuum mechanics consequently fails to capture the physical phenomena at the scale of a few lattice spacings. In our opinion, the present results demonstrate instead that an incompatible, nonlocal continuum description involving fields of displacement and crystal defect densities that are smooth at this scale can be adequate for the purpose of describing such phenomena, in complement to atomistic representations. Indeed, the continuous disclination and dislocation theory proposed here was satisfactorily used to model the core structure of grain boundaries as well as the resulting elastic fields and excess energies. It was also shown to predict the complex interactions between disclinations/dislocations at work in the generation of plastic rotations/strains during shear-coupled boundary migration.

A continuous description of lattice defects may be more attractive than discrete approaches for various reasons. First, smoothness is desirable from the point of view of mathematical analysis and numerical computation, and because it allows dealing with defects' core properties. When viewed at a small scale, lattice defects and the corresponding distributions of elastic strain and energy are certainly better described by smooth density fields than by singularities. Building on this point of view, recent work [16,70,71,74–76] has been devoted to the construction of field representations of the lattice deformation from the atomic displacements resulting from molecular dynamics simulations.

Secondly, such descriptions do not have to resolve atomic vibrations in the manner of atomistic simulations. The kinetic energy of atomic vibrations is time-averaged over periods of microseconds and rendered as dissipation, which results in higher computational efficiency. As a practical result, finite element or spectral FFT numerical approximations may allow considering the dynamics of crystal defect ensembles over time scales in the μs or more, under realistic loading rates and stresses.

Finally, continuous approaches of crystal defects at nanoscale offer a convenient basis for seamless coarse-graining by using appropriate averaging techniques, which allows building mesoscale continuous theories of grain boundary-mediated plasticity in grain aggregates [56]. This ability could allow investigating alternative disclination-based mechanisms for plasticity, e.g., triple junction motion or emission/absorption of dislocations at grain boundaries, in media where dislocation-based plasticity is limited, either by scarcity of the slip systems, as in olivine, or because dislocation glide is restricted, as in polynanocrystals.

Author Contributions: Writing–review and editing, C.F. and V.T. All authors have read and agreed to the published version of the manuscript.

Funding: This research received no external funding.

Conflicts of Interest: The authors declare no conflict of interest.

References

1. Volterra, V. Sur l'équilibre des corps élastiques multiplement connexes. *Ann. Sci. Ecol. Norm. Supér. III* **1907**, *24*, 401–517. [CrossRef]
2. De Wit, R. Linear theory of static disclinations. In *Fundamental Aspects of Dislocation Theory*; Simmons, J.A., de Wit, R., Bullough, R., Eds.; National Bureau of Standards Special Publication: Washington, DC, USA, 1970; Volume 317, pp. 651–680.
3. Friedel, J. *Dislocations*; Pergamon Press: Oxford, UK, 1964.
4. Romanov, A.E.; Vladimirov, V.I. Disclinations in crystalline solids. In *Dislocations in Solids*; Nabarro, F.R.N., Ed.; Elsevier: Amsterdam, The Netherlans, 1992; Volume 9, p. 191.
5. Romanov, A.E.; Kolesnikova, A.L. Application of disclination concept to solid structures. *Prog. Mater. Sci.* **2009**, *54*, 740–769. [CrossRef]
6. Beausir, B.; Fressengeas, C. Disclination densities from EBSD orientation mapping. *Int. J. Solids Struct.* **2013**, *50*, 137–146. [CrossRef]
7. Fressengeas, C.; Beausir, B. Tangential continuity of the curvature tensor at grain boundaries underpins disclination density determination from spatially mapped orientation data. *Int. J. Solids Struct.* **2019**, *156–157*, 210–215. [CrossRef]
8. Bozhko, S.I.; Taupin, V.; Lebyodkin, M.; Fressengeas, C.; Levchenko, E.A.; Radikan, K.; Lübben, O.; Semenov, V.N.; Shvets, I.V. Disclinations in C_{60} molecular layers on $WO_2/W(110)$ surfaces. *Phys. Rev. B* **2014**, *90*, 214106. [CrossRef]

9. Gertsman, V.Y.; Nazarov, A.A.; Romanov, A.E.; Valiev, R.Z.; Vladimirov, V.I. Disclination-structural unit model of grain boundaries. *Philos. Mag. A* **1989**, *59*, 1113–1118. [CrossRef]

10. Hurtado, J.A.; Eliott, B.R.; Shodja, H.M.; Gorelikov, D.V.; Campbell, C.E.; Lippard, H.E.; Isabell, T.C.; Weertmann, J. Disclination grain boundary model with plastic deformation by dislocations. *Mater. Sci. Eng. A* **1995**, *190*, 1. [CrossRef]

11. Li, J.C.M. Disclination model of high angle grain boundaries. *Surf. Sci.* **1972**, *31*, 12. [CrossRef]

12. Shih, K.K.; Li, J.C.M. Energy of grain boundaries between cusp misorientations. *Surf. Sci.* **1975**, *50*, 109–124. [CrossRef]

13. Berbenni, S.; Paliwal, B.; Cherkaoui, M. A micromechanical-based model for shear-coupled grain boundary migration in bicrystals. *Int. J. Plast.* **2013**, *44*, 68–94. [CrossRef]

14. Cahn, J.W.; Mishin, Y.; Suzuki, A. Coupling grain boundary motion to shear deformation. *Acta Mater.* **2006**, *54*, 4953–4975. [CrossRef]

15. Farkas, D.; Froseth, A.; Van Swygenhoven, H. Grain boundary migration during room temperature deformation of nanocrystalline Ni. *Scr. Mater.* **2006**, *55*, 695–698. [CrossRef]

16. Tucker, G.J.; Zimmerman, J.A.; McDowell, D.L. Shear deformation kinematics of bicrystalline grain boundaries in atomistic simulations. *Model. Simul. Mater. Sci. Eng.* **2010**, *18*, 015002. [CrossRef]

17. Taupin, V.; Capolungo, L.; Fressengeas, C. Disclination mediated plasticity in shear-coupled boundary migration. *Int. J. Plast.* **2014**, *53*, 179–192. [CrossRef]

18. Fressengeas, C.; Taupin, V.; Capolungo, L. An elasto-plastic theory of dislocation and disclination fields. *Int. J. Solids Struct.* **2011**, *48*, 3499–3509. [CrossRef]

19. Acharya, A.; Fressengeas, C. Continuum mechanics of the interactions between phase boundaries and dislocations in solids. In *Differential Geometry and Continuum Mechanics*; Chen, G.Q., Grinfeld, M., Knops, R.J., Eds.; Springer: Berlin/Heidelberg, Germany, 2015; Volume 137, pp. 125–168.

20. Kröner, E. *Kontinuumstheorie der Versetzungen und Eigenspannungen (Ergebnisse der Angewandten Mathematik, No. 5)*; Collatz, L., Lösch, F., Eds.; Springer: Berlin/Heidelberg, Germany, 1958.

21. Kröner, E. Continuum theory of defects. In *Physics of Defects*; North Holland: Amsterdam, The Netherlands, 1980; pp. 218–314.

22. Acharya, A. A model of crystal plasticity based on the theory of continuously distributed dislocations. *J. Mech. Phys. Solids* **2001**, *49*, 761. [CrossRef]

23. Acharya, A. Driving forces and boundary conditions in continuum dislocation mechanics. *Proc. R. Soc. Lond.* **2003**, *A459*, 1343–1363. [CrossRef]

24. Puri, S.; Das, A.; Acharya, A. Mechanical response of multicrystalline thin films in mesoscale field dislocation mechanics. *J. Mech. Phys. Solids* **2011**, *59*, 2400–2417. [CrossRef]

25. Taupin, V.; Varadhan, S.; Chevy, J.; Fressengeas, C.; Beaudoin, A.J.; Montagnat, M.; Duval, P. Effects of size on the dynamics of dislocations in ice single crystals. *Phys. Rev. Lett.* **2007**, *99*, 155507. [CrossRef]

26. Taupin, V.; Varadhan, S.; Fressengeas, C.; Beaudoin, A.J. Directionality of yield point in strain-aged steels: The role of polar dislocations. *Acta Mater.* **2008**, *56*, 3002–3010. [CrossRef]

27. Varadhan, S.; Beaudoin, A.J.; Fressengeas, C. Lattice incompatibility and strain-aging in single crystals. *J. Mech. Phys. Solids* **2009**, *57*, 1733–1748. [CrossRef]

28. Gbemou, K.; Taupin, V.; Raulot, J.M.; Fressengeas, C. Building compact dislocation cores in an elasto-plastic model of dislocation fields. *Int. J. Plast.* **2016**, *82*, 241–259. [CrossRef]

29. Zhang, X.; Acharya, A.; Walkington, N.J.; Bielak, J. A single theory for some quasi-static, supersonic, atomic, and tectonic scale applications of dislocations. *J. Mech. Phys. Solids* **2015**, *84*, 145–195. [CrossRef]

30. Chevy, J.; Fressengeas, C.; Lebyodkin, M.; Taupin, V.; Bastie, P.; Duval, P. Characterizing short range vs. long range spatial correlations in dislocation distributions. *Acta Mater.* **2010**, *58*, 1837–1849. [CrossRef]

31. Fressengeas, C.; Beaudoin, A.J.; Entemeyer, D.; Lebedkina, T.; Lebyodkin, M.; Taupin, V. Dislocation transport and intermittency in the plasticity of crystalline solids. *Phys. Rev. B* **2009**, *79*, 014108. [CrossRef]

32. Kossecka, E.; de Wit, R. Disclination kinematics. *Arch. Mech.* **1977**, *29*, 633–651.

33. Kossecka, E.; de Wit, R. Disclination dynamics. *Arch. Mech.* **1977**, *29*, 749–767.

34. Das, E.S.P.; Marcinkowski, M.J.; Armstrong, R.W.; de Wit, R. The movement of Volterra disclinations and the associated mechanical forces. *Philos. Mag.* **1973**, *27*, 369–391. [CrossRef]

35. Pauli, W. *Handbuch der Physik, Band XXIV, Teil 1, Vol. V, Part 1*; Springer: Berlin, Germany, 1933.

36. Schrödinger, E. The energy-impulse hypothesis of material waves. *Ann. Phys.* **1927**, *82*, 265. [CrossRef]

37. Nielsen, O.H.; Martin, R.M. Quantum-mechanical theory of stress and force. *Phys. Rev. B* **1985**, *32*, 3780–3791. [CrossRef]

38. Jiang, B. *The Least-Squares Finite Element Method*; Springer: Berlin/Heidelberg, Germany, 1998.

39. Acharya, A.; Roy, A. Size effects and idealized dislocation microstructure at small scales: Predictions of a phenomenological model of Mesoscopic Field Dislocation Mechanics: Part I. *J. Mech. Phys. Solids* **2006**, *54*, 1687–1710. [CrossRef]

40. Weingarten, J. Sulle superficie di discontinuità nella teoria della elasticità dei corpi solidi. *Rend. R. Accad. Lincei* **1901**, *10*, 57–60.

41. Mindlin, R.D.; Tiersten, H.F. Effects of couple-stresses in linear elasticity. *Arch. Ratlat. Mech. Anal.* **1962**, *11*, 415–448. [CrossRef]

42. Eringen, A.C. Theory of micropolar elasticity. In *Fracture*; Liebowitz, H., Ed.; Academic Press: New York, NY, USA, 1968; Volume 2, pp. 621–729.

43. Koiter, W.T. Couple stresses in the theory of elasticity. *I II Proc. Ned. Akad. Wet.* **1964**, *B*, 17–44.

44. Fressengeas, C.; Sun, X. On the theory of dislocations and generalized disclinations and its application to straight and stepped symmetrical tilt boundaries. *J. Mech. Phys. Solids* **2020**, *143*, 104092. [CrossRef]

45. Eringen, A.C. A unified theory of thermomechanical materials. *Int. J. Eng. Sci.* **1966**, *4*, 179–202. [CrossRef]

46. Eringen, A.C.; Edelen, D.G.B. Nonlocal elasticity. *Int. J. Eng. Sci.* **1972**, *10*, 233–248. [CrossRef]

47. Eringen, A.C.; Kim, B.S. Relation between nonlocal elasticity and lattice dynamics. *Cryst. Lattice Defects* **1977**, *7*, 51–57.

48. Eringen, A.C. *Continuum Mechanics at the Atomic Scale*; No. 77-SM-1; Princeton University: Princeton, NJ, USA, 1977.

49. Eringen, A.C. *Non Local Continuum Field Theories*; Springer: New York, NY, USA, 2002.

50. Dingreville, R.; Hallil, A.; Berbenni, S. From coherent to incoherent mismatched interfaces: A generalized continuum formulation of surface stresses. *J. Mech. Phys. Solids* **2014**, *72*, 40–60. [CrossRef]

51. Spearot, D.E.; Jacob, K.I.; McDowell, D.L. Dislocation nucleation from bicrystal interfaces with dissociated structure. *Int. J. Plast.* **2007**, *23*, 143–160. [CrossRef]

52. Spearot, D.E.; Capolungo, L.; Qu, J.; Cherkaoui, M. On the elastic tensile deformation of <100> bicrystal interfaces in copper. *Comput. Mater. Sci.* **2008**, *42*, 57–67.

53. Tschopp, M.A.; Tucker, G.J.; McDowell, D.L. Atomistic simulations of tension-compression asymmetry in dislocation nucleation for copper grain boundaries. *Comput. Mater. Sci.* **2008**, *44*, 351–362. [CrossRef]

54. Upadhyay, M.; Capolungo, L.; Taupin, V.; Fressengeas, C. Elastic constitutive laws for incompatible crystalline media: The contributions of dislocations, disclinations and G-disclinations. *Philos. Mag.* **2013**, *93*, 794–832. [CrossRef]

55. Taupin, V.; Gbemou, K.; Fressengeas, C.; Capolungo, L. Nonlocal elasticity tensors in dislocation and disclination cores. *J. Mech. Phys. Solids* **2017**, *100*, 62–84. [CrossRef]

56. Taupin, V.; Capolungo, L.; Fressengeas, C.; Upadhyay, M.; Beausir, B. A mesoscopic theory of dislocation and disclination fields for grain boundary-mediated crystal plasticity. *Int. J. Solids Struct.* **2015**, *71*, 277–290. [CrossRef]

57. Mura, T. Continuous distributions of moving dislocations. *Philos. Mag.* **1963**, *89*, 843–857. [CrossRef]

58. Kleman, M.; Friedel, J. Disclinations, dislocations, and continuous defects: A reappraisal. *Rev. Mod. Phys.* **2008**, *80*, 61–115. [CrossRef]

59. Djaka, K.S.; Taupin, V.; Berbenni, S.; Fressengeas, C. A numerical spectral approach to solve the dislocation density transport equation. *Model. Simul. Mater. Sci. Eng.* **2015**, *23*, 065008. [CrossRef]

60. Roy, A.; Acharya, A. Finite element approximation of field dislocation mechanics. *J. Mech. Phys. Solids* **2005**, *53*, 143–170. [CrossRef]

61. Varadhan, S.; Beaudoin, A.J.; Acharya, A.; Fressengeas, C. Dislocation transport using an explicit Galerkin/least-squares formulation. *Model. Simul. Mater. Sci. Eng.* **2006**, *14*, 1245–1270. [CrossRef]

62. Coleman, B.D.; Gurtin, M.E. Thermodynamics with internal state variables. *J. Chem. Phys.* **1967**, *47*, 597–613. [CrossRef]

63. Djaka, K.S.; Villani, A.; Taupin, V.; Capolungo, L.; Berbenni, S. Field Dislocation Mechanics for heterogeneous elastic materials: A numerical spectral approach. *Comput. Methods Appl. Mech. Eng.* **2017**, *315*, 921–942. [CrossRef]

64. Fressengeas, C.; Taupin, V.; Capolungo, L. Continuous modeling of the structure of symmetric tilt boundaries. *Int. J. Solids Struct.* **2014**, *51*, 1434–1441. [CrossRef]

65. Hasson, G.; Boos, J.Y.; Herbeuval, I.; Biscondi, M.; Goux, C. Theoretical and experimental determinations of grain-boundary structures and energies—Correlations with various experimental results. *Surf. Sci.* **1972**, *31*, 115–137. [CrossRef]

66. Tschopp, M.A.; Spearot, D.E.; McDowell, D.L. Influence of grain boundary structure on dislocation nucleation in FCC metals. In *Dislocations in Solids*; Hirth, J.P., Ed.; North Holland Publishing: Amsterdam, The Netherlands, 2008; Volume 14, pp. 43–139.

67. Bachurin, D.V.; Murzaev, R.T.; Nazarov, A.A. Atomistic computer and disclination simulation of [001] tilt boundaries in nickel and copper. *Phys. Met. Metallogr.* **2003**, *96*, 555–561.

68. Gorkaya, T.; Molodov, D.A.; Gottstein, G. Stress-driven migration of symmetric <001> tilt grain boundaries in Al bicrystals. *Acta Mater.* **2009**, *57*, 5396–5405.

69. Mompiou, F.; Caillard, D.; Legros, M. Grain boundary shear-migration coupling—I. In situ TEM straining experiments in Al polycrystals. *Acta Mater.* **2009**, *57*, 2198–2209. [CrossRef]

70. Tucker, G.J.; Tiwari, S.; Zimmerman, J.A.; McDowell, D.L. Investigating the deformation of nanocrystalline copper with microscale kinematic metrics and molecular dynamics. *J. Mech. Phys. Solids* **2012**, *60*, 471–486. [CrossRef]

71. Zimmerman, J.A.; Bammann, D.J.; Gao, H. Deformation gradients for continuum mechanical analysis of atomistic simulations. *Int. J. Solids Struct.* **2009**, *46*, 238–253. [CrossRef]

72. Adjaoud, O.; Marquardt, K.; Jahn, S. Atomic structures and energies of grain boundaries in Mg_2SiO_4 forsterite from atomistic modeling. *Phys. Chem. Miner.* **2012**, *39*, 749–760. [CrossRef]

73. Cordier, P.; Demouchy, S.; Beausir, B.; Taupin, V.; Barou, F.; Fressengeas, C. Disclinations provide the missing mechanism for deforming olivine-rich rocks in the mantle. *Nature* **2014**, *507*, 51–56. [CrossRef] [PubMed]

74. Sun, X.Y.; Taupin, V.; Fressengeas, C.; Cordier, P. Continuous description of the atomic structure of a grain boundary using dislocations and generalized disclination density fields. *Int. J. Plast.* **2016**, *77*, 75–89. [CrossRef]

75. Sun, X.Y.; Cordier, P.; Taupin, V.; Fressengeas, C.; Jahn, S. Continuous description of a grain boundary in forsterite from atomic scale simulations: The role of disclinations. *Philos. Mag. A* **2016**, *96*, 1757–1772. [CrossRef]

76. Sun, X.Y.; Fressengeas, C.; Taupin, V.; Cordier, P.; Combe, N. Disconnections, dislocations and generalized disclinations in grain boundary ledges. *Int. J. Plast.* **2018**, *104*, 134–146. [CrossRef]

Publisher's Note: MDPI stays neutral with regard to jurisdictional claims in published maps and institutional affiliations.

Article

Experimental and Computational Approach to Fatigue Behavior of Polycrystalline Tantalum

Damien Colas [1,2,†], **Eric Finot** [3], **Sylvain Flouriot** [2], **Samuel Forest** [1,*], **Matthieu Mazière** [1] and **Thomas Paris** [2]

1   Centre des Matériaux (CMAT), MINES ParisTech, PSL University, CNRS UMR 7633, BP 87 91003 Evry, France; colasdam@gmail.com (D.C.); matthieu.maziere@mines-paristech.fr (M.M.)
2   CEA Valduc, 21120 Is-sur-Tille, France; sylvain.flouriot@cea.fr (S.F.); thomas.paris@cea.fr (T.P.)
3   Laboratoire Interdisciplinaire Carnot de Bourgogne, UMR 5209 CNRS, Université de Bourgogne Franche Comté, 9 Avenue Alain Savary, BP 17870, 21078 Dijon CEDEX, France; Eric.Finot@u-bourgogne.fr
*   Correspondence: samuel.forest@mines-paristech.fr
†   Current address: Varinor SA Rue St Georges 7, 2800 Delémont, Switzerland.

Abstract: This work provides an experimental and computational analysis of low cycle fatigue of a tantalum polycrystalline aggregate. The experimental results include strain field and lattice rotation field measurements at the free surface of a tension–compression test sample after 100, 1000, 2000, and 3000 cycles at ±0.2% overall strain. They reveal the development of strong heterogeneites of strain, plastic slip activity, and surface roughness during cycling. Intergranular and transgranular cracks are observed after 5000 cycles. The Crystal Plasticity Finite Element simulation recording more than 1000 cycles confirms the large strain dispersion at the free surface and shows evidence of strong local ratcheting phenomena occurring in particular at some grain boundaries. The amount of ratcheting plastic strain at each cycle is used as the main ingredient of a new local fatigue crack initiation criterion.

Keywords: tantalum; crystal plasticity; slip lines; fatigue; polycrystal; ratcheting; crack initiation; finite element

Citation: Colas, D.; Finot, E.; Flouriot, S.; Forest, S.; Mazière, M.; Paris, T. Experimental and Computational Approach to Fatigue Behavior of Polycrystalline Tantalum. *Metals* **2021**, *11*, 416.
https://doi.org/10.3390/met11030416

Academic Editor: Mikhaïl A. Lebyodkin

Received: 23 January 2021
Accepted: 21 February 2021
Published: 3 March 2021

Publisher's Note: MDPI stays neutral with regard to jurisdictional claims in published maps and institutional affiliations.

1. Introduction

Polycrystalline tantalum, a refractory material used in several industrial applications such as nuclear, biomedical, and chemical engineering, has been the subject of intensive research dealing with the understanding of plastic deformation modes, mainly at large strain and strain rates [1–3]. Modern experimental and computational techniques have been applied to body centered cubic (BCC) metals at the grain level [4]. They include Electron Back-Scatter Diffraction (EBSD) analysis of crystal lattice orientation and its evolution during deformation, Digital Image Correlation (DIC) analysis for the determination of strain fields [5], as well as Crystal Plasticity Finite Element simulations (CPFEM) at the continuum level relying on appropriate constitutive equations [6], and molecular dynamics simulations [7]. These techniques have been applied to tantalum polycrystals and oligocrystals in [8,9]. They give access to plastic slip mechanisms by identification of the traces of slip lines [10], to the lattice curvature and identification of Geometrically Necessary Dislocations (GND) densities [11], and to damage mechanisms [12].

All previous referenced works deal with the monotonic behavior of tantalum. In contrast, the cyclic behavior of tantalum has attracted only limited interest from the scientific community. Fatigue life assessment of polycrystalline tantalum is of the utmost importance in biomedical applications [13]. Reported results on High Cycle Fatigue (HCF) and Low Cycle Fatigue (LCF) behavior of tantalum seem to be limited to those in [14,15], respectively. In the latter reference, a macroscopic constitutive model was proposed to simulate the hysteresis stress–strain loops obtained experimentally. The shown experiments and proposed model include specific strain aging effects exhibited by pure tantalum in the form of a peak stress at the early plastic flow, due to interactions between dislocations and solute atoms like oxygen. An important feature of the cyclic response of metallic alloys is the ratcheting behavior which results from locally or globally non-symmetric stress loading conditions [16]. The authors of [17] provide unique computational results evaluating the amount of ratcheting inside the grains of a tantalum polycrystalline aggregate. The approach is based on the identification of tantalum single crystal model from single crystal monotonic and cyclic curves. An alternative single crystal model incorporating the effect of GNDs was proposed recently [18]. A major result in [17] is the clear evidence of a free surface effect in the ratcheting behavior which has implications in the initiation of fatigue cracks. Crystal plasticity-based fatigue initiation criteria make use of appropriate indicators involving the effects of resolved shear stress on the slip systems, cumulative slip and effect of normal to slip, or cleavage planes [19–23]. Statistical probabilistic fatigue criteria can then be proposed based on these indicators [24,25]. As an example, a fatigue crack nucleation criterion was developed for titanium Ti-6242 using accelerated crystal plasticity FEM simulations in [26].

The objectives of the present work is (*i*) to provide the first experimental results on plasticity and cracking at the free surface of a tantalum polycrystal by combining EBSD, DIC, and CPFEM, and (*ii*) to propose a novel fatigue crack initiation model at the continuum level of crystal plasticity. For that purpose, the results of an interrupted cyclic tensile–compression test are documented in terms of plastic slip activity, strain, and lattice rotation field measurements and detection of intergranular and transgranular cracks at the free surface of a tantalum polycrystalline aggregate. These results are compared with a unique CPFEM simulation of more than 1000 cycles on a semi-periodic polycrystalline aggregate. The experimentally observed and computed polycrystalline aggregates are distinct. The authors of [27,28] show that the full 3D grain structure is needed to make meaningful predictions of the plasticity at the surface by CPFEM. The experimental analysis is limited here to the surface description. No FIB image of the sample was available.

The original results of the present contribution deal with (*i*) evolution of the surface roughness, correlated with local EBSD and strain fields evolutions; (*ii*) dominance of intergranular or transgranular crack initiation observed after 5000 cycles; (*iii*) numerical estimation of strain heterogeneities at the surface and in the bulk of a polycrystalline aggregate and ratcheting effects at the free surface; and (*iv*) design of a new fatigue crack initiation criterion at the mesoscale of the polycrystalline aggregate.

The paper is organized as follows. The experimental methods and results are presented in Section 2. Section 3 deals with the description of the FE simulation of a polycrystalline aggregate and the design of the fatigue crack initiation criterion. Experimental and computational results are discussed and compared in Section 4.

The index notation is used throughout the work to denote vectors and tensors: Cartesian coordinates x_i of the material point and of the strain and stress tensors $\varepsilon_{ij}, \sigma_{ij}$, within the small strain framework.

2. Experimental Approach

2.1. Material and Experimental Methods

Tantalum is a refractory material used in several industries such as nuclear, capacitors, lighting, biomedical, and chemical processing. The material studied is a commercially pure tantalum (99.95% w.) coming from Cabot Performance Materials (USA). The sheet used has been recrystallized during 2 h between 1000 and 1200 °C and under 10^{-4} to 10^{-5} mbar. After recrystallization, the mean grain size is close to 120 µm and the initial dislocation density is rather high, between $\rho_d \approx 10^{13}$ m^{-2} and $\rho_d \approx 10^{14}$ m^{-2} [29–33]. These values are in agreement with measurements of GND densities deduced from EBSD data for recrystallized tantalum in the references [9,34].

The tantalum purity is characterized by the following composition of solute atoms: C: 10, N: <10, O: 19, H: 6 ppm. Some of these elements are responsible for strain aging effects studied in [15].

The EBSD map over a large area on Figure 1 shows quasi-equiaxial grains without marked crystallographic texture. The fact that no predominant direction is observed is due to cross rolling of the sheets. Some curved grain boundaries are present, probably due to sub-optimal recrystallization conditions. The pole figure analysis on raw data does not show a significant crystallographic texture (cf. Figure 2). Thus, the hypothesis of an isotropic texture is adopted in the following work, particularly for the input data of the polycrystalline aggregate generation.

The macroscopic cyclic behavior of the considered polycrystalline tantalum has been investigated for various amplitudes and strain rates in another article by the same authors [15]. In the present paper, the attention is focused on the observation of local plasticity events at the surface of one single sample. An interrupted cyclic tension–compression test has been performed on a thick enough flat sample to avoid potential buckling. The dogbone specimen had a gauge part with dimensions $15 \times 10 \times 8$ mm^3, see Figure 1b in [15] for the detailed sample geometry. The specimen geometry is not a standard one but allows for DIC and microstructural observations on the flat surfaces. The strain rate was $\dot{\varepsilon} = 10^{-3}$ s^{-1} for a strain amplitude of $\Delta\varepsilon/2 = 0.2\%$.

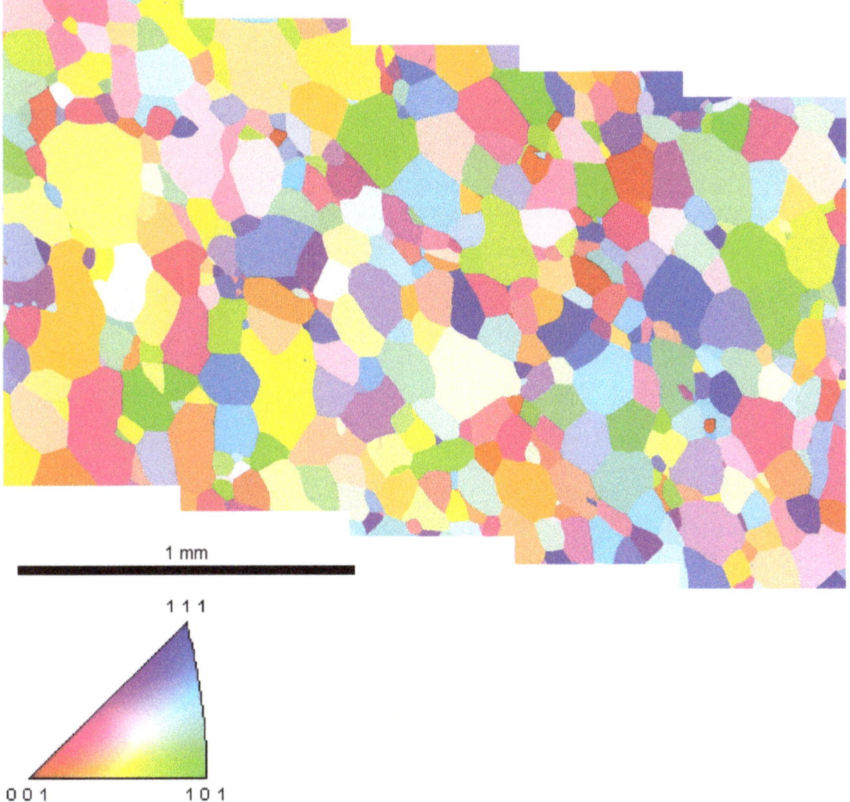

Figure 1. EBSD map of tantalum polycrystal over a large area in the center of a sample.

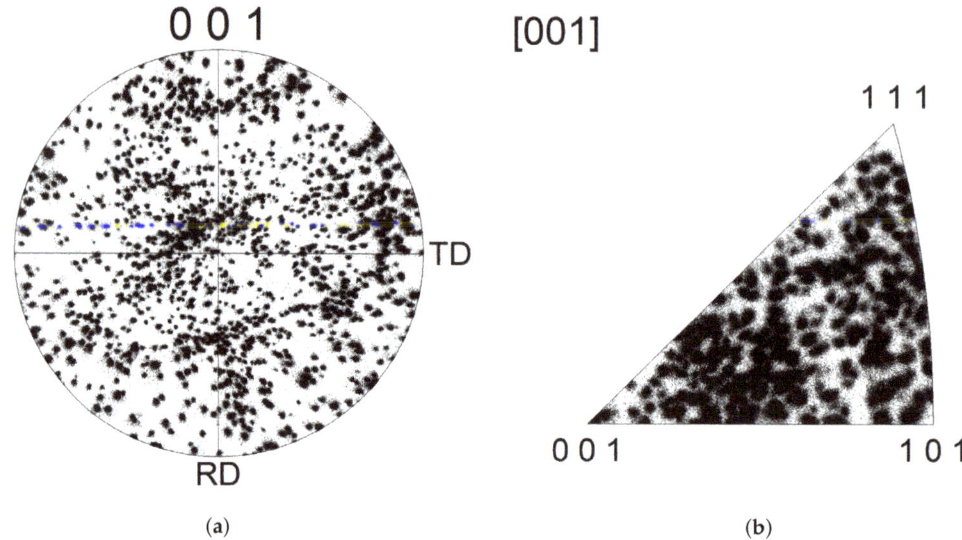

Figure 2. (**a**) Direct <100> pole figure and (**b**) inverse pole figure of tantalum over a large zone in the center of a sample.

The sample was polished before Electron Beam Lithography deposition. Several grids have been spread along the sample, consisting in 500 nm markers, with a 5 μm pitch on a 400×400 μm^2 zone.

The cyclic test has been interrupted after 100, 1000, 2000, and 3000 cycles. Each time, the sample was unloaded from the test machine and placed into a SEM for secondary, backscattered, and EBSD analyses of each grid. The testing machine used is a servo-hydraulic Instron 8800. All the SEM observations have been done using the same beam parameters and adjustments and the same magnification.

The shape of the hysteresis loops does not evolve significantly during the test, with a marked Bauschinger effect. The isotropic hardening of about 20 MPa reaches a saturation point after ≈50 cycles. The sample is unloaded at zero force, with an increment of macroscopic plastic strain about $\langle \varepsilon_{22} \rangle = 0.001$ after each sequence.

2.2. Local Evolution of the Surface during Cycling

At each step, the load was decreased to zero in order to observe the specimen ex situ in a SEM. The SEM pictures give some information about the surface evolution.

The pictures of Figure 3 show the roughness evolution after 100, 1000, 2000, and 3000 cycles. The roughness gradually increases, with a stronger effect close to grain boundaries. Grains encircled in blue, for instance, are subjected to the progressive formation of a step along grain boundaries. This can induce some stress concentration or notch effect, and lead to microstrucural ratcheting, as discussed later [35]. The images also show numerous curved slip lines associated with the easy cross slip characteristic of BCC materials [36]. Such sinuous or wavy slip lines are also referred to as pencil glide [37].

Moreover, most of the grains exhibit slip bands that look like Persistent Slip Bands (PSB). They progressively intensify with increasing cycle numbers. PSBs are stopped at grain boundaries, except for the area circled in red, where PSBs spread on both sides of the grain boundary, probably because of the weak misorientation between the two grains. This has been observed in previous work [38].

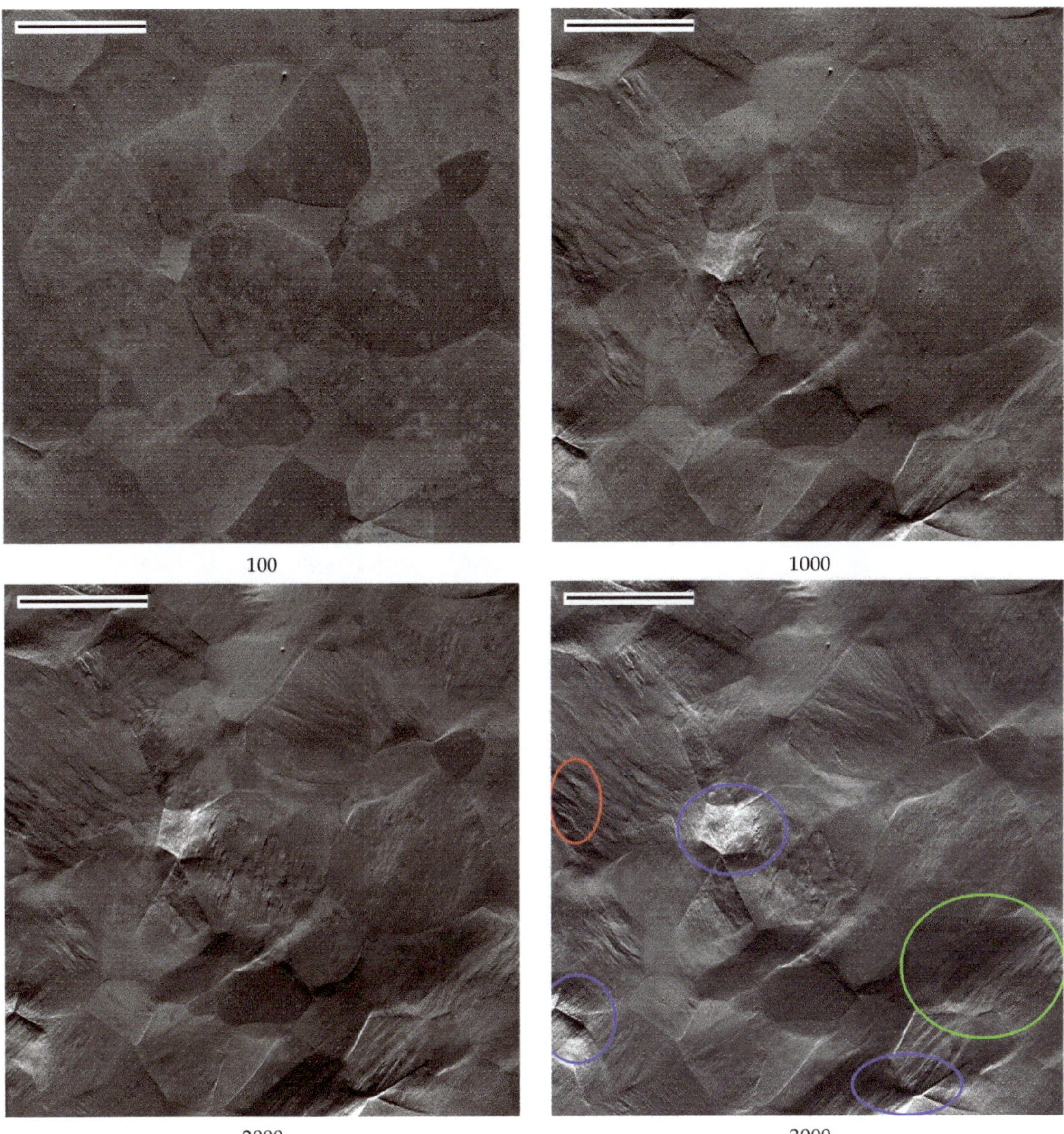

Figure 3. Roughening of a $400 \times 400 \ \mu m^2$ zone at the surface for several cycles during a fatigue test at $\pm 0.2\%$ and $\dot{\varepsilon} = 10^{-3} \ s^{-1}$. Loading direction is horizontal. The small dots of the deposited grid are visible on the SEM pictures. The upper left bar is 100 μm long.

The green circle on Figure 3 denotes a zone of a large grain that progressively becomes concave. This grain also presents PSB-like structures.

Thus, the surface activity is very intense. It can be characterized by local strain fields measurements provided by Digital Image Correlation (DIC) and also by lattice rotation fields given by EBSD.

2.3. Strain Field Measurements

The strain control loading is characterized by the half-amplitude $\Delta\varepsilon/2$ and the loading ratio $R = -1$. Four successive sequences of cyclic loading are carried out: The first sequence between 0 and 100 cycles, the second until 1000 cycles, the third until 2000, and the fourth until 3000 cycles. After each loading sequence, the sample is unmounted in order to observe the specimen ex situ in a SEM equipped with an EBSD detector. SEM data are used in order to build the displacement field of an area between two loading sequences using Vic-2D digital image correlation (DIC) software in the sequential mode (the first picture is used as a reference in any case). The spatial resolution of the acquired SEM images about 100 nm per pixel.

The ZNSSD method proposed by the Vic-2D 5 software (Zero-mean Normalized Sum of Squared Differences) has been used to handle the images because it accounts for both the lightening scale factor and the gray level offset [39]. This method is especially suitable for SEM pictures with significant gray level variations.

The measurement uncertainty associated with the strain field calculation from SEM pictures is due to three main factors: the magnification, the scanning, and the drift of the beam. In any case, the maximum cumulated error is about 1 pixel [40–43]. Moreover, the out-of-plane displacements and the system alignment compared to the surface can induce significant errors that are not taken into account. A quantification and a correction procedure have been proposed by Sutton [44]. However, the correction levels are smaller than our resolution and are thus not taken into account here.

The grid of nano-markers is built in six successive steps, as described in previous works [45,46]. The markers are made of nickel in order to have a sufficient contrast with tantalum. Several grids of 400×400 μm^2 have been spread along the sample gauge length. The markers have a 500 nm diameter and a 5 μm periodic pitch. The Electron Beam Lithography method is quite time consuming for large scale patterns, and thus a compromise between insolation time and covered surface versus spatial resolution has been chosen. The grids are not directly used for the DIC, but they are useful to mark the area of interest. Instead, the DIC is based on the gray level variations of the surface (including the variations due to the presence of the grids). DIC from natural contrast associated with varying gray levels due to surface roughness for example in SEM images has been used extensively in the past during in situ testing, see, for instance, the work in [47].

Experimental strain fields ε_{11}, ε_{12} and ε_{22} for several sequences are presented in Figure 4, where the loading direction is 2. They have been computed using Vic-2D from back-scattered electrons pictures, with a correlation window 100×100 pixels and a 50 pixel grid spacing. When applied to a strain value of 5%, this gives a displacement of 2.5 pixel. Correlation errors in our measurements range from 0.1 pixel to 1 pixel maximum, which leads to errors from 0.1% to 1% in absolute strain values.

Strongly heterogeneous strain fields are observed with local strain values reaching a few percents in spite of 0.2% macroscopic prescribed half-amplitude. This strain heterogenity increases with the cycle number culminating at the final observed stage of 3000 cycles. Strain discontinuities are visible at some grain boundaries resulting from strain incompatibilities from grain to grain due to different crystallographic orientations. The red circle points out a zone of strong strain discontinuities on both sides of several grain boundaries. Zones of rather homogeneous strain are observed that encompass several grains. Their size corresponds to grain clusters of 150–200 μm.

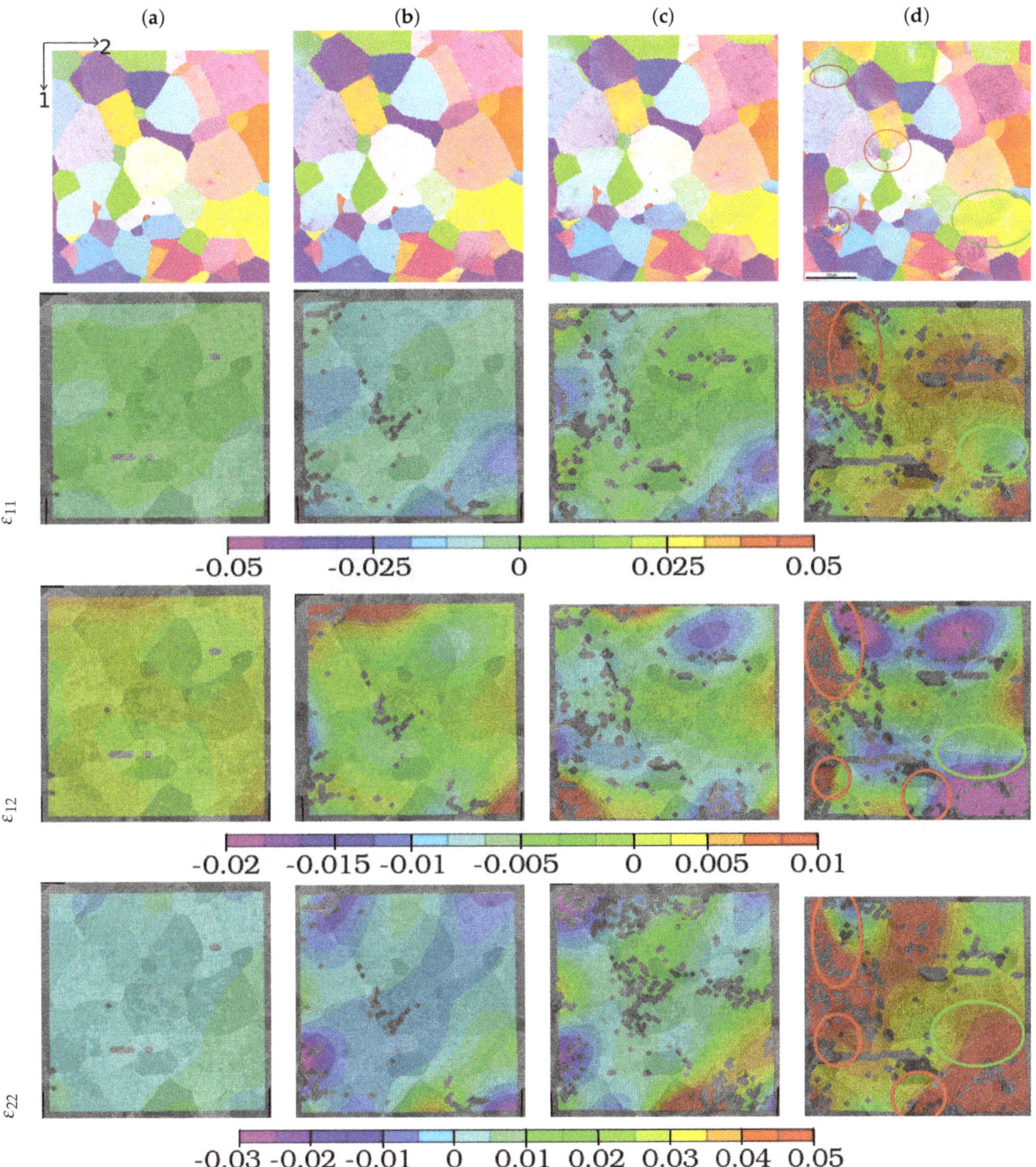

Figure 4. Local residual strain evolution during a fatigue test at $\pm 0.2\%$ and $\dot{\varepsilon} = 10^{-3}\,\text{s}^{-1}$: (**a**) After 100 cycles, (**b**) 1000 cycles, (**c**) 2000 cycles, and (**d**) 3000 cycles. Loading direction is horizontal (direction 2). The picture size is $400 \times 400\,\mu\text{m}^2$.

The green circle indicates an area of strong intragranular strain in a zone of two large grains with low misorientation, as is indicated on the EBSD maps of the top pictures of Figure 4. The EBSD maps show the evolution of lattice orientation during cycling. Lattice curvature clearly develops inside the grain as a result of heterogeneous plastic strain [48]. The building-up of rotation gradient inside the grains is particularly visible in the green circled region. The evolution of lattice orientation changes inside the grains is illustrated in Figure 5. The grain orientation spread (GOS) is computed in each grain at each cycle number, as the averaged misorientation range inside each grain. Orientation spreading clearly increases with cycle numbers reaching values of a few degrees. After 3000 cycles, lattice spreading exceeds 5° in some regions circled in the black and white zones. As a rule of thumb, crystal plasticity teaches us that one degree of lattice rotation is associated with a plastic strain of typically 0.01. These suggested local strain levels are confirmed by the strain field measurements. On the other hand, a misorientation of typically 5° over 100 micron can be associated to a GND density in the range $10^{12\text{--}13}$ m^{-2} which is expected in annealed tantalum. Note, however, that statistically stored dislocations (SSD) also contribute to the overall dislocation density and cannot be derived solely from EBSD maps.

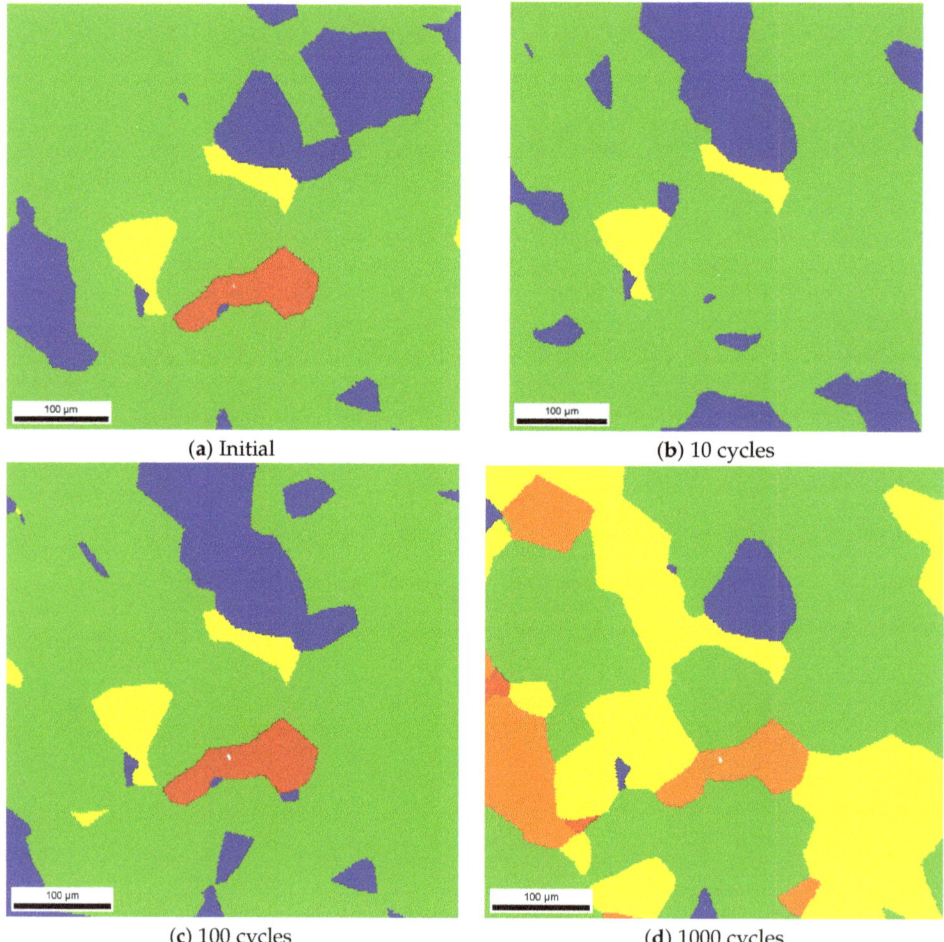

(**a**) Initial   (**b**) 10 cycles

(**c**) 100 cycles   (**d**) 1000 cycles

Figure 5. *Cont.*

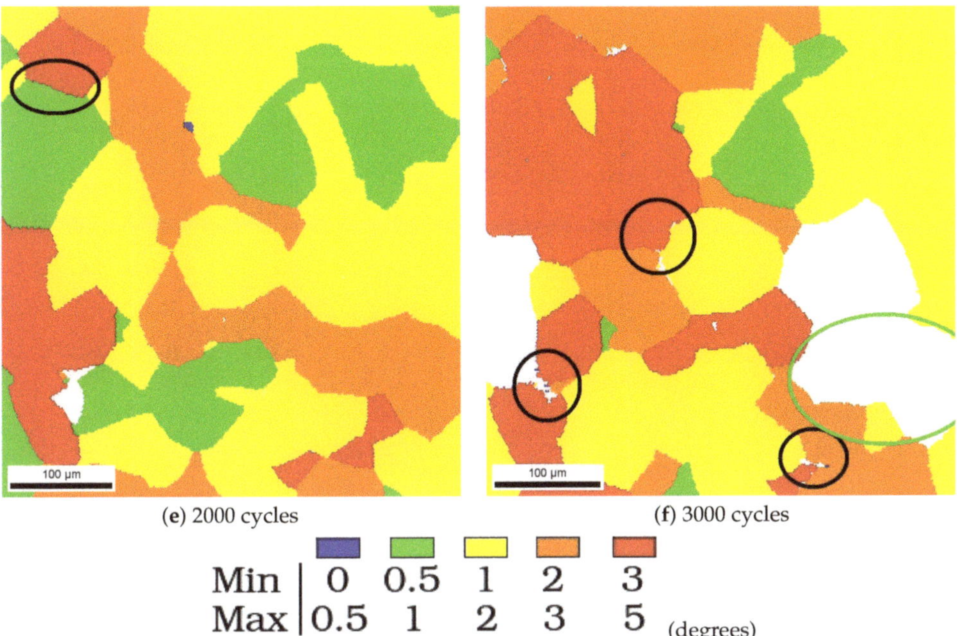

(**e**) 2000 cycles   (**f**) 3000 cycles

| Min | 0 | 0.5 | 1 | 2 | 3 |
| Max | 0.5 | 1 | 2 | 3 | 5 |

(degrees)

Figure 5. Evolution of the lattice orientation spreading (GOS) per grain with increasing cycle numbers. The loading direction is horizontal.

Histograms are provided for each strain components at 4 cycle numbers in Figure 6. The strain distributions correspond to the zone observed in Figure 4. A spreading of the strain distribution is observed for increasing cycle numbers. This illustrates quantitatively the strain heterogeneities discussed in the previous paragraphs. A significant part of the distribution is located at high values of surface strains, in spite of the low applied global amplitude. This corresponds to the visible roughening of the surface, a well-known effect in fatigue [49,50]. Significant errors in the measured strain values are expected due to the out of plane displacement component which limits the validity of the 2D correlation. This is especially the case after 3000 cycles where surface roughening is very pronounced. A remarkable feature of the ε_{22} and ε_{11} histograms is the shift toward larger mean values after 2000 and 3000 cycles. This indicates that considered region made of about 60 grains behaves as a whole differently from macroscopic straining conditions. This means that the representative volume element size for fatigue loading conditions must be larger than this observed zone. Maximal values for the ε_{11} and ε_{12} components are greater than 5% up to 11% after 3000 cycles. After 3000 cycles, the computed ε_{11} and ε_{22} strain values are strongly affected by errors induced by surface roughing. These high values are in agreement with the literature, where similar strains are calculated from experimental data [51] in stainless steels. Similar maximal local strains have been calculated in the literature for different loading conditions from the present case [45,52,53] in Nickel-based superalloys.

For the ε_{22} component after 3000 cycles, the distribution becomes highly scattered around a mean strain close to 3%. The zones of maximum strain correspond to areas including sliding grain boundaries (see the red circles in Figure 4 and blue circles in Figure 3). Clear evidence of grain boundary sliding (GBS) can be seen in Figure 3 after 1000 and 2000 cycles in the form of an inclined white line in the lower right part of the pictures indicating out of plane sliding of one grain boundary.

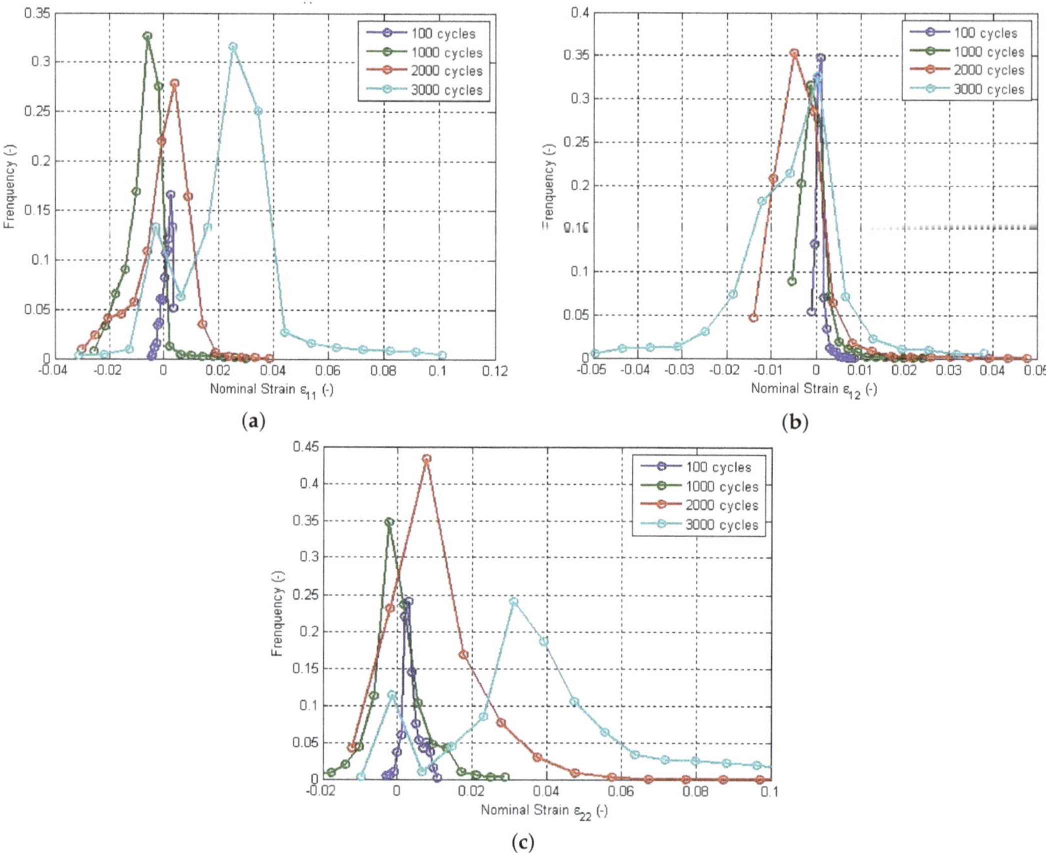

Figure 6. Strain histograms for a fatigue test at ±0.2%. (**a**) ε_{11}, (**b**) ε_{12}, and (**c**) ε_{22} at the surface. The histograms are given after 10, 1000, 2000, and 3000 cycles.

2.4. Fatigue Crack Initiation

Crack initiation is observed mainly at grain boundaries, thus indicating an intergranular cracking mechanism. Intergranular crack branching is observed at triple junctions in Figure 7. Transgranular cracking is also observed as a competing mechanism, see the red circle in Figure 7. Severe extrusion–intrusion zones are visible in the blue circle of the same figure. They are associated with numerous intense slip lines visible in some grains. Crack branching from a grain boundary to some PSB is found to take place in a grain circled in red in Figure 7. As the fatigue test has been interrupted after 5000 cycles, it is not known whether the final fracture is due to inter- or trans-granular cracks. In previous studies at CEA (unpublished), a combined intergranular–transgranular fracture has been reported [38,54]. After 5000 cycles, the minimum crack size observed is about ≈10 μm and the maximal one is ≈120 μm.

The presented cyclic test was carried out until the appearance of the first stage I crack at the surface, as defined by Forsyth [55]. In that case, the cracks have a length similar to the grain size and can be both intergranular or transgranular. This corresponds to micro-crack initiation for which a criterion will be proposed in the following.

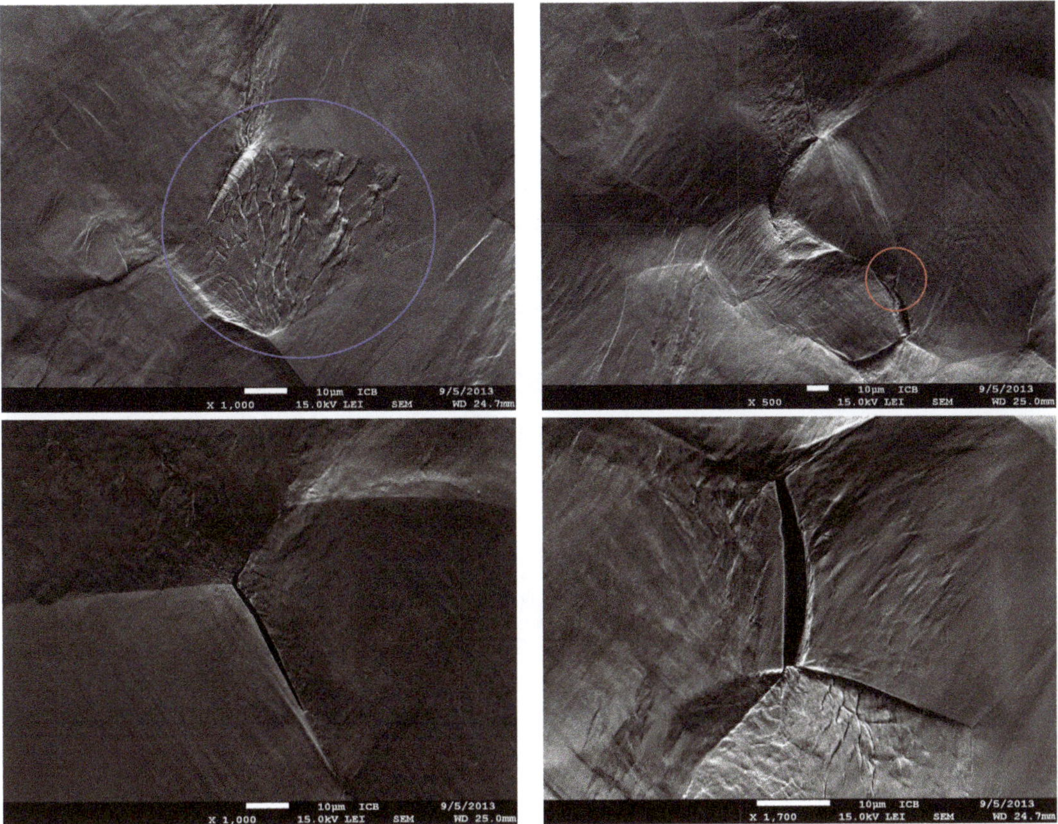

Figure 7. Pictures of the surface after 5000 cycles at $\pm 0.2\%$ and $\dot{\varepsilon} = 10^{-3}$ s^{-1}. They illustrate the occurrence of grain boundary sliding (**top left**), intragranular cracks (**top**), and intergranular cracks (**bottom**).

3. Finite Element Simulation

The Finite Element simulation described in the following has been introduced in the reference [17]. The computational method and the constitutive model are only briefly recalled. The remainder of this section concentrates on results not presented in [17] that can be compared to the experimental data explored in the previous section.

3.1. Description of Semi-Periodic Polycrystalline Aggregates

The simulation is based on a synthetic polycrystalline aggregate because a one-to-one mapping to the experimentally observed zone was not possible due to the ignorance of the sub-surface grain morphology. The polycrystalline aggregates considered in this work is characterized by two parallel flat surfaces perpendicular to the space direction 3, and four lateral surface displaying periodicity of grain morphology along the directions 1 and 2, as shown in Figure 8. The through-thickness view on Figure 8 shows that the grain morphology of the parallel flat surfaces are not periodic. The aggregate on Figure 8 contains 250 grains with 4 to 5 grains through the thickness in average. The construction of such semi-periodic polycrystalline aggregates starting from a classical Voronoi tesselation follows the strategy proposed by J. Guilhem in [56]. By construction, grain boundaries are flat surfaces ensuring continuity of displacement and reaction forces in finite element simulations. This is a simplification compared to the real curved grain boundaries visible in Figure 1.

The three-dimensional finite element mesh is made of quadratic tetrahedral elements with full integration. The mesh of Figure 8 contains 55,089 nodes and 37,595 quadratic elements. The authors of [27,28] address the question of suitable mesh size for proper representation of surface field. We have used these recommendations in the current work, even though the computational cost lead us to use rather coarse elements.

The number of grains in the polycrystalline volume and the mesh refinement have been chosen as a compromise between computational cost for cycling testing and sufficient representativity of the polycrystalline response according to previous studies of Representative Volume Element size for f.c.c. crystals [57–59]. In particular, the thickness was chosen following the works in [27,28] that show that the strain field at a free surface of a cubic polycrystal is affected mainly by the three layers of grains below the surface. A discussion on the Representative Volume Element size for polycrystals under cyclic loading conditions can be found in [17] based on the few works in this field [58,60] where polycrystalline aggregates made of 100 grains were considered. Finally, it must be noted that only one Finite Element simulation was carried out in this work due to the fact that more than 1000 cycles were simulated leading to computation times not compatible with a statistical analysis for several realizations of the microstructures.

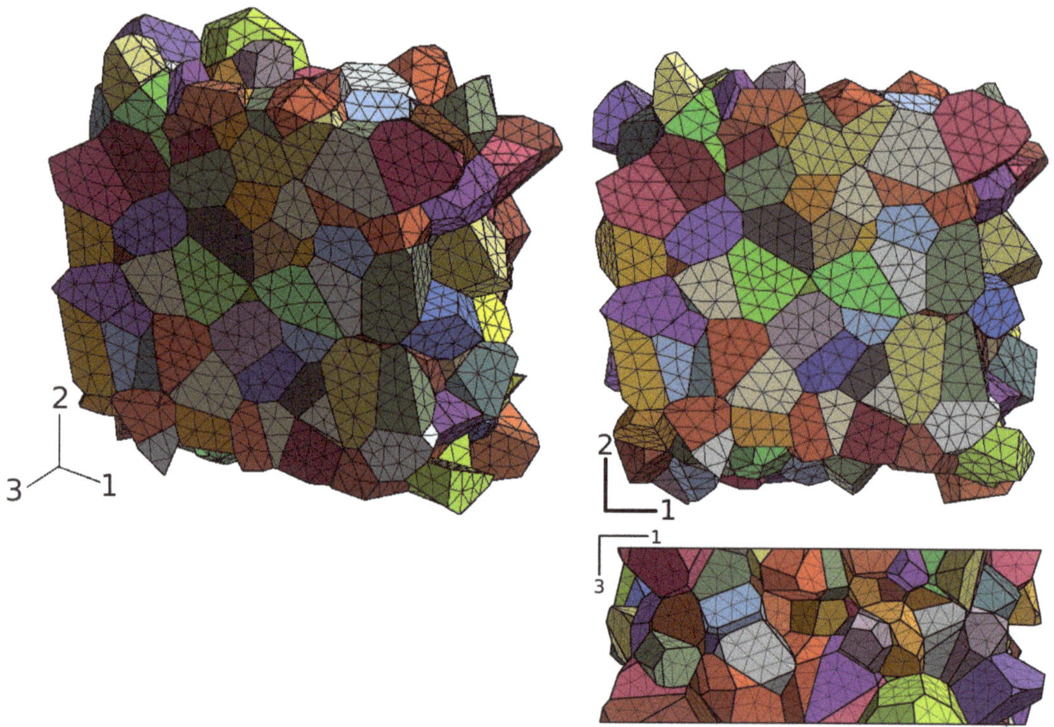

Figure 8. Synthetic semi-periodic aggregate made of 250 grains: general view (**top left**), free surface (**top right**), and view through the thickness (**bottom**), after the work in [17].

The detailed boundary conditions are described in [17]. They amount to prescribing the mean strain in the tension–compression direction 2 with vanishing mean stress components (except $< \sigma_{22} >$). The two flat surfaces of the polycrystalline aggregate of Figure 8 are called $z = 0$ and $z = H$. The surface $z = 0$ is subjected to the Dirichlet boundary condition $u_3 = 0$ where u_3 is the displacement component along the direction 3, see the coordinate frame in Figure 8. The remaining boundary conditions for this surface $z = 0$ are vanishing traction components along 1 and 2. The surface $z = H$ is traction-free in all 3 directions, corresponding to a free surface, as in the experiment. The strain fields on this free surface will be analyzed and compared, in a statistical sense, with the experimental results.

Periodicity conditions are prescribed for the remaining four lateral surfaces. Such conditions are known to limit boundary layer effects in the simulation of material volume elements in contrast to homogeneous Dirichlet or Neumann conditions, see the works in [58,61–63].

The average stress and strain components over the whole polycrystalline volume element V are computed as

$$\Sigma_{ij} = \, < \sigma_{ij} > \, = \frac{1}{V}\int_V \sigma_{ij}\, dV, \quad E_{ij} = \, < \varepsilon_{ij} > \, = \frac{1}{V}\int_V \varepsilon_{ij}\, dV \tag{1}$$

The macroscopic strain component $E_{22} = \pm 0.2\%$ is imposed. The loading conditions therefore correspond to a simple tension–compression test with one free surface, four periodic surfaces, and one flat surface.

The results presented in this work correspond to the simulation of more than 1000 cycles on the previously described polycrystalline aggregate. They represent 11 months of computation time on a 12 core Intel Xeon 3 GHz processor with 25 GB RAM. 1250 time steps were saved for 70 variables (stress, strain, and plastic strain tensor components; slip amounts; accumulated slip amounts; kinematic hardening variables; accumulated plastic strain ...) at each integration point, which amounts to 250 GB disk space.

3.2. Crystal Plasticity Model and Identification of Material Parameters

The constitutive equations of the crystal plasticity model used for tantalum are now given. The Cailletaud crystal plasticity model [64] is extended here to include static strain aging effects as displayed by tantalum crystals [15]. The strain tensor is the sum of the elastic and plastic contributions:

$$\varepsilon_{ij} = \varepsilon_{ij}^e + \varepsilon_{ij}^p \tag{2}$$

The local elastic behavior is cubic, with an elasticity tensor defined by the three independent elastic moduli: $C_{11} = 267$ MPa, $C_{12} = 159$ MPa, and $C_{44} = 83$ MPa, after the work in [65].

The plastic strain tensor results from plastic slip processes with respect to all slip systems. In the present work, 12 slip systems are considered on {110} slip planes with slip directions <111> corresponding to the b.c.c. structure. This choice represents a simplification as more slip planes are known to be available in b.c.c. crystals [66]. An alternative approach would be to introduce the pencil glide mechanism as done in [37]. Motivations for the present choice can be found in [67] where the selection of {110} vs. {112} slip planes is discussed. The viscoplastic strain rate tensor is written as

$$\dot{\varepsilon}_{ij}^p = \sum_s \dot{\gamma}^s m_{ij}^s, \quad \text{with} \quad m_{ij}^s = \frac{1}{2}(\ell_i^s n_j^s + n_i^s \ell_j^s) \tag{3}$$

where n_i^s are the components of the normal vector to the slip plane and ℓ_j^s that of the slip direction vector, for the slip system s. The slip rate for each slip system is evaluated using the following model equation:

$$\dot{\gamma}^s = \dot{v}_0 \sinh\left(\frac{<|\tau^s - x^s| - r^s - r_a>}{\sigma_0}\right)\text{sign}\,(\tau^s - x^s), \quad \dot{v}^s = |\dot{\gamma}^s| \tag{4}$$

where $\tau^s = \sigma_{ij} m_{ij}^s$ is the resolved shear stress for slip system s and v^s is the cumulative slip variable. The parameters $\dot{v}_0$ and σ characterize the viscosity effects. The Macaulay brackets $< \bullet > = \text{Max}(\bullet, 0)$ were used. The isotropic hardening law describing the evolution of the critical resolved shear stress (CRSS) follows a nonlinear evolution law:

$$r^s = r_0 + Q \sum_r h^{sr}(1 - \exp(-bv^r)) \tag{5}$$

where r_0 is the initial CRSS. Q and b are material parameters; h^{sr} is the interaction matrix which characterizes both self-hardening and cross(latent)-hardening between the different slip systems.

The kinematic hardening term is the main ingredient for the description of internal stresses building up inside the grains, for instance due to dislocation pile–ups and dislocation structure formation. It is decomposed into two contributions:

$$x^s = x_1^s + x_2^s \tag{6}$$

The evolution equation of each component is

$$\dot{x}_i^s = c_i \dot{\gamma}^s - d_i x_i^s \dot{v}^s, \quad i \in \{1, 2\} \tag{7}$$

where c_i and d_i are kinematic hardening material parameters. The reason for introducing two components x_1^s and x_2^s is a better description of the hysteresis loops at various strain amplitudes.

The previous crystal plasticity model was enhanced in [17] by the addition of a resistance term associated with static strain aging, namely, the component r_a in Equation (4). This strain aging effect is not discussed here as it concerns mainly the first hysteresis loops, whereas the focus of the present work is crack initiation at larger cycle numbers.

The values of the model parameters were identified from tensile and cyclic tests carried out on tantalum single crystals. The identification procedure was detailed in [17]. Table 1 provides the list of parameter values found in the latter reference and used for the present simulation.

Table 1. Parameter values of the single crystal model identified for tantalum at room temperature, after the work in [17].

$\dot{v}_0$	$5 \times 10^{-5}\ \mathrm{s}^{-1}$	c_1	$360 \times 10^3\ \mathrm{MPa}$
σ_0	5 MPa	D_1	8000
r_0	0 MPa	C_2	250 MPa
Q	1 MPa	D_2	1.5
b	1.5	h^{rs}	1

The single crystal model has been implemented in the implicit finite element code Zset [68], following the procedure presented in [69]. Global equilibrium is solved using a Newton–Raphson algorithm and the constitutive equations are integrated using a second order Runge–Kutta method with adaptive time stepping.

3.3. Simulation of the Mechanical Fields at the Free Surface

The overall stress–strain loops, i.e., the $< \sigma_{22} >$–$< \varepsilon_{22} >$ curves, of the studied cyclic test on the considered polycrystalline aggregate are given in Figure 9 for various cycle numbers from the CPFEM simulation. The stress and strain amplitudes are clearly visible. The loops at 66, 466, and 1066 are almost identical showing that no macroscopic ratcheting takes place for the considered loading. The first loop is strongly different due to the initial peak stress associated with static strain aging discussed in [15,17]. Comparison between experimental initial and stabilized loops can be found in the latter reference. It is not reported here since the present work concentrates on cycle numbers from 100 to 5000.

The evolution of several mechanical variables on the free surface of the polycrystal is now discussed with respect to the number of cycles. All surface fields are provided for a vanishing value of the mean axial stress $\Sigma_{22} = \langle \sigma_{22} \rangle = 0$ MPa in accordance with the experimental observation conditions.

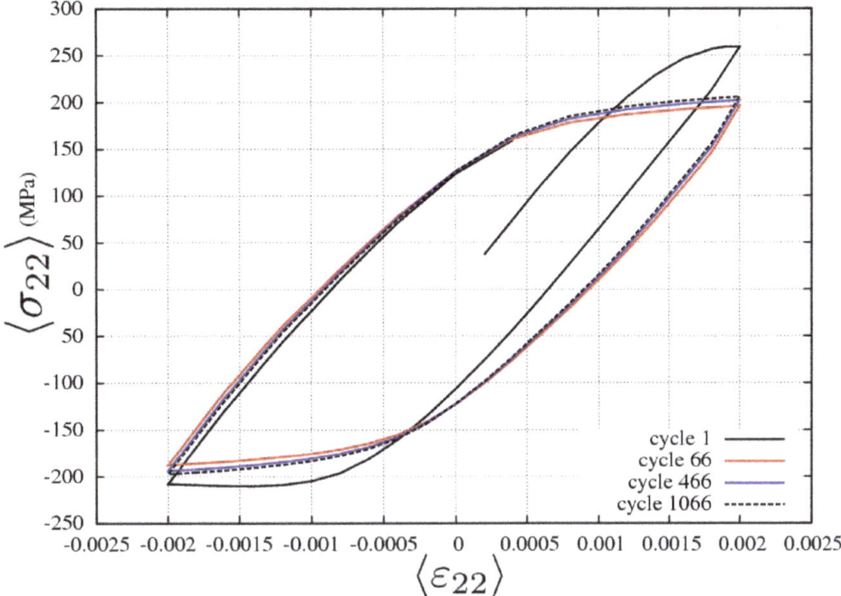

Figure 9. Macroscopic stress-strain loops of the cyclic test on the studied polycrystalline aggregate, as predicted by the CPFEM simulation.

3.3.1. Surface Strain Field

The fields of the strain components $\varepsilon_{11}, \varepsilon_{22}$, and ε_{12} at the free surface of the polycrystalline aggregate are shown in Figures 10–12, respectively. Three snapshots are presented at the cycle numbers 66, 666, and 1066 for each component. The residual strain field is strongly heterogeneous with values from -0.003 to more than 0.003, to be compared to the loading amplitude ± 0.002. The deformation structures extend over several grains as observed under monotonic conditions [57,59]. This heterogeneity is found to increase significantly between $N_{cycle} = 66$ and $N_{cycle} = 666$, with certain grains localizing more deformation, especially at some grain boundaries. The localization remains the same after 400 additional cycles but the contrast intensifies between tensile and compressive zones. The heterogeneity culminates for the shear component. Moreover, some areas close to grain boundaries and triple junctions display discontinuous strain values due to change in crystallographic orientation. This strain localization close to grain boundaries is shown to increase with the number of cycles. The dramatic development of strain heterogeneities under cyclic loading is in qualitative agreement with the experimental results. However, the local strain values predicted after 1000 cycles significantly underestimate the corresponding strain field measurements of Figure 4b. There are several reasons to explain such a discrepancy, which will be discussed later in this work.

Figures 10–12 show the strain components predicted at the free surface but also at the mid-section parallel to the flat surfaces and also in a cross-section perpendicular to the loading axis. They show a significant free surface effect as strain values are considerably more heterogeneous at the free surface than in the bulk. Note that strain values shown at the free surface correspond to values computed at the integration (Gauss) points close of the surface whereas values at mid-sections are interpolated which may lead to some smoothing effect due to the relatively coarse mesh. This surface effect has dramatic consequences on ratcheting phenomena discussed in [15].

Significant roughening of the surface is predicted in qualitative agreement with the observations of Figure 3, and with previous cyclic behavior simulations of copper in [70]. This is illustrated by Figure 13 showing the out of plane displacement field at the free surface for 6 cycle numbers. After 466 cycles, specific zones of raising or sinking-in material points are observed. Their location, mainly at triple junctions and grain boundaries but also in the interior of some grains, does not change until 1000 cycles. Increasing the cycle numbers leads to further raising or sinking of the material points, as observed experimentally. The existence, general location, characteristic size and wave length roughness development at the free surface are in qualitative agreement with experimental observations of Figure 3. However statistical analysis of the results would require several simulations and experiments, which are not available due to the high cost of both computation and experiment. Compared with the experimental observations, the predicted absolute values of the roughness induced by heterogeneous straining of the free surface are significantly lower. The reasons are the same as for the underestimation of strain values predicted by the FE simulation compared to DIC results. They will be discussed later in this work. The presented experimental and computational results presented in this work are the first of this kind for low cycle fatigue in tantalum. Future work is needed to consolidate them and add measurements like roughness and out of plane displacement.

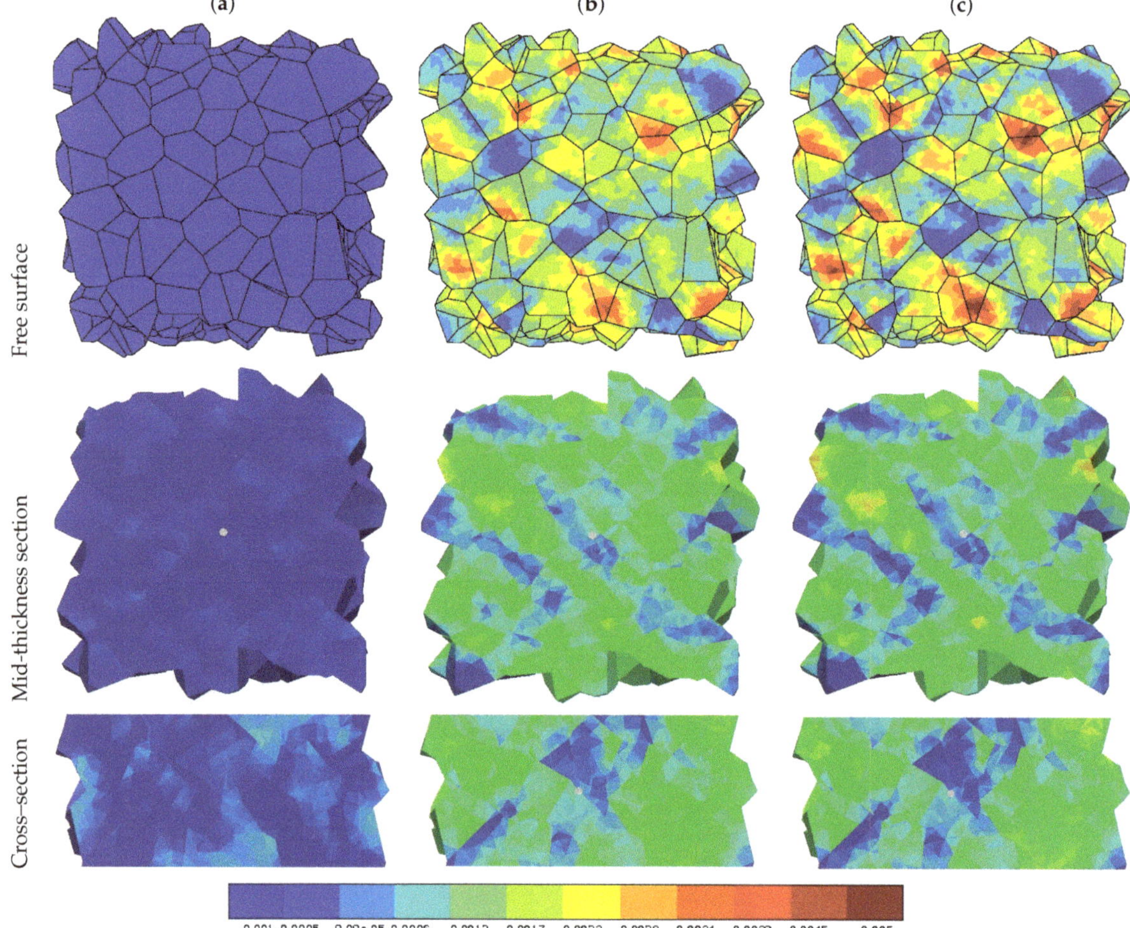

Figure 10. Simulation of a cyclic test at $\pm 0.2\%$. Evolution of the strain component ε_{22} at (**a**) cycle 66, (**b**) cycle 666, and (**c**) cycle 1066 at the free surface (top) and at two mid-sections perpendicular to directions 3 (middle) and 2 (bottom). The loading direction 2 is vertical for the top and middle pictures.

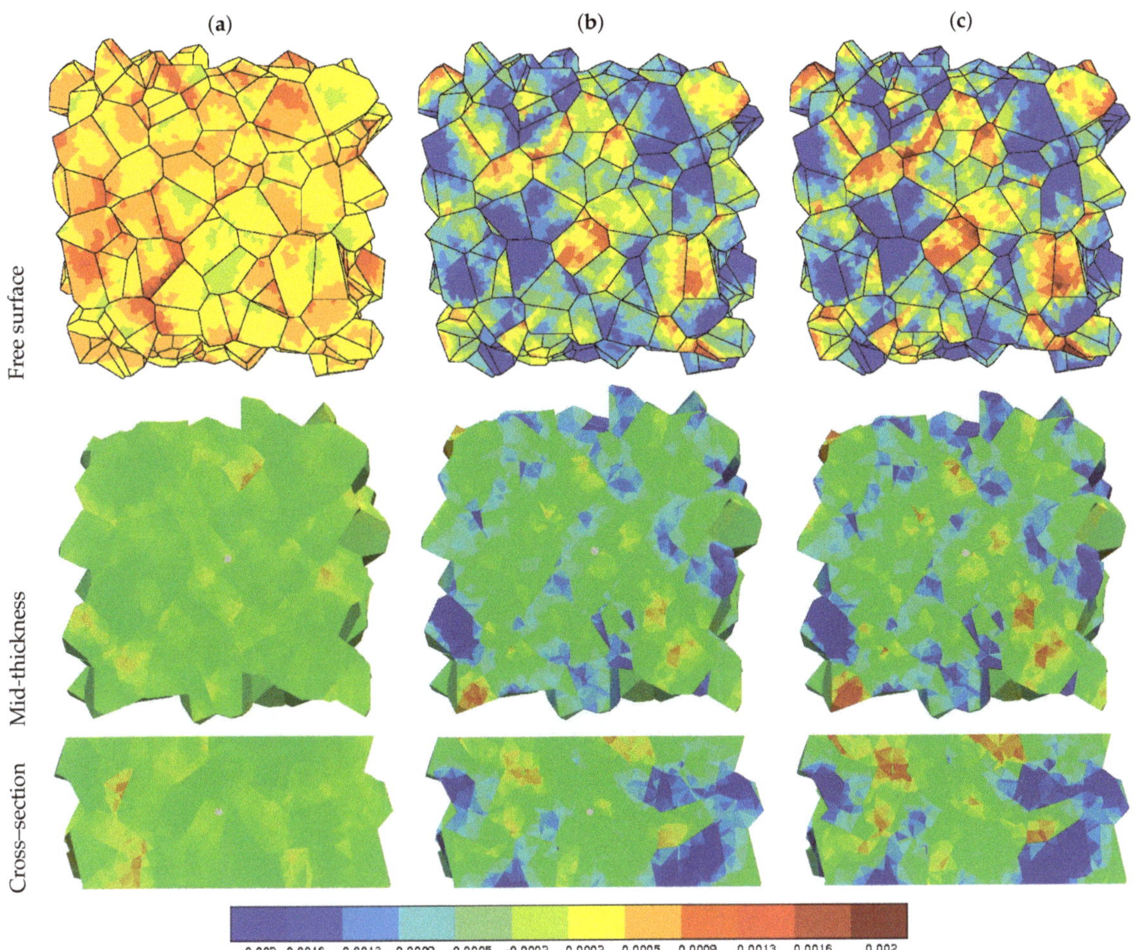

Figure 11. Simulation of a cyclic test at $\pm 0.2\%$. Evolution of the strain component ε_{11} at (**a**) cycle 66, (**b**) cycle 666, and (**c**) cycle 1066 at the free surface (top) and at two mid-sections perpendicular to directions 3 (middle) and 2 (bottom). The loading direction 2 is vertical for the top and middle pictures.

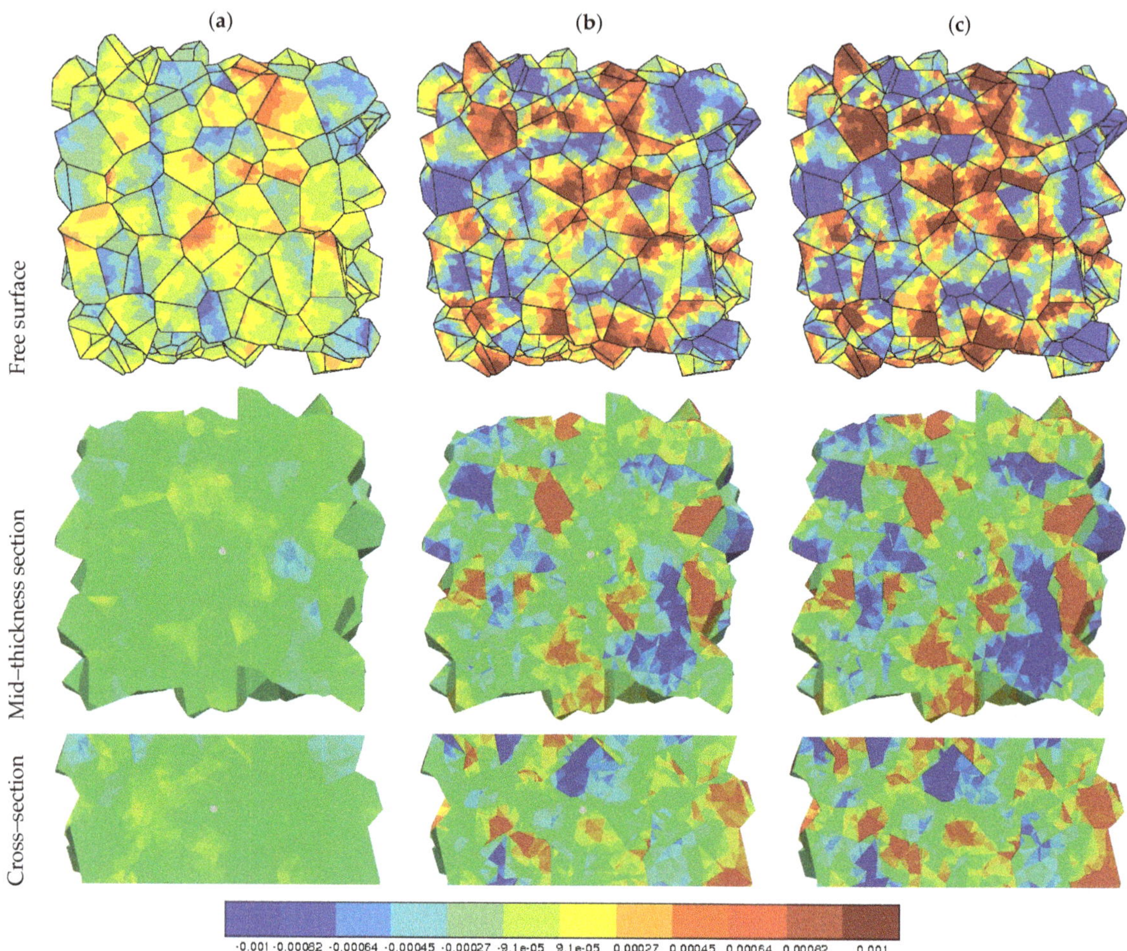

Figure 12. Simulation of a cyclic test at ±0.2%. Evolution of the strain component ε_{12} at (**a**) cycle 66, (**b**) cycle 666, and (**c**) cycle 1066 at the free surface (top) and at two mid-sections perpendicular to directions 3 (middle) and 2 (bottom). The loading direction 2 is vertical for the top and middle pictures.

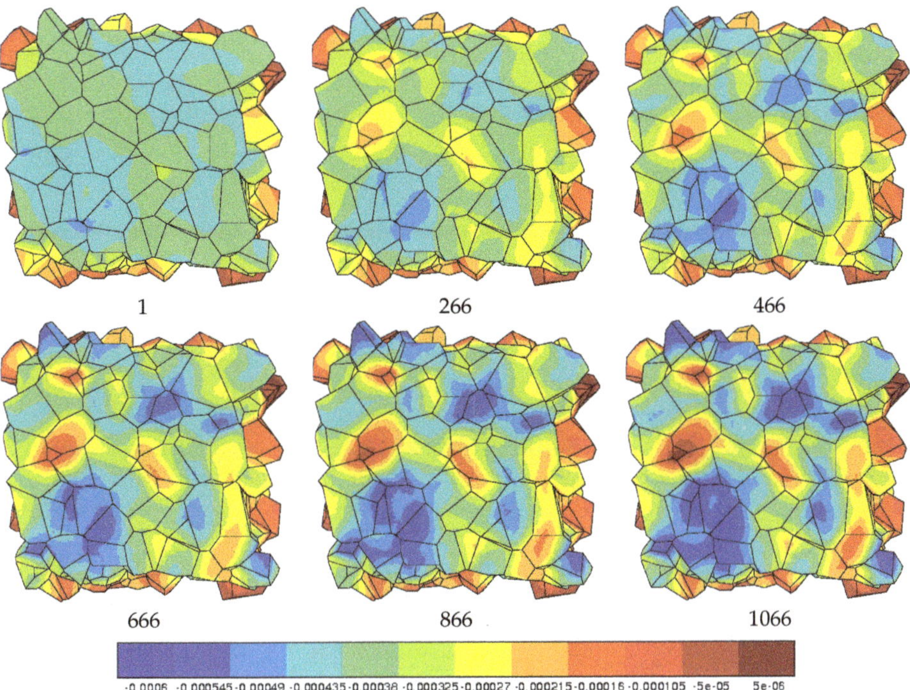

Figure 13. Fields of the displacement component in the direction 3 at the free surface showing the evolution of surface rougness with cycle numbers. Displacement in mm, to be compared to the specimen thickness used in the FE simulation, namely, 0.5 mm.

3.3.2. Plastic Slip Evolution

The sum of plastic slip contributions, $\sum_s |\gamma^s|$, is a variable characterizing the intensity of plastic slip activity inside the grains. The corresponding field is shown in Figure 14 at three different cycle numbers. According to the proposed model, the slip variables γ^s can take positive or negative values depending on the sign of the effective stress $\tau^s - x^s$. This variable must be distinguished from cumulative slip v^s which always increases. Blue zones indicate some unfavorably oriented grains where no (or low) plastic slip activity is observed after 1000 cycles. In contrast, red zones point out regions where plastic slip reaches high values, sometimes more than 0.03. Strong heterogeneities from grain to grain are circled by black lines in the top right of Figure 14. They are in qualitative agreement with the experimentally observed zones of strong plastic slip gradients in Figure 3. The slip activity is found to increase drastically over the more than thousand cycles. Significant evolution in plastic slip activity is observed between the cycles 666 and 1066. This is in qualitative agreement with the experimental observations of Figure 3, for instance between 1000 and 2000 cycles. The effect is stronger than for the strain fields of Figures 11 and 12. When the number of cycles increases, the slip increments decrease but remain positive and strongly heterogeneous. These features of plastic slip accumulation are discussed in the next section.

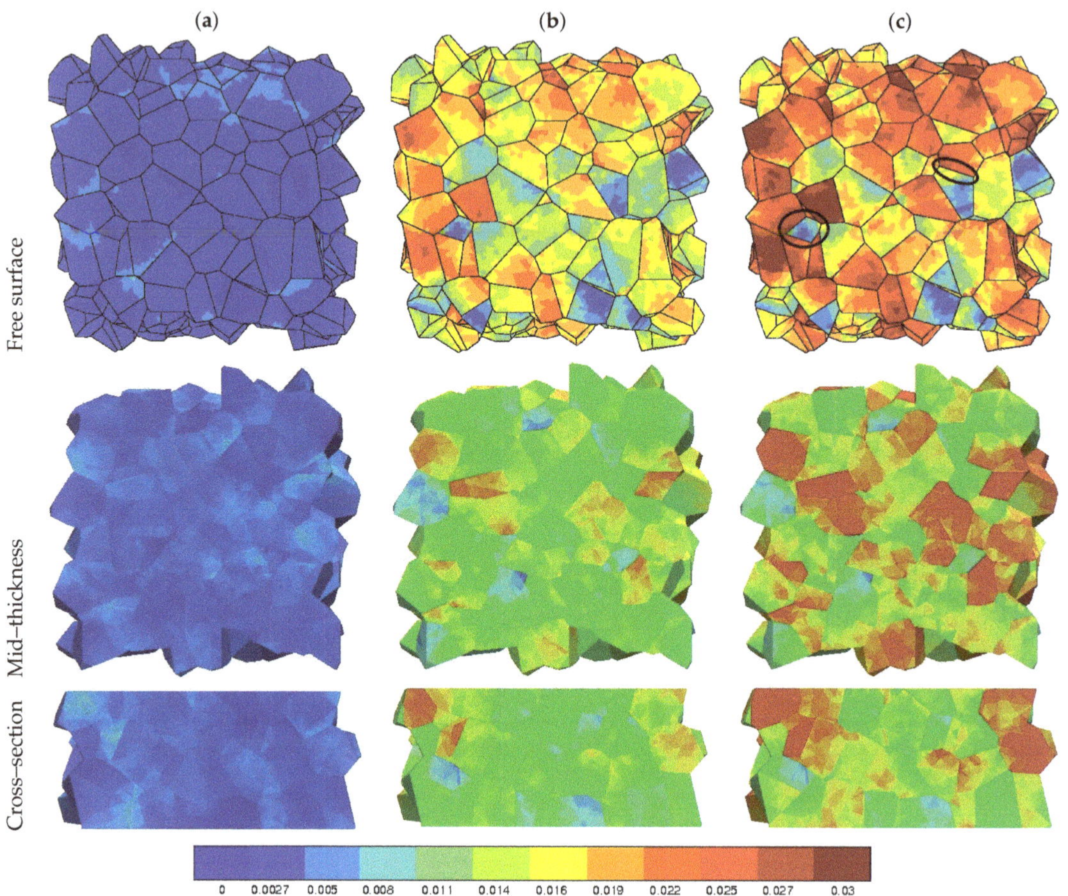

Figure 14. Total slip evolution $\sum_s |\gamma^s|$ for several cycles during a fatigue test simulation at $\pm0.2\%$.

4. Discussion

The salient features of the computational results are now discussed, namely, comparison between experimental and simulation results, local ratcheting phenomena at the free surface, and the development of a fatigue crack initiation criterion.

4.1. The FE Model Underestimates Strain Heterogeneities and Roughness in Fatigue

Experiment and simulation agree in showing strong strain heterogeneities developing at the free surface of a tantalum polycrystaline aggregate after a few thousands of cycles, even for a moderate cyclic total strain amplitude of 0.2%. Strain localization and roughness evolution are observed at grain boundaries, triple junctions but can also cross several grains. In particular the FE model predicts highly contrasted fields of plastic activity in some grains where high values of plastic slip are found, more than ten times the prescribed strain. However, FE simulations quantitatively underpredicts the strain levels at the free surface. Several reasons can be put forward. First, the proposed model assumes continuity of the displacement components at grain boundaries where strain and roughness effects tend to localize as shown in the experiment and simulations. The experiment shows that significant grain boundary sliding (GBS) takes place in some grains as noticed in several observations of polycrystalline plasticity [71]. GBS is an important deformation and damage mechanism which has been observed in the present work but not sufficiently documented. Instead, we provide more evidence of grain boundary cracking after 5000 cycles, see Figure 7. Second, sharp slip lines and intense intrusion and extrusion bands are observed in the experiments that cannot be accounted for by the proposed continuum mechanics approach. These discontinuities are characteristic of crystal plasticity physical mechanisms and can lead to strong local strain levels. Alternative modeling approaches are needed to capture these mechanisms like discrete dislocation dynamics simulations [72]. Third, the local strain levels also depend on the mesh resolution, which is definitely too coarse in the presented simulations in order to spare computational time and make the simulation possible over a large number of cycles. Finally, the experiment reveals that some cluster of 50 to 100 grains can undergo average strain levels that significantly exceed the prescribed overall strain level. In contrast, the overall mean strain is prescribed to the simulated volume of 250 grains, which strongly limits the development of strain heterogeneities from grain to grain. This pleads for the consideration of larger Volume Element sizes than the one considered in this work, which represents a challenging task for fatigue simulations. Strain heterogeneities and roughness development are closely related to ratcheting phenomena occurring close to and at the free surfaces, as evidenced and discussed in [17]. Only 1000 cycles were simulated here but extrapolations are possible to 3000 cycles based on local established ratcheting rates.

4.2. Local Ratcheting Behavior

The accumulation of plastic slip at some specific locations of the aggregate plays a major role in crack initiation. This accumulation is enhanced by local ratcheting phenomena, i.e., the existence of a strain component increment of given sign and magnitude after each cycle [73]. The evidence of such ratcheting effects has been demonstrated using the present simulation results in [17]. A major result of this analysis was that ratcheting values are significantly higher at the free surface than in the bulk, which is an essential feature of fatigue crack initiation correctly reproduced by the analysis. Some further results are presented here that serve for the definition of a fatigue criterion in the next subsection.

From the previous observation of the free surface, 6 nodes have been selected in order to document the local cyclic response of material points, see Figure 15. Among the 6 nodes, 3 are located in plastically active zones (nodes 1–3), whereas the three others are situated in less active areas (nodes 4–6). This can be inferred from the cyclic stress–strain σ_{22}–ε_{22} loops of Figure 16. All nodes display largely open cyclic loops revealing significant plastic deformation. The first loop is characterized in each case by the peak stress associated with static strain aging typical of this material [15]. Then, the loops become symmetric with respect to the stress, due to relaxation of the mean stress. In contrast, the minimal and maximal strain levels are not symmetric. The loops are not saturated after 1000 cycles and strain accumulates either in tension or compression, which corresponds to local ratcheting phenomena. Significant tensile ratcheting is found for the nodes 1–3, meaning that the mean strain is positive and increases monotonically, whereas more limited compressive ratcheting is observed for the nodes 4–6. The local stress–strain loops strongly differ from the macroscopic loop of Figure 9 which is fully symmetric. Local stress levels σ_{22} depend on the crystallographic orientation of each grain and of its neighbors. They are generally lower than the macroscopic stress (150 vs. 200 MPa), except for node 5 (300 MPa).

The ratcheting phenomenon (also called cyclic creep) occurs when the plastic strain increment is not fully reversed for a cyclic loading [74]. One-dimensional ratcheting occurs for instance for non-vanishing mean stress loadings: a positive (resp. negative) ratcheting occurs for a positive (resp. negative) mean stress [75–77]. However, the mean stress is found here to relax to zero in all the plotted loops after less than 100 cycles. The origin of the continuing ratcheting is the multiaxial loading experienced by the material points. The σ_{11}–ε_{11} loops in Figure 17 show that the transverse stress component is significant for the six selected nodes. Transverse strain ratcheting is observed for nodes 1–3. Similar observation was made for the shear component so that it can be concluded that the material points experience multiaxial ratcheting, responsible for plastic strain accumulation and ultimately crack initiation. The multiaxial stress–strain state is induced by the complex plastic strain incompatibilities from grain to grain.

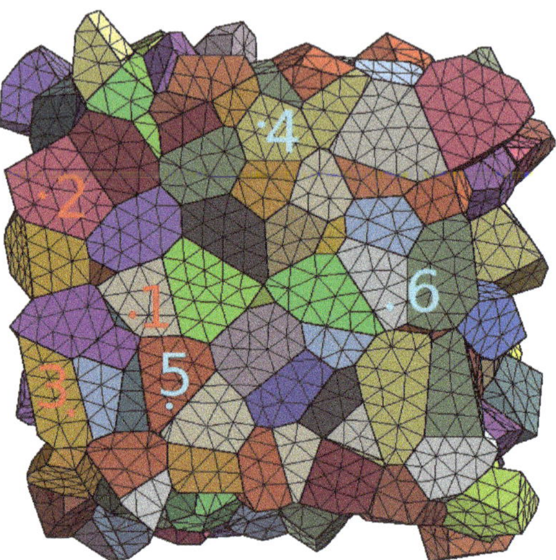

Figure 15. Selected nodes for the local study, after the work in [17].

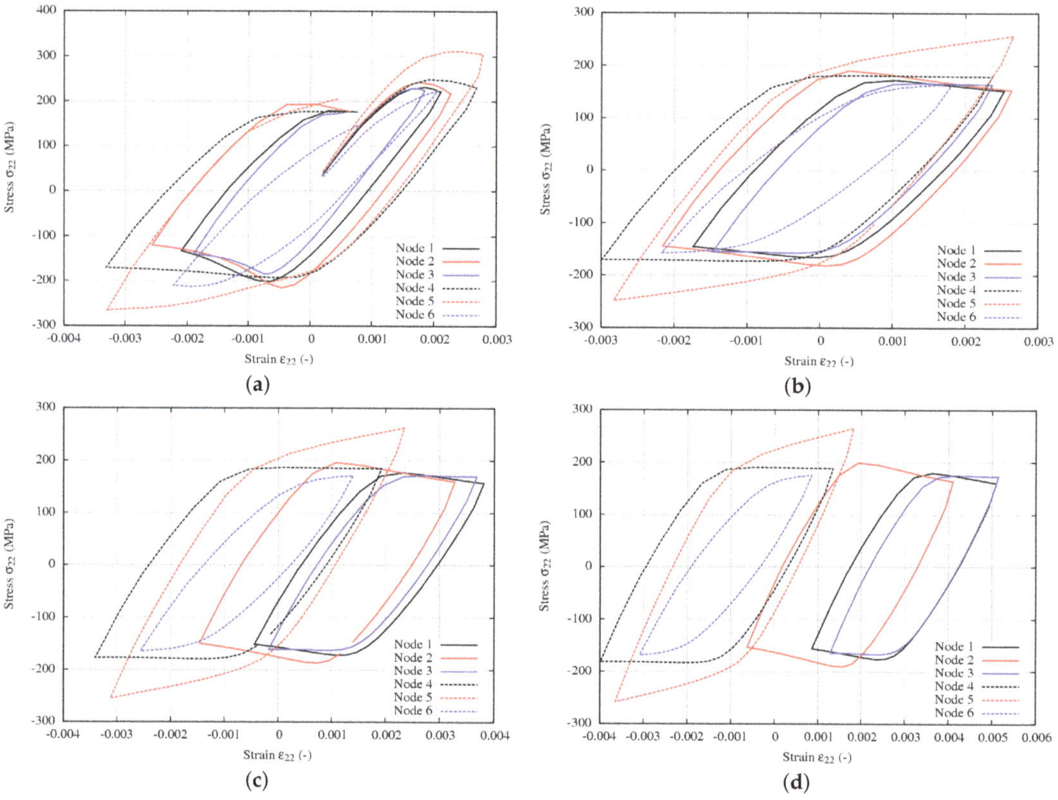

Figure 16. Strain loops σ_{22}–ε_{22} for the six nodes defined in Figure 15 at several cycle numbers: (**a**) cycle 1, (**b**) cycle 66, (**c**) cycle 466, and (**d**) cycle 1066.

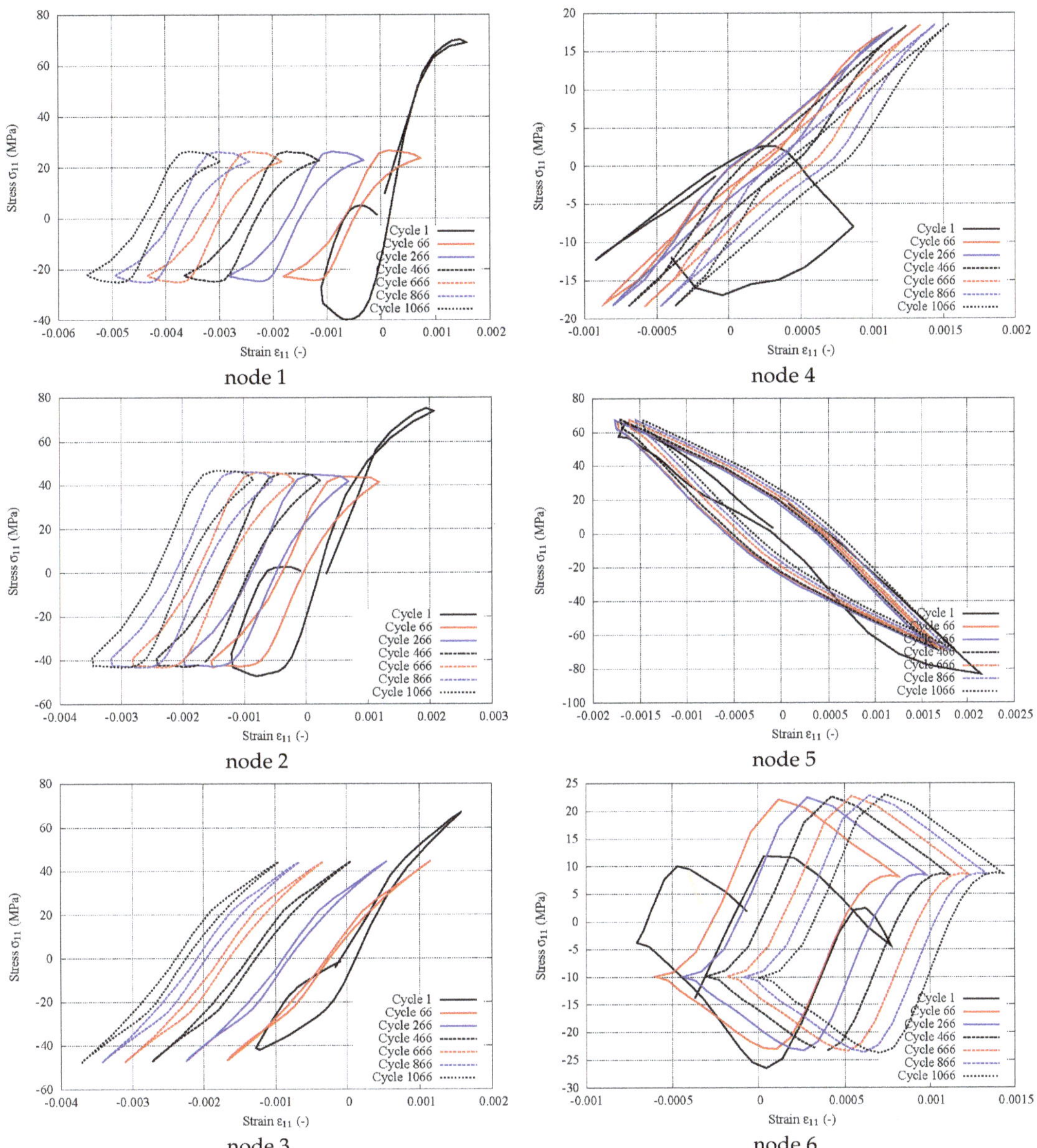

Figure 17. Local curves σ_{11}–ε_{11} for several cycles at 6 different finite element nodes, after the work in [17].

The previous results are used to define a

Ratcheting Indicator

The progressive accumulation of plastic strain at the surface during strain-controlled cycling is regarded in this work as the relevant physical mechanism for fatigue crack initiation in the grains of a polycrystal. It has been reported that it can lead to crack nucleation and finally to the global failure of a structure (see the works in [35,73,74,77]), and is in accordance with the experimental observations of the present work. A measure of local ratcheting has been proposed in the reference [77] for the analysis of plasticity induced by fretting contact. For that purpose, the local plastic strain increment $(\Delta\varepsilon_{ij}^{p})_{ratch}$ is defined at each material point for each component of the plastic strain tensor as

$$(\Delta\varepsilon_{ij}^{p})_{ratch} = \varepsilon_{ij}^{p}\big|_{end\ of\ the\ cycle} - \varepsilon_{ij}^{p}\big|_{start\ of\ the\ cycle} \tag{8}$$

The effective plastic strain increment is then computed as

$$\Delta\varepsilon_{ratch,eff}^{p} = \sqrt{\frac{2}{3}(\Delta\varepsilon_{ij}^{p})_{ratch}(\Delta\varepsilon_{ij}^{p})_{ratch}} \tag{9}$$

where repeated indices are summed up. It is computed for each individual cycle at each material point. It is regarded as an indicator of local ratcheting as departure of $\Delta\varepsilon_{ratch,eff}^{p}$ from zero is a signature of accumulation of plastic staining. Fields of this variable were provided at various cycle numbers in [15] to demonstrate the significance of ratcheting at the free surface of the polycrystal. The highest ratcheting values are reached mainly close to grain boundaries and at triple junctions.

4.3. Fatigue Criterion

According to the micrographs of Figure 7, about 5000 cycles are required for the nucleation of the first cracks under the considered loading conditions. These cracks are found to be mainly intergranular. The experimental observations of Section 2.2 also show the strong local strain heterogeneity, with intense surface roughening occurring at some grain boundaries with corresponding peaks and sinks. The previous finite element simulations have shown that strain gradients and crystal misorientations develop on both sides of some grain boundaries and that they are leading to local ratcheting at several places. These experimental and numerical observations are in accordance with grain boundary behavior in BCC metals reported in [35].

Local fatigue criteria at the grain scale have been proposed and discussed in [19,21] and in the references quoted therein. They take into account the crystallographic deformation processes, the accumulation and the amplitude of plastic slip, mean resolved shear and normal stresses. The Dang–Van criterion applied to slip systems is based on the hydrostatic stress and the resolved shear stress, and thus predicts crack initiation in the bulk and not at the surface [78]. However, in the present case, it has been shown that crack initiation occurs at the surface, see Figure 7. Thus, such criteria are inappropriate for our application.

The accumulated slip at the local level in polycrystals was used by [23,79,80] to predict microcrack propagation, including the evaluation of the crack length increment and its direction. In the present work, it is proposed to use the previously defined ratcheting indicator as the main ingredient of the fatigue crack initiation criterion. Using the local plastic strain increment $\Delta\varepsilon_{ratch,eff}^{p}(x_i, t)$ at material point x_i at time t allows to determine the most active plastic zones. In addition, based on the local data, the local equivalent plastic strain is defined at each material point located at x_i at time t as

$$\varepsilon_{eq}^{p}(x_i, t) = \sqrt{\frac{2}{3}\varepsilon_{ij}^{p}(x_i, t)\varepsilon_{ij}^{p}(x_i, t)} \tag{10}$$

A critical strain threshold ε_c^{p} is defined, as proposed in the literature on tantalum as a first indicator [81].

The plastic strain increment $\Delta\varepsilon_{ratch,eff}(x_i, t)$ at every material point x_i is assumed to have reached a constant value after 1000 cycles, but the local heterogeneity in the aggregate is preserved.

A crack is then assumed to initiate at the material point x_i at cycle number N as soon as

$$\varepsilon_{eq}^{p}(x_i, N) = \varepsilon_c^{p} \tag{11}$$

where $\varepsilon_{eq}^{p}(x_i, N)$ must be extrapolated from the value at the 1000th cycle. Considering that the plastic strain increment is constant at that time, i.e., $\Delta\varepsilon_{ratch,eff}(x_i, N > 1000) \approx \Delta\varepsilon_{ratch,eff}(x_i, 1000)$, the local equivalent plastic strain value is estimated as

$$\varepsilon_{eq}^{p}(x_i, N) = \varepsilon_{eq}^{p}(x_i, 1000) + \Delta\varepsilon_{ratch,eff}^{p}(x_i, 1000)(N - 1000) \tag{12}$$

It follows that the number of cycles for crack initiation at the material point x_i is

$$N_{initiation}(x_i) = 1000 + \frac{\varepsilon_c^{p} - \varepsilon_{eq}^{p}(x_i, 1000)}{\Delta\varepsilon_{ratch,eff}^{p}(x_i, 1000)} \tag{13}$$

A postprocessing of the FE results has been applied in order to calculate an average value of the fatigue criterion around each integration point in a set of elements. This volume average value is then assigned to each integration point. A value of 10% of the grain size is chosen as the radius of the spherical averaging domain, as proposed in the literature for crack initiation, for instance, in [82]. Half a sphere is used for points lying at the free surface. This local averaging procedure is used to avoid artifacts associated with local extremal values that may be due to poor element shape or size. This also allows for the consideration of process zone size depending on the fatigue crack initiation mechanisms.

In [83] it is proposed to consider that the critical local strain in fatigue ε_c^{p} at a material point in the aggregate is chosen identical to the overall ultimate strain for a monotonic tensile stress. Thus, the local strain needed for the nucleation of a crack is the same as the macroscopic ductility. This is a rough estimation that needs to be refined if sufficient experimental data is available. According to a recent study by Nadal [81] on tantalum, the macroscopic critical strain has been assessed for several strain amplitudes as $\varepsilon_f = \varepsilon_c^{p} = 0.07$. This critical value has been selected for the evaluation of the present fatigue crack initiation criterion.

Figure 18 presents the prediction of the number of cycles required for crack initiation as defined by Equation (13) using $\varepsilon_c^{p} = 0.07$. Crack initiation occurs in areas studied earlier for nodes 1–3, especially for node 3 where the ratcheting was very intense. This figure can be compared to the maps of effective plastic strain increments at the free surface and mid-section in Figure 6 from the work in [17]. The lowest values of cycle numbers are listed in red in Figure 18 indicates the locations where fatigue cracking is expected to start. They correspond to the locations of maximal ratcheting. Some areas present a strong strain discontinuity on both sides of grain boundaries and are preferential crack initiation sites. The predominant crack initiation close to grain boundaries predicted by the simulation is in agreement with the previous experimental observations. Moreover, the comparison of Figure 18a,b demonstrates that crack initiate earlier at the free surface than in the bulk, which is a well-known feature of fatigue in pure metals. This is related to the fact that ratcheting phenomena were also found to be dominant at the free surface [17].

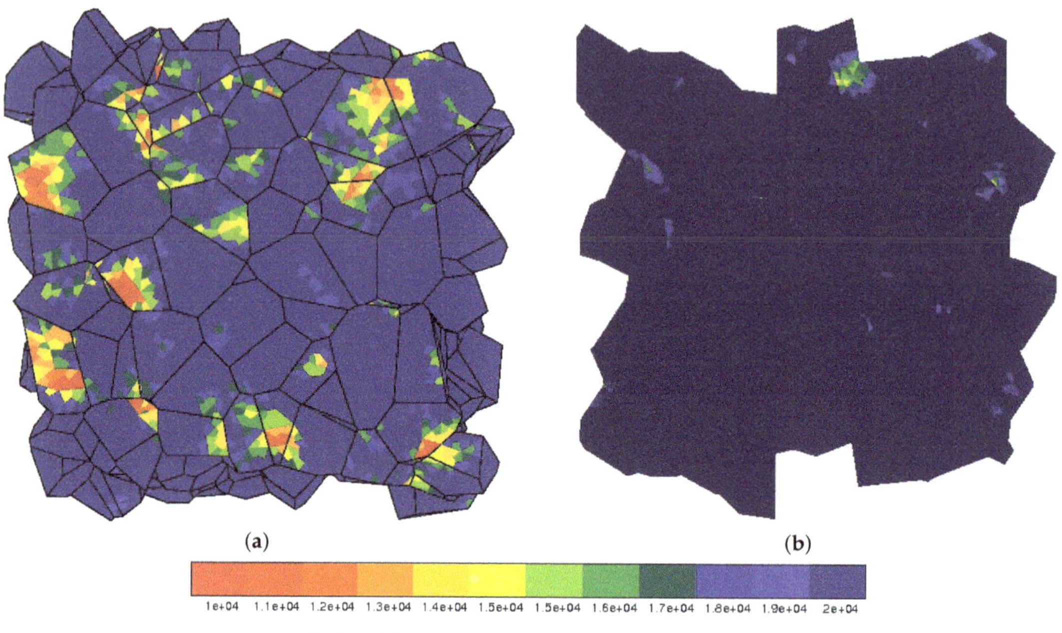

Figure 18. Prediction of cycle numbers to crack initiation per element for the fatigue test simulation at $\pm0.2\%$: (**a**) surface prediction and (**b**) half thickness prediction.

Figure 19 shows the histogram of the number of cycles to crack initiation for each node of the aggregate. The smallest values are reached at the surface close to node 3 identified on Figure 15. The analysis of this histogram shows that the lowest value is about 5000 cycles, and that the maximal value is 4.2×10^5 cycles. The mean value is 3.7×10^4 cycles and the standard deviation is 1.65×10^4 cycles. The found lowest value is in good agreement with the experimental observations of first cracks in Figure 7 already observed after 5000 cycles for stage I fatigue crack initiation. More accurate determination of crack initiation cycle numbers requires the choice of two additional quantities: a threshold for the statistical definition of crack initiation in the histogram of Figure 19 and the definition of a crack length for macro-initiation cracking, for instance, 300 µm, which are not reached after 5000 cycles in our experiment.

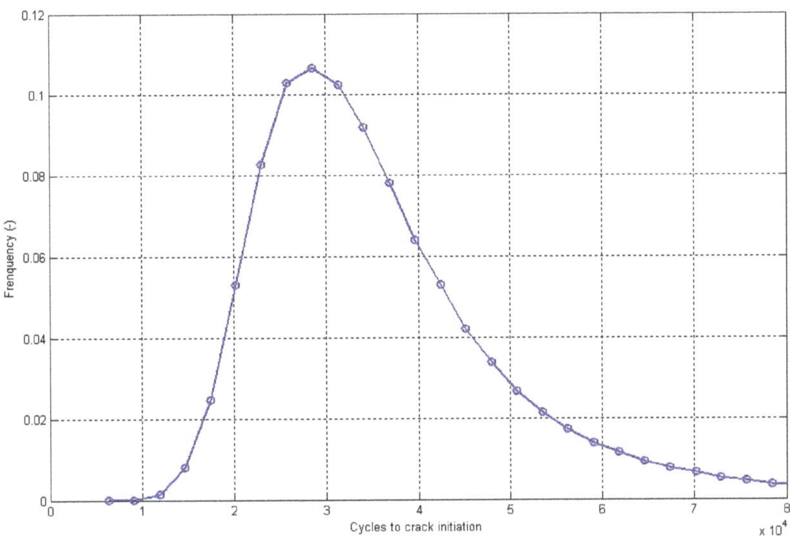

Figure 19. Histogram of cycle numbers for crack initiation for the whole aggregate at ±0.2%.

5. Conclusions

The main findings of the present work are the following:

- Unique experimental results from an interrupted cyclic tension–compression test have been provided. They show the development of plastic activity in the form of sinuous slip lines and intense slip bands. The development of lattice curvature resulting from inhomogeneous plastic slip inside the grains was illustrated after 100, 1000, 2000, and 3000 cycles.
- Strain field measurements reveal local strain values of a few percents even though the overall loading was ±0.2%. These values are compatible with observed local amounts of lattice rotation. This results in significant roughening of the free surface during cycling.
- Fatigue cracks were shown to initiate between 3000 and 5000 cycles mainly in the form of intergranular cracks.
- Unique FE simulation results of crystal plasticity in a semi-periodic polycrystalline aggregate were provided for more than 1000 cycles. They confirm but underestimate the strain heterogeneity that develops at the free surface.
- An original fatigue criterion was proposed for stage I crack initiation at the grain level. It is based on the local determination of a multiaxial ratcheting indicator to be compared with a critical plastic equivalent strain value. The criterion indirectly incorporates the effects of resolved shear stresses and mean stress through the induced multiaxial plastic slip.

In this contribution, the hypothesis that local ratcheting phenomena in the surface grains of a tantalum polycrystal are responsible for fatigue crack initiation has been tested quantitavely. This is a well-known phenomenon contributing to the formation of extrusion/intrusion bands at the surface of many cyclically loaded metals and alloys. The ratcheting values deduced from the full-field simulations were used to make lifetime assessments in terms of number of cycle to initiation of fatigue cracks of given length. The experimentally observed surface roughness induced by cyclic loading, especially at some grain boundaries, corroborates the ratcheting phenomenon as a plausible fatigue crack initiation mechanism. More detailed microscopic analyses remain however to be performed to investigate the dislocation structures in such zones of accumulated plastic strain.

In this first step, it was not possible to have a one to one mapping between the experimentally investigated zone and the polycrystalline aggregate used in the FE simulation. Such a correspondence requires a precise knowledge of the 3D grain structure that can be obtained using Focused Ion Beam techniques. This remains to be done in the future for a more detailed analysis of the plastic behavior of tantalum. Guidelines to perform such one to one mapping can be found in [84]. Note that only one sample was experimentally tested and only one polycrystalline aggregate was simulated, so that a more complete statistical analysis of the results should be the objective of future work in this field. In particular, a larger number of cracks should be considered to confirm the predominance of intergranular fatigue cracking. Such additional analysis is necessary to draw definitive conclusions on the cracking mechanisms and on the validity of the proposed fatigue crack initiation model.

Improvements of the used crystal plasticity model are possible to include more sophisticated dislocation evolution equations for cyclic plasticity [85]. It must be acknowledged that the used modeling framework cannot account for grain boundary sliding (or cracking) or PSB formation. Validation of the presented fatigue crack initiation model is necessary for other loading conditions including non-vanishing mean stress loading and various loading ratios. The proposed fatigue crack initiation criterion could be combined with a more specific intergranular cracking model based on the alternating traction vector acting locally at grain boundaries. It is not sure that the consideration of a competition between transgranular and intergragranular crack initiation will give a better description of the experimental results as the proposed model predicts, in any case, crack initiation close to grain boundaries and triple junctions. More importantly, coupled fatigue damage–plasticity models could then be used to describe the local propagation of such initiated cracks. In that case, intergranular and transgranular cracks can be distinguished. Such a coupled crystal plasticity and damage model was proposed and applied to crack initiation and propagation in single crystal superalloys in [86]. This includes a gradient damage term which is necessary to obtain mesh-independent crack simulation results. Such a regularization was used in [87,88].

Author Contributions: D.C. performed the experiments and the finite element analysis. He also prepared the figures and the first draft of the paper; E.F. supervised the experimental work; S.F. (Sylvain Flouriot) and T.P. supervised the experimental and computational work and did the project administration; M.M. and S.F. (Samuel Forest) supervised the modeling and simulation part. All authors have read and agreed to the published version of the manuscript.

Funding: This research received no external funding.

Data Availability Statement: Not applicable.

Conflicts of Interest: The authors declare no conflict of interest.

References

1. Nemat-Nasser, S.; Isaacs, J.; Liu, M. Microstructure of high-strain, high-strain-rate deformed tantalum. *Acta Mater.* **1998**, *46*, 1307–1325. [CrossRef]
2. Nemat-Nasser, S.; Okinaka, T.; Ni, L. A physically-based constitutive model for BCC crystals with application to polycrystalline tantalum. *J. Mech. Phys. Solids* **1998**, *46*, 1009–1038. [CrossRef]
3. Chen, Y.; Meyers, M.; Nesterenko, V. Spontaneous and forced shear localization in high-strain-rate deformation of tantalum. *Mater. Sci. Eng. A* **1999**, *268*, 70–82. [CrossRef]
4. Diehl, M.; Niehuesbernd, J.; Bruder, E. On the PLC Effect in a Particle Reinforced AA2017 Alloy. *Metals* **2019**, *9*, 1252. [CrossRef]
5. Bornert, M.; Brémand, F.; Doumalin, P.; Dupré, J.C.; Fazzini, M.; Grédiac, M.; Hild, F.; Mistou, S.; Molimard, J.; Orteu, J.J.; et al. Assessment of Digital Image Correlation Measurement Errors: Methodology and Results. *Exp. Mech.* **2009**, *49*, 353–370. [CrossRef]
6. Cho, H.; Bronkhorst, C.A.; Mourad, H.M.; Mayeur, J.R.; Luscher, D.J. Anomalous plasticity of body-centered-cubic crystals with non-Schmid effect. *Int. J. Solids Struct.* **2018**, *139–140*, 138–149. [CrossRef]
7. Alleman, C.; Ghosh, S.; Luscher, D.J.; Bronkhorst, C.A. Evaluating the effects of loading parameters on single-crystal slip in tantalum using molecular mechanics. *Philos. Mag.* **2014**, *94*, 92–116. [CrossRef]
8. Lim, H.; Carroll, J.D.; Battaile, C.C.; Boyce, B.L.; Weinberger, C.R. Quantitative comparison between experimental measurements and CP-FEM predictions of plastic deformation in a tantalum oligocrystal. *Int. J. Mech. Sci.* **2015**, *92*, 98–108. [CrossRef]

9. Moussa, C.; Bernacki, M.; Besnard, R.; Bozzolo, N. Statistical analysis of dislocations and dislocation boundaries from EBSD data. *Ultramicroscopy* **2017**, *179*, 63–72. [CrossRef]
10. Lim, H.; Carroll, J.D.; Michael, J.R.; Battaile, C.C.; Chen, S.R.; Lane, J.M.D. Investigating active slip planes in tantalum under compressive load: Crystal plasticity and slip trace analyses of single crystals. *Acta Mater.* **2020**, *185*, 1–12. [CrossRef]
11. Hansen, L.T.; Fullwood, D.T.; Homer, E.R.; Wagoner, R.H.; Lim, H.; Carroll, J.D.; Zhou, G.; Bong, H.J. An investigation of geometrically necessary dislocations and back stress in large grained tantalum via EBSD and CPFEM. *Mater. Sci. Eng. A* **2020**, *772*, 138704. [CrossRef]
12. Chen, J.; Hahn, E.N.; Dongare, A.M.; Fensin, S.J. Understanding and predicting damage and failure at grain boundaries in BCC Ta. *J. Appl. Phys.* **2019**, *126*, 165902. [CrossRef]
13. Wauthle, R.; van der Stok, J.; Yavari, S.A.; Van Humbeeck, J.; Kruth, J.P.; Zadpoor, A.A.; Weinans, H.; Mulier, M.; Schrooten, J. Additively manufactured porous tantalum implants. *Acta Biomater.* **2015**, *14*, 217–225. [CrossRef] [PubMed]
14. Marechal, D.; Saintier, N.; Palin-Luc, T.; Nadal, F. High-Cycle Fatigue Behaviour of Pure Tantalum under Multiaxial and Variable Amplitude Loadings. *Adv. Mater. Res.* **2014**, *891–892*, 1341–1346. [CrossRef]
15. Colas, D.; Finot, E.; Forest, S.; Flouriot, S.; Mazière, M.; Paris, T. Investigation and modelling of the anomalous yield point phenomenon in pure Tantalum. *Mater. Sci. Eng.* **2014**, *A615*, 283–295. [CrossRef]
16. Taleb, L.; Keller, C. Experimental contribution for better understanding of ratcheting in 304LSS. *Int. J. Mech. Sci.* **2018**, *105*, 527–535. [CrossRef]
17. Colas, D.; Finot, E.; Flouriot, S.; Forest, S.; Mazière, M.; Paris, T. Local Ratcheting Phenomena in the Cyclic Behavior of Polycrystalline Tantalum. *JOM J. Miner. Met. Mater. Soc.* **2019**, *71*, 2586–2599. [CrossRef]
18. Zhou, G.; Jeong, W.; Homer, E.R.; Fullwood, D.T.; Lee, M.G.; Kim, J.H.; Lim, H.; Zbib, H.; Wagoner, R.H. A predictive strain-gradient model with no undetermined constants or length scales. *J. Mech. Phys. Solids* **2020**, *145*, 104178. [CrossRef]
19. McDowell, D. Simulation-based strategies for microstructure-sensitive fatigue modeling. *Mater. Sci. Eng.* **2007**, *A468*, 4–14. [CrossRef]
20. Li, Y.; Aubin, V.; Rey, C.; Bompard, P. Polycrystalline numerical simulation of variable amplitude loading effects on cyclic plasticity and microcrack initiation in austenitic steel 304L. *Int. J. Fatigue* **2012**, *42*, 71–81. [CrossRef]
21. Dunne, F. Fatigue crack nucleation: Mechanistic modelling across the length scales. *Curr. Opin. Solid State Mater. Sci.* **2014**, *18*, 170–179. [CrossRef]
22. Lu, J.; Sun, W.; Becker, A. Material characterisation and finite element modelling of cyclic plasticity behaviour for 304 stainless steel using a crystal plasticity model. *Int. J. Mech. Sci.* **2016**, *105*, 315–329. [CrossRef]
23. Proudhon, H.; Li, J.; Ludwig, W.; Roos, A.; Forest, S. Simulation of short fatigue crack propagation in a 3D experimental microstructure. *Adv. Eng. Mater.* **2017**. [CrossRef]
24. Hor, A.; Saintier, N.; Robert, C.; Palin-Luc, T.; Morel, F. Statistical assessment of multiaxial HCF criteria at the grain scale. *Int. J. Fatigue* **2014**, *67*, 151–158. [CrossRef]
25. Le, V.; Morel, F.; Bellett, D.; Saintier, N.; Osmond, P. Multiaxial high cycle fatigue damage mechanisms associated with the different microstructural heterogeneities of cast aluminium alloys. *Mater. Sci. Eng. A* **2016**, *649*, 426–440. [CrossRef]
26. Ghosh, S.; Chakraborty, P. Microstructure and load sensitive fatigue crack nucleation in Ti-6242 using accelerated crystal plastificty FEM simulations. *Int. J. Fatigue* **2013**, *48*, 231–246. [CrossRef]
27. Zeghadi, A.; Nguyen, F.; Forest, S.; Gourgues, A.F.; Bouaziz, O. Ensemble averaging stress–strain fields in polycrystalline aggregates with a constrained surface microstructure—Part 1: Anisotropic elastic behaviour. *Philos. Mag.* **2007**, *87*, 1401–1424. [CrossRef]
28. Zeghadi, A.; Forest, S.; Gourgues, A.F.; Bouaziz, O. Ensemble averaging stress–strain fields in polycrystalline aggregates with a constrained surface microstructure—Part 2: Crystal plasticity. *Philos. Mag.* **2007**, *87*, 1425–1446. [CrossRef]
29. Wasserbach, W. *Work-Hardening and Dislocation Behaviour of Tantalum and Tantalum Alloys*; The Minerals, Metals and Materials Society: Pittsburgh, PA, USA, 1996.
30. Hosseini, E.; Kazeminezhad, M. Dislocation structure and strength evolution of heavily deformed tantalum. *Int. J. Refract. Met. Hard Mater.* **2009**, *27*, 605–610. [CrossRef]
31. Frenois, S.; Munier, E.; Feaugas, X.; Pilvin, P. A polycrystalline model for stress-strain behaviour of tantalum at 300cK. *J. De Phys. IV* **2001**, *11*, 301–308.
32. Norlain, M. Comportement Mécanique du Tantale, Texture et Recristallisation. Ph.D. Thesis, Mines ParisTech, Paris, France, 1999.
33. Kerisit, C.; Logé, R.E.; Jacomet, S.; Llorca, V.; Bozzolo, N. EBSD coupled to SEM in situ annealing for assessing recrystallization and grain growth mechanisms in pure tantalum. *J. Microsc.* **2013**, *250*, 189–199. [CrossRef] [PubMed]
34. Moussa, C.; Bernacki, M.; Besnard, R.; Bozzolo, N. About quantitative EBSD analysis of deformation and recovery substructures in pure Tantalum. *IOP Conf. Ser. Mater. Sci. Eng.* **2015**, *89*, 012038. [CrossRef]
35. Priester, L. *Grain Boundaries: From Theory to Engineering*, 1st ed.; Springer Series in Materials Science 172; Springer: Dordrecht, The Netherlands, 2013.
36. Magnin, T.; Driver, J.; Lepinoux, J.; Kubin, L. Aspects microstructuraux de la deformation cyclique dans les metaux et alliages C.C. et C.F.C. - I. Consolidation cyclique. *Rev. Phys. Appl.* **1984**, *19*, 467–482. [CrossRef]
37. Le, L.T.; Ammar, K.; Forest, S. Efficient simulation of single and poly–crystal plasticity based on the pencil glide mechanism. *C. R. Méc.* **2020**. [CrossRef]

38. Guillaumain, J. *Etude du Comportement du Tantale en Fatigue à Grand Nombre de Cycles*; Document Interne CEA; 2009.
39. VIC-3D. Correlated Solutions. 2010. Available online: www.correlatedsolutions.com (accessed on 3 March 2021).
40. Allais, L.; Bornert, M.; Bretheau, T.; Caldemaison, D. Experimental characterization of the local strain field in a heterogeneous elastoplastic material. *Acta Metall. Mater.* **1994**, *42*, 3865–3880. [CrossRef]
41. Doumalin, P. Microextensométrie Locale par Corrélation d'Images Numériques: Application Aux études Micromécaniques par Microscopie Électronique à Balayage. Ph.D. Thesis, Ecole Polytechnique, Palaiseau, France, 2000.
42. Racine, A.; Bornert, M.; Sainte Catherine, C.; Caldemaison, D. Étude expérimentale des micro-mécanismes d'endommagement et de rupture des zircaloy hydrurés. *J. Phys. IV Fr.* **2003**, *106*, 109–118.:20030221. [CrossRef]
43. Bodelot, L.; Charkaluk, E.; Sabatier, L.; Dufrénoy, P. Experimental study of heterogeneities in strain and temperature fields at the microstructural level of polycrytalline metals through fully-coupled full-field measurements by Digital Image Correlation and Infrared Thermography. *Mech. Mater.* **2011**, *43*, 654–670. [CrossRef]
44. Sutton, M.; Yan, J.; Tiwari, V.; Schreier, H.; Orteu, J. The effect of out-of-plane motion on 2D and 3D digital image correlation measurements. *Opt. Lasers Eng.* **2008**, *46*, 746–757. [CrossRef]
45. Clair, A. Caractérisation Expérimentale des Propriétés Micromécaniques et Micromorphologiques des Alliages Base Nickel Contraints par la Croissance d'une Couche d'oxydes formée Dans le Milieu Primaire d'une Centrale Nucléaire. Ph.D. Thesis, Université de Bourgogne, Dijon, France, 2011.
46. Vignal, V.; Finot, E.; Oltra, R.; Lacroute, Y.; Bourillot, E.; Dereux, A. Mapping the 3D-surface strain field of patterned tensile stainless steels using atomic force microscopy. *Ultramicroscopy* **2005**, *103*, 183–189. [CrossRef]
47. Shi, Q.; Roux, S.; Latourte, F.; Hild, F.; Loisnard, D.; Brynaert, N. On the use of SEM correlative tools for in situ mechanical tests. *Ultramicroscopy* **2018**, *184*, 71–87. [CrossRef] [PubMed]
48. Proudhon, H.; Guéninchault, N.; Forest, S.; Ludwig, W. Incipient Bulk Polycrystal Plasticity Observed by Synchrotron In-Situ Topotomography. *Materials* **2018**, *11*, 2018. [CrossRef]
49. Pineau, A.; McDowell, D.L.; Busso, E.P.; Antolovich, S.D. Failure of metals II: Fatigue. *Acta Mater.* **2016**, *107*, 484–507. [CrossRef]
50. François, F.; Pineau, A.; Zaoui, A. *Mechanical Behaviour of Materials. Volume 2: Fracture Mechanics and Damage*; Solid Mechanics and its Applications; Springer: Berlin/Heidelberg, Germany, 2013; Volume 191.
51. El Bartali, A.; Aubin, V.; Degallaix, S. Surface observation and measurement techniques to study the fatigue damage micromechanisms in a duplex stainless steel. *Int. J. Fatigue* **2009**, *31*, 2049–2055. [CrossRef]
52. Kuo, J.C.; Chen, D.; Tung, S.H.; Shih, M.H. Prediction of the orientation spread in an aluminum bicrystal during plane strain compression using a DIC-based Taylor model. *Comput. Mater. Sci.* **2008**, *42*, 564–569. [CrossRef]
53. Clair, A.; Foucault, M.; Calonne, O. Strain mapping near a triple junction in strained Ni-based alloy using EBSD and biaxial nanogauges. *Acta Mater.* **2011**, *59*, 3116–3123. [CrossRef]
54. Helstroffer, A. *Etude en Fatigue du Tantale*; Rapport de Stage Document Interne CEA; 2010.
55. Forsyth, P. *Fatigue Behaviour and Its Dependence on Microstructure*; Colloque de Métallurgie de Saclay: Saclay, France, 1972.
56. Guilhem, Y.; Basseville, S.; Curtit, F.; Stephan, J.; Cailletaud, G. Numerical investigations of the free surface effect in three-dimensional polycrystalline aggregates. *Comput. Mater. Sci.* **2013**, *70*, 150–162. [CrossRef]
57. Barbe, F.; Quey, R.; Musienko, A.; Cailletaud, G. Three-dimensional characterization of strain localization bands in high-resolution elastoplastic polycrystals. *Mech. Res. Commun.* **2009**, *36*, 762–768. [CrossRef]
58. Bouchedjra, M.; Kanit, T.; Boulemia, C.; Amrouche, A.; Belouchrani, M.E.A. Determination of the RVE size for polycrystal metals to predict monotonic and cyclic elastoplastic behavior: Statistical and numerical approach with new criteria. *Eur. J. Mech. A Solids* **2018**, *72*, 1–15. [CrossRef]
59. Barbe, F.; Forest, S.; Cailletaud, G. Intergranular and intragranular behavior of polycrystalline aggregates. Part 2: Results. *Int. J. Plast.* **2001**, *17*, 537–563. [CrossRef]
60. Zhang, K.S.; Ju, J.W.; Bai, Y.L.; Brocks, W. Micromechanics based fatigue life prediction of a polycrystalline metal applying crystal plasticity. *Mech. Mater.* **2015**, *85*, 16–37. [CrossRef]
61. Kanit, T.; Forest, S.; Galliet, I.; Mounoury, V.; Jeulin, D. Determination of the size of the Representative Volume Element for random composites: Statistical and numerical approach. *Int. J. Solids Struct.* **2003**, *40*, 3647–3679. [CrossRef]
62. Gérard, C.; Bacroix, B.; Bornert, M.; Cailletaud, G.; Crépin, J.; Leclercq, S. Hardening description for FCC materials under complex loading paths. *Comput. Mater. Sci.* **2009**, *45*, 751–755. [CrossRef]
63. Guilhem, Y.; Basseville, S.; Curtit, F.; Stephan, J.; Cailletaud, G. Investigations of the effect of grain clusters on fatigue crack initiation in plycrystals. *Int. J. Fatigue* **2010**, *32*, 1748–1763. [CrossRef]
64. Méric, L.; Cailletaud, G. Single crystal modeling for structural calculations. Part 2: Finite element implementation. *J. Eng. Mater. Technol.* **1991**, *113*, 171–182. [CrossRef]
65. Eyraud, V.; Nadal, M.H.; Gondard, C. Texture measurement of shaped material by impulse acoustic microscopy. *Ultrasonics* **2000**, *38*, 438–442. [CrossRef]
66. Hoc, T.; Crépin, J.; Gélébart, L.; Zaoui, A. A procedure for identifying the plastic behavior of single crystals from the local response of polycrystals. *Acta Mater.* **2003**, *51*, 5477–5488. [CrossRef]
67. Lim, H.; Carroll, J.D.; Battaile, C.C.; Buchheit, T.E.; Boyce, B.L.; Weinberger, C.R. Grain-scale experimental validation of crystal plasticity finite element simulations of tantalum oligocrystals. *Int. J. Plast.* **2015**, *60*, 1–18. [CrossRef]

68. Z–Set Package. Non-Linear Material & Structure Analysis Suite. 2013. Available online: www.zset-software.com (accessed on 3 March 2021).
69. Besson, J.; Cailletaud, G.; Chaboche, J.L.; Forest, S.; Blétry, M. *Non–Linear Mechanics of Materials*; Solid Mechanics and Its Applications 167; Springer-Verlag: Berlin/Heidelberg, Germany, 2009.
70. Šiška, F.; Forest, S.; Gumbsch, P.; Weygand, D. Finite element simulations of the cyclic elastoplastic behavior of copper thin films. *Model. Simul. Mater. Sci. Eng.* **2007**, *15*, S217–S238. [CrossRef]
71. Linne, M.A.; Venkataraman, A.; Sangid, M.D.; Daly, S. Grain Boundary Sliding and Slip Transmission in High Purity Aluminum. *Exp. Mech.* **2019**, *59*, 643–658. [CrossRef]
72. Déprés, C.; Robertson, C.F.; Fivel, M.C. Low-strain fatigue in AISI 316L steel surface grains: A three-dimensional discrete dislocation dynamics modelling of the early cycles I. Dislocation microstructures and mechanical behaviour. *Philos. Mag.* **2004**, *84*, 2257–2275. [CrossRef]
73. Lemaitre, J.; Chaboche, J.L. *Mechanics of Solid Materials*; University Press: Cambridge, UK, 1994.
74. Suresh, S. *Fatigue of Materials*; Cambridge University Press: Cambridge, UK, 1998.
75. Bower, A.; Johnson, K. The influence of strain hardening on cumulative plastic deformation in rolling and sliding contact. *J. Mech. Phys. Solids* **1989**, *37*, 471–493. [CrossRef]
76. McDowell, D. Stress state dependence of cyclic ratchetting behavior of two rail steels. *Int. J. Plast.* **1995**, *11*, 397–421. [CrossRef]
77. Zhang, M.; Neu, R.; McDowell, D. Microstructure-sensitive modelling: Application to fretting contacts. *Int. J. Fatigue* **2009**, *31*, 1397–1406. [CrossRef]
78. Agbessi, K. Approches Expérimentales et Multi-Échelles des Processus d'amorÇage des Fissures de Fatigue sous Chargements Complexes. Ph.D. Thesis, Ecole Nationale Supérieure d'Arts et Métiers, Bordeaux, France, 2013.
79. Li, J.; Proudhon, H.; Roos, A.; Chiaruttini, V.; Forest, S. Crystal plasticity finite element simulation of crack growth in single crystals. *Comput. Mater. Sci.* **2014**, *90*, 191–197. [CrossRef]
80. Proudhon, H.; Li, J.; Wang, F.; Roos, A.; Chiaruttini, V.; Forest, S. 3D simulation of short fatigue crack propagation by finite element crystal plasticity and remeshing. *Int. J. Fatigue* **2016**, *82*, 238–246. [CrossRef]
81. Nadal, F. *Résistance en Fatigue du Tantale*; Document Interne CEA; 2011.
82. Fleury, E.; Rémy, L. Low cycle fatigue damage in nickel-base superalloy single crystals at elevated temperature. *Mater. Sci. Eng. A* **1993**, *167*, 23–30. [CrossRef]
83. Bathias, C.; Pineau, A. *La Fatigue des Matériaux et des Structures*, 1st ed.; Hermès Science: Paris, France, 2008.
84. Lebensohn, R.A.; Brenner, R.; Castelnau, O.; Rollett, A.D. Orientation image-based micromechanical modelling of subgrain texture evolution in polycrystalline copper. *Acta Mater.* **2008**, *56*, 3914–3926. [CrossRef]
85. Ren, X.; Yang, S.; Wen, G.; Zhao, W. A Crystal-Plasticity Cyclic Constitutive Model for the Ratchetting of Polycrystalline Material Considering Dislocation Substructures. *Acta Mech. Sin.* **2020**, *33*, 268–280. [CrossRef]
86. Aslan, O.; Quilici, S.; Forest, S. Numerical modeling of fatigue crack growth in single crystals based on microdamage theory. *Int. J. Damage Mech.* **2011**, *20*, 681–705. [CrossRef]
87. Lindroos, M.; Laukkanen, A.; Andersson, T.; Vaara, J.; Mäntylä, A.; Frondelius, T. Micromechanical modeling of short crack nucleation and growth in high cycle fatigue of martensitic microstructures. *Comput. Mater. Sci.* **2019**, *170*, 109185. [CrossRef]
88. Mareau, C. A non-local damage model for the fatigue behaviour of metallic polycrystals. *Philos. Mag.* **2020**, *100*, 955–981. [CrossRef]

Article

Modeling the Strain-Range Dependent Cyclic Hardening of SS304 and 08Ch18N10T Stainless Steel with a Memory Surface

Radim Halama [1,*], Jaromír Fumfera [2], Petr Gál [1,3], Tadbhagya Kumar [4] and Alexandros Markopoulos [1]

[1] Department of Applied Mechanics, Faculty of Mechanical Engineering, VSB-Technical University of Ostrava, 17.listopadu 2172/15, 70800 Ostrava, Czech Republic

[2] Department of Mechanics, Biomechanics and Mechatronics, Faculty of Mechanical Engineering, Center of Advanced Aerospace Technology, Czech Technical University in Prague, Technicka Street 4, 16607 Prague 6, Czech Republic

[3] ÚJV Řež, a. s., Hlavní 130, Řež, 250 68 Husinec, Czech Republic

[4] Mechanical and Aerospace Engineering Department, University of Florida, Gainesville, FL 32611, USA

* Correspondence: radim.halama@vsb.cz; Tel.: +420-597-321-288; Fax: +420-596-916-490

Received: 23 June 2019; Accepted: 22 July 2019; Published: 26 July 2019

Abstract: This paper deals with the development of a cyclic plasticity model suitable for predicting the strain range-dependent behavior of austenitic steels. The proposed cyclic plasticity model uses the virtual back-stress variable corresponding to a cyclically-stable material under strain control. This new internal variable is defined by means of a memory surface introduced in the stress space. The linear isotropic hardening rule is also superposed. First, the proposed model was validated on experimental data published for the SS304 material (Kang et al., Constitutive modeling of strain range dependent cyclic hardening. Int J Plast 19 (2003) 1801–1819). Subsequently, the proposed cyclic plasticity model was applied to our own experimental data from uniaxial tests realized on 08Ch18N10T at room temperature. The new cyclic plasticity model can be calibrated by the relatively simple fitting procedure that is described in the paper. A comparison between the results of a numerical simulation and the results of real experiments demonstrates the robustness of the proposed approach.

Keywords: cyclic plasticity; cyclic hardening; finite element method; austenitic steel 08Ch18N10T; stainless steel 304

---

1. Introduction

The SS304 material, which includes 18 percent chromium and eight percent nickel, is the most widely-used austenitic stainless steel. It has good drawability and welding properties together with strong corrosion resistance. Austenitic stainless steel 08Ch18N10T is a chrome-nickel steel that is stabilized by titanium. This steel is widely used in the nuclear industry for piping systems and reactor internals in the Russian-designed VVERwater-water power reactors for nuclear power plants (NPP). Reactor internals are the part of an NPP that provides support, guidance, and protection for the reactor core and for the control elements. The block of guided tubes, the core barrel, the core barrel bottom, and the core shroud are some of the internal components that are exposed to very harsh operating regimes. The operating regime, e.g., heating and shut-downs, has a significant influence on the service life of the components. The vibration and pressure pulsation of the water pumps also have to be taken into account. These regimes expose the reactor internals to cyclic loading.

In practice, cyclic loading of structural parts can lead to the formation and propagation of cracks through the process referred to as fatigue. In all areas of industry, the operational safety of machinery depends on an appropriate design process, which includes an analysis of all possible critical states. In the low-cycle fatigue domain, seismic analysis and the simulation of operational tests of the piping systems of NPPs may be used as an example. In these cases, it is crucial to have an accurate description of the stress–strain behavior of the material that is being considered.

Phenomenological models [1] are the most widely-used models in practical applications. Their goal is to provide an as accurate as possible description of the stress–strain behavior of the material, which is found on the basis of experiments [2]. The stress–strain behavior of structural materials under cyclic loading is very diverse, and a case-by-case approach is required [3].

The most progressive group of cyclic plasticity models, which are commonly encountered in commercial finite element method programs, is the single yield surface models based on differential equations. Their development is closely linked to the creation of a nonlinear kinematic hardening rule with a memory term, introduced by Armstrong and Frederick in 1966 for the evolution of back-stress [4] and the discovery by Chaboche [5] of the vast possibilities offered by the superposition of several back-stress parts.

Developments in the field of non-linear kinematic hardening rules were mapped in detail in [1]. In the current paper, we will mention only the most important theories. In 1993, Ohno and Wang [6] proposed two nonlinear kinematic hardening rules. For both models, it was considered that each part of the back-stress had a certain critical state of dynamic recovery. Ohno–Wang Model I leads to plastic shakedown under uniaxial loading with a nonzero mean axial stress value (no ratcheting), and under multiaxial loading, it gives lower accumulated plastic deformation values than have been observed in experiments. The memory term of Ohno–Wang Model II [6] is partially active before reaching the critical state of dynamic recovery, which allows a good prediction of ratcheting under uniaxial loading and also under multiaxial loading. The Abdel-Karim–Ohno nonlinear kinematic hardening rule [7] was published in 2000. This rule is in fact a superposition of the Ohno–Wang I and Armstrong–Frederick rules. The proposed model was designed to predict the behavior of materials that exhibit a constant increment of plastic deformation during ratcheting. Other modifications to this kinematic hardening rule, leading to a better prediction of uniaxial ratcheting and also multiaxial ratcheting, were proposed by one of the authors of the paper [8]. In order to capture the additional effects of cyclic plasticity, the concept of kinematic and isotropic hardening has been further modified. Basically, the available theories can be divided into two approaches. The first approach is related to the actual distortion of the yield surface [9–11], while the second approach is related to the memory effect of the material [5,12]. The effect of cyclic hardening as a function of the size of the strain amplitude is usually assumed in the second approach.

The first comprehensive model of cyclic plasticity with a memory surface was proposed by Chaboche and co-authors in [5]. Chaboche's memory surface was established in the principal plastic strain space and captures the influence of plastic strain amplitude and also the mean value of the plastic strain. The memory surface is associated with a non-hardening strain region in a material point, as was explained by Ohno [12] for the general case of variable amplitude loading. Memory surfaces established in the stress space have also been developed. Their main advantage is that they enable more accurate ratcheting strain prediction to be achieved, as presented by Jiang and Sehitoglu in their robust cyclic plasticity model [13].

It should be mentioned that both of these memory surface concepts lead to an increase in the number of material parameters and in the number of evolution equations, which complicates their use in engineering practice.

The original application of the memory surface, introduced by Jiang and Sehitoglu, was extended by some authors of the present paper to capture the memory effect of ST52 material, in [14]. Uniaxial experiments indicated that, in the case of a cyclically-softening/hardening material, larger strain amplitudes cause a significant change in the shape of the hysteresis loops. Only a very small number of researchers in the field of cyclic plasticity have investigated the influence of strain amplitude on the cyclic hardening effect. Good agreement with experiments has been achieved in the case of steel SS304 [15], but at the cost of defining more than 70 material parameters.

Some of the material models have been used to capture cyclic material behavior. To describe the cyclic behavior of SAE4150 martensitic steel [16], Schäfer et al. considered three kinematic hardening models, i.e., the Chaboche [5], Armstrong-Frederick [4] and Ohno–Wang [6] models. They used these kinematic approaches to simulate the micromechanical behavior of the selected material. Moeini et al. [17] used the Chaboche model [5] to predict the low cyclic behavior of dual-phase steel. The selected kinematic hardening model provided good agreement with experimental results. Msolli used the unified viscoplastic model [18] developed by Chaboche when modeling the elastoviscoplastic behavior of JLF-1 steel at higher temperatures (400 °C and 600 °C). In this study, the Chaboche model was slightly modified to capture cyclic hardening followed by cyclic softening. The material model also falls into the category of coupled damage models. The material model showed good agreement with the experimental results. The effect of torsional pre-strain on low cycle fatigue performance of SS304 was studied in [19]. Kang et al. [20] used the viscoplastic constitutive model with the extended Abdel-Karim–Ohno nonlinear kinematic hardening rule with some temperature-dependent terms. This constitutive model was verified on uniaxial and non-proportional multiaxial ratcheting experimental results at room temperature and at elevated temperatures. Another viscoplastic constitutive model was used by Kang, Gao, and Yang [21] in their study to simulate uniaxial and multiaxial ratcheting of cyclically-hardening materials. They used the Ohno–Wang kinematic hardening rule with the critical state of dynamic recovery. The effect of loading history was also considered by introducing a fading memorization function for the maximum plastic strain amplitude.

This paper shows the advantages of using the memory surface established by Jiang and Sehitoglu in 1996 [13] to treat the impact of the strain amplitude on the material stress response. The new theory is shown on the kinematic hardening rule based on Chaboche's model with three back-stress parts, but it can also easily be applied to the Abdel-Karim–Ohno model or its modified version with promised ratcheting prediction [8]. Recently, an approach was introduced that takes into account a new internal variable referred to as virtual back-stress, corresponding to a cyclically-stable material. This provides an easy way to identify the parameters and to use fewer material parameters than in earlier models, for example [21]. New experimental results from uniaxial fatigue tests realized on 08Ch18N10T at room temperature are presented and subsequently used for the validation of the new cyclic plasticity model.

2. New Constitutive Model

In this paper, isothermal conditions are considered, and the influence of the strain rate is neglected. However, the model can be extended by standard techniques for use in the area of viscoplasticity [7].

2.1. Yield Surface and Flow Rule

In this work, the concept of a single yield surface for metallic materials is used, based on the von Mises yield function, which can be expressed for the general mixed hardening model as:

$$f = \sqrt{\frac{2}{3}(\boldsymbol{s} - \boldsymbol{a}) : (\boldsymbol{s} - \boldsymbol{a})} - Y = 0, \tag{1}$$

where $\boldsymbol{s}$ is the deviatoric part of stress tensor $\boldsymbol{\sigma}$, $\boldsymbol{a}$ is the deviatoric part of back-stress $\boldsymbol{\alpha}$, and the current size of yield surface (or the actual yield stress) Y is defined as the sum of the isotropic variable R and the initial size of the yield surface σ_y (the yield strength) by the equation:

$$Y = \sigma_y + R. \tag{2}$$

It should be mentioned that the colon between the second-order tensors in Equation (1) denotes their inner product $x : y = x_{ij}y_{ij}$ (considering the Einstein summation convention).

The associative plasticity is considered, so the normality flow rule is considered in the case of active loading:

$$de^p = d\lambda \frac{\partial f}{\partial \sigma}. \tag{3}$$

This expresses mathematically that the plastic strain increment de^p is collinear with the exterior normal to the yield surface for the current stress state. In associative plasticity, the scalar multiplier $d\lambda$ is equal to the accumulated plastic strain increment dp, which is defined as:

$$dp = \sqrt{\frac{2}{3}de^p : de^p}. \tag{4}$$

2.2. Virtual Back-Stress

A new internal variable is established to provide an easy way to calibrate the model. The variable is the back-stress of a cyclically-stable material corresponding to the response of the material investigated under a large strain range. It will be referred to as the virtual back-stress. The Chaboche superposition of the back stress parts is used in the following form:

$$\alpha_{virt} = \sum_{i=1}^{M} \alpha_{virt}^i \tag{5}$$

taking into consideration the nonlinear kinematic hardening rule of Armstrong and Frederick [4] for each part:

$$d\alpha_{virt}^i = \frac{2}{3}C_i de_p - \gamma_i \alpha_{virt}^i dp, \tag{6}$$

where C_i and γ_i are material parameters. For all calculations in this paper, the superposition of three kinematic hardening rules ($M = 3$) will be used.

It should be mentioned that the virtual back-stress is used only in the definition of the memory surface, which will be described in the next section. Zero components of the virtual back-stress are considered in the initial state. The increment of the virtual back-stress is calculated according to Equations (5) and (6) assuming the current increment of accumulated plastic strain dp and the current increment of plastic strain tensor de_p in each iteration of the local problem. Further details of the implementation algorithm that is used can be found in [22], where a more complex model with the memory surface of Jiang and Sehitoglu [13] was considered.

2.3. Memory Surface

To provide a correct description of the cyclic hardening for various strain ranges, a memory surface in the stress space is established. The concept is analogous to the theory of Jiang and Sehitoglu [13]. A scalar function is introduced to represent the memory surface in the deviatoric stress space:

$$g = \|\alpha_{virt}\| - R_M \leq 0, \tag{7}$$

where R_M is the size of the memory surface and $\|\alpha_{virt}\|$ is the magnitude of the total virtual back-stress, which is defined as $\|\alpha_{virt}\| = \sqrt{\alpha_{virt} : \alpha_{virt}}$. The evolution equation ensuring the possibility of memory surface expansion, Figure 1, is therefore:

$$dR_M = H(g) \langle L : d\alpha_{virt} \rangle, \tag{8}$$

where

$$L = \frac{\boldsymbol{\alpha}_{virt}}{\|\boldsymbol{\alpha}_{virt}\|}.$$

(9)

Contraction of the memory surface is not allowed in this paper. It can be implemented according to the stress space-based memory surface concept of Jiang and Sehitoglu [13].

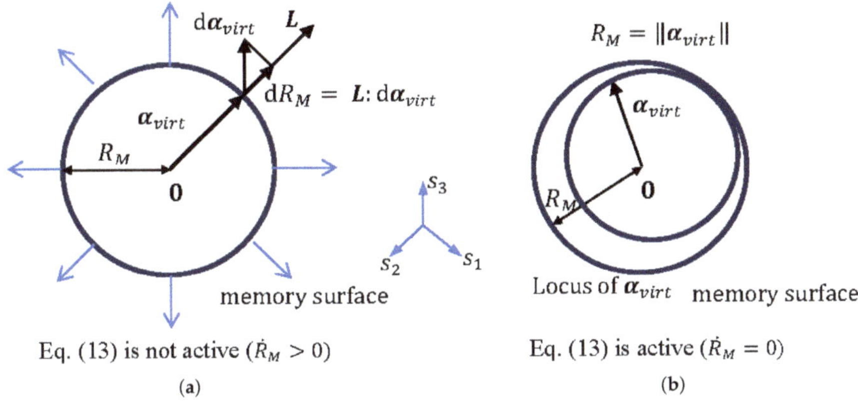

Figure 1. Expansion of the memory surface and the stabilized memory surface. (**a**) Equation (13) is not active; (**b**) Equation (13) is active.

2.4. Kinematic Hardening Rule

Consistent with the previous sections, the back-stress is composed of M parts:

$$\boldsymbol{\alpha} = \sum_{i=1}^{M} \boldsymbol{\alpha}_i,$$

(10)

but the memory term is dependent on the size of memory surface R_M and accumulated plastic strain p; thus:

$$d\boldsymbol{\alpha}_i = \frac{2}{3}C_i d\boldsymbol{\epsilon}_p - \gamma_i \phi(p, R_M)\boldsymbol{\alpha}_i dp,$$

(11)

where C_i and γ_i are the same as in Equation (6). The multiplier ϕ of parameters γ_i is composed of a static part and a cyclic part:

$$\phi(p, R_M) = \phi_0 + \phi_{cyc}(p, R_M),$$

(12)

where ϕ_0 has the meaning of a material parameter, while the cyclic part is variable and can change only in the case of $\dot{R}_M = 0$. In this case, the evolution equation is defined in the following way:

$$d\phi_{cyc} = \omega(R_M) \cdot \left(\phi_\infty + \phi_{cyc}(p, R_M)\right) dp.$$

(13)

$$\phi_\infty(R_M) = A_\infty R_M^4 + B_\infty R_M^3 + C_\infty R_M^2 + D_\infty R_M + F_\infty,$$

(14)

$$\omega(R_M) = A_\omega + B_\omega R_M^{-C_\omega} \text{ for } R_M \geq R_{M\omega},$$

(15)

$$\omega(R_M) = A_\omega + B_\omega R_{M\omega}^{-C_\omega} \text{ otherwise,} \tag{16}$$

where A_∞, B_∞, C_∞, D_∞, F_∞, A_ω, B_ω, C_ω, $R_{M\omega}$, and R_{M0} are additional parameters to Chaboche's material parameters C_i and γ_i. The evolution parameter ω directs the rate of cyclic hardening behavior according to the current size of memory surface R_M.

2.5. Isotropic Hardening Rule

Continuous cyclic hardening has been observed for austenitic stainless steels for a large strain range under uniaxial loading [15]. To capture this behavior, we introduce the linear isotropic hardening rule:

$$dR = R_0(R_M)dp, \tag{17}$$

where parameter R_0 is dependent on the size of the memory surface:

$$R_0(R_M) = A_R R_M^2 + B R_M + C_R \text{ for } R_M \geq R_{M0}, \tag{18}$$

$$R_0(R_M) = A_R R_{M0}^2 + B R_{M0} + C_R \text{ otherwise,} \tag{19}$$

because of the strong dependence on the strain range observed in the experiments [15].

3. Identification of Material Parameters and Model Verification on SS304 Data

The cyclic plasticity model was implemented in the ANSYS FE code, using the algorithm described in [22]. The methodology for calibrating the proposed material model will be explained according to the classical Chaboche material model, which requires the following parameters to be identified: σ_y, E, μ, C_1, C_2, C_3, γ_1, γ_2, and γ_3, where E is the Young modulus and μ is the Poisson ratio.

Generally, 14 additional parameters have to be specified for the proposed cyclic plasticity model: ϕ_0, A_∞, B_∞, C_∞, D_∞, F_∞, A_ω, B_ω, C_ω, R_{M0}, $R_{M\omega}$, A_R, B_R, C_R.

It is customary to determine the Young modulus E from cyclic curves rather than from a tensile test. The Poisson ratio μ can be determined by standard procedures. The material parameter E was established from the elastic region of the largest available hysteresis loop, using a linear regression. The initial yield strength σ_y was chosen to get the best possible description of the static stress–strain curve using Equation (23).

A sequence of steps is retained that should be applied in the following description of the calibration of the proposed model. The sections are named according to the required experimental data. The parameters are identified on the basis of the experimental set of stainless steel SS304, available at [15].

3.1. Uniaxial Large Hysteresis Loop

It is well known that the material parameters of the Chaboche model can be determined (under cyclic loading) from the cyclic strain curve or from the large uniaxial hysteresis loop [23]. Figure 2 shows us the results for identifying the parameter in the case of stainless steel 304. According to [24], it is possible to use a relation that defines the loading part of the stabilized hysteresis loop in the stress—plastic strain diagram:

$$\sigma_x = \sigma_y + \alpha_{virt}, \tag{20}$$

$$\sigma_x = \sigma_y + \frac{C_1}{\gamma_1}\left(1 - 2e^{-\gamma_1(\epsilon_p - (-\epsilon_{pL}))}\right) + \frac{C_2}{\gamma_2}\left(1 - 2e^{-\gamma_2(\epsilon_p - (-\epsilon_{pL}))}\right) + C_3\epsilon_p, \tag{21}$$

where ϵ_{pl} is the plastic strain corresponding to the compressive peak strain and σ_x is the axial stress. Relation (21) is valid in the case of $\gamma_3 = 0$ and for a large hysteresis loop.

If the Chaboche model is calibrated using the large hysteresis loop ($\Delta\epsilon = 6\%$), it predicts a higher stress amplitude for small strain amplitudes than was observed in the experiments. This phenomenon is shown for 1% strain amplitude in Figure 2. For this reason, a new cyclic plasticity model is needed. Parameters C_i and γ_i can be obtained for the newly-proposed model after applying Relation (21) to data from the largest available hysteresis loop, while parameter $\gamma_3 = 0$. In our case, we have used the Levenberg–Marquardt algorithm of the nonlinear least squares method. It is now clear that material parameters C_i and γ_i in the new model can be estimated on the basis of a single uniaxial hysteresis loop. The applicability of the new model to different strain ranges is given by the multiplier ϕ in Equation (11). Its evolution depends on memory surface size R_M and accumulated plastic strain p. How the necessary parameters are identified will be explained below.

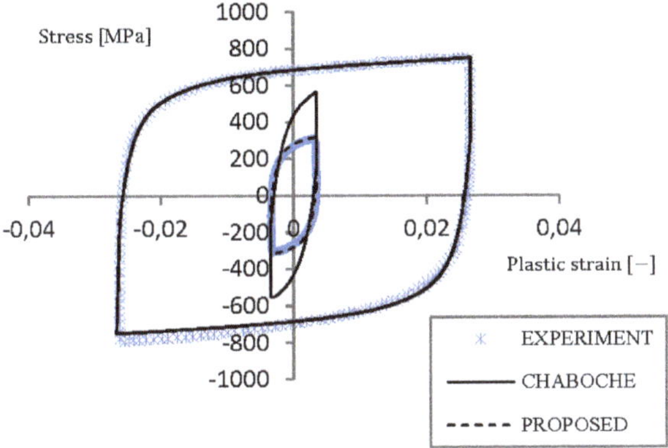

Figure 2. Prediction of two uniaxial hysteresis loops of SS304 (experimental data were taken from [15]).

3.2. Static Strain Curve

Under monotonic loading, the kinematic hardening rule of the proposed model is reduced to:

$$d\alpha_i = \frac{2}{3}C_i d\epsilon_p - \gamma_i \phi_0 \alpha_i dp. \tag{22}$$

If isotropic hardening is neglected, the material parameter ϕ_0 can be determined by a constitutive relation commonly used for the Chaboche model:

$$\sigma = \sigma_y + \frac{C_1}{\gamma_1 \phi_0}\left(1 - e^{-\gamma_1 \phi_0 \epsilon_p}\right) + \frac{C_2}{\gamma_2 \phi_0}\left(1 - e^{-\gamma_2 \phi_0 \epsilon_p}\right) + C_3 \epsilon_p. \tag{23}$$

For stainless steel SS304, the value of the parameter is $\phi_0 = 4.5$. The material model prediction of the static stress–strain curve is depicted in Figure 3.

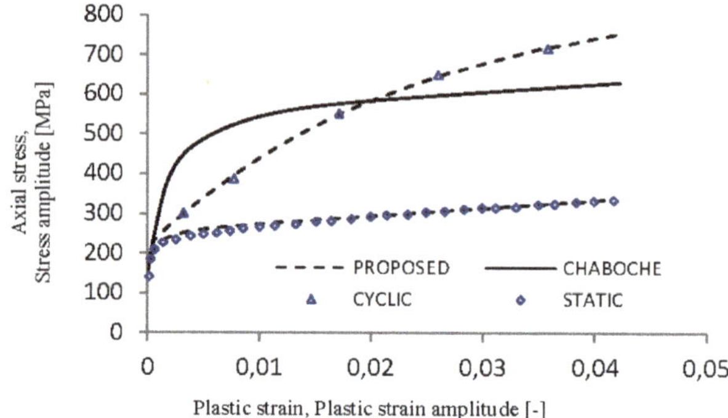

Figure 3. Prediction of static and cyclic uniaxial stress–strain curves; the experimental data for SS304 were reproduced from [15].

3.3. Cyclic Stress–Strain Curve

For the Chaboche model, the relation between stress amplitude and plastic strain amplitude can be derived, in accordance with [23], in the form:

$$\Delta\sigma/2 = \sigma_y + \frac{C_1}{\gamma_1} \tanh\left(\gamma_1 \Delta\epsilon_p/2\right) + \frac{C_2}{\gamma_2} \tanh\left(\gamma_2 \Delta\epsilon_p/2\right) + C_3 \Delta\epsilon_p/2. \tag{24}$$

By analogy, the relation for the cyclic hardening curve can be obtained for the proposed constitutive model. Neglecting the isotropic part, we can write,

$$\Delta\sigma/2 = \sigma_y + \frac{C_1}{\gamma_1 \phi_\infty} \tanh\left(\gamma_1 \phi_\infty \Delta\epsilon_p/2\right)$$
$$+ \frac{C_2}{\gamma_2 \phi_\infty} \tanh\left(\gamma_2 \phi_\infty \Delta\epsilon_p/2\right) + C_3 \Delta\epsilon_p/2. \tag{25}$$

This is a scalar nonlinear equation, which can be solved for selected experimental points, for example by the successive substitution method. Afterwards, the ϕ_∞ values for each peak of the hysteresis loop are fitted by the approximate function (13).

The experimental cyclic stress–strain curve data considered without linear isotropic hardening (published in [15]) and the predicted data corresponding to the proposed model are also shown in Figure 3. The cyclic stress–strain curve of the classic Chaboche model, calibrated using the large hysteresis loop ($\Delta\epsilon = 6\%$), is also presented in Figure 3. It is again clear that a more robust model is needed.

3.4. Cyclic Hardening Curves

In order to provide a good description of the cyclic hardening properties for a wide range of strain amplitudes, it is necessary to identify the isotropic hardening and kinematic hardening functions.

The remaining parameters A_ω, B_ω, C_ω, $R_{M\omega}$, R_{M0}, A_R, B_R, and C_R were estimated by a fitting procedure, using the nonlinear relations between the peak stress and the accumulated plastic strain p for all available cases of constant strain amplitude tests for the particular type of stainless steel ($\Delta\epsilon = 1, 2, 4, 6,$ and 8%).

The isotropic hardening parameters were determined from the slope of each cyclic hardening/softening curve in a saturated state. More precisely, a unique value for each case of a strain range was obtained, which was afterwards used for approximation by (18) and (19). The estimated material parameters of the proposed cyclic plasticity model are stated in Table 1.

Note that parameter γ_3 was equal to 10, which corresponds more to the behavior of metallic materials, e.g., if ratcheting occurs under stress-controlled loading with a nonzero mean axial stress value.

Table 1. Material parameters of the proposed model for SS304.

E [MPa]	v	σ_y [MPa]	C_1 [MPa]	γ_1	C_2 [MPa]
196,000	0.3	150	150,000	622	19,827
γ_2	C_3 [MPa]	γ_3	A_∞	B_∞	C_∞
128	2000	10	0	1.15×10^{-7}	-1.23×10^{-4}
D_∞	F_∞	A_R [MPa^{-1}]	B_R	C_R [MPa]	R_{M0} [MPa]
0.032	-3.6	0.000915	-0.5	60.7	305
A_ω	B_ω	C_ω	$R_{M\omega}$ [MPa]	ϕ_0	
0	4.02×10^{-17}	6.424	344	4.5	

3.5. Prediction Results for SS304

For implementation in ANSYS, the user subroutine called USERMAT1D.F, which was originally distributed for bilinear isotropic hardening. It was necessary to modify the user subroutine according to the used radial return algorithm [22]. A single LINK180 element was used for all analyses in ANSYS, because only uniaxial loading cases are considered in this paper. Material SS304, which was used as an example to explain the calibration procedure, exhibited very strong cyclic hardening at larger amplitudes of plastic strain. All investigated cases corresponded to the uniaxial experiments published by Kang et al. [15]. Predictions of cyclic hardening, corresponding to the proposed model, are shown together with the results of experiments in the form of peak tensile stress values as a function of the number of cycles (Figure 4).

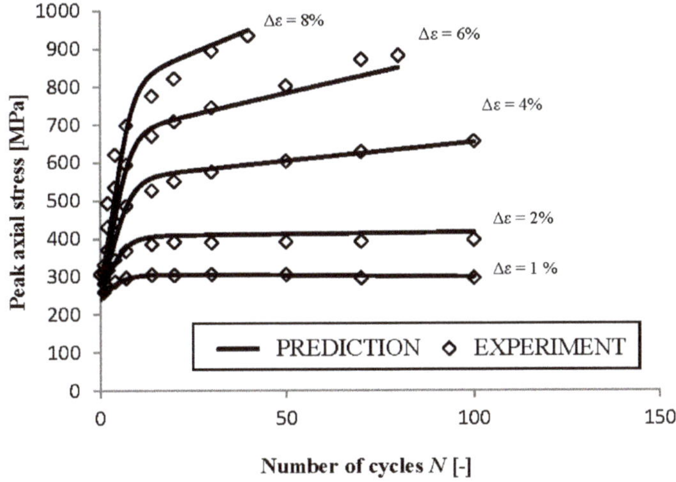

Figure 4. A comparison of simulations and experiments in the form of tensile peak stress variation; uniaxial strain controlled tests (the experiment was taken from [15]).

The results of the prediction of the transition behavior of the SS304 steel material during uniaxial cyclic loading (Figure 4) were supplemented by hysteresis loops for strain amplitudes of 1% and 6%, respectively (Figures 5 and 6). These results can be compared with the experimentally-obtained hysteresis loops published in [15].

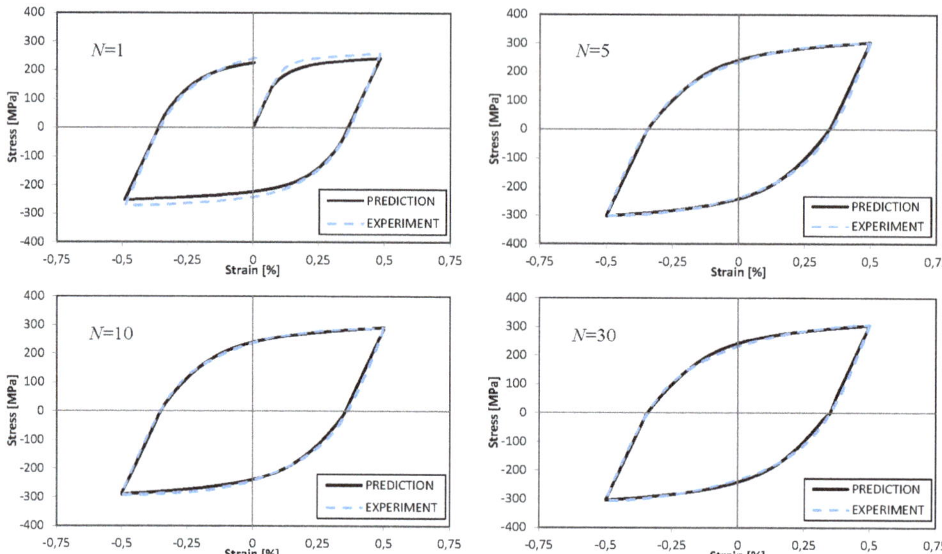

Figure 5. Prediction of the uniaxial test with $\Delta\epsilon = 1\%$ (experimental data were taken from [15]).

Another simulation was of a cyclic test with a linearly-increased/-decreased strain amplitude composed of five identical blocks. In each block, the strain range was increased within 20 cycles to a value of 5%, and subsequently, it was reduced, with the same increment. The resulting stress–total strain dependence for the strain range increasing stage is shown in Figure 7. The increasing amplitude of the stress was more progressive in the prediction than in the experiment, as can also be seen in Figure 8.

Figure 8 presents a comparison with an experiment, in which the peaks of the hysteresis loops are plotted for the first and fifth loading block. The relative error of the prediction was about 12.3% in the first block (for amplitude of strain 2.5%), but it was reduced to 6.7% in the fifth loading block.

The proposed model was able to capture the static and cyclic stress–strain curve for SS304 correctly. It also simulated well the shapes of the stress–strain hysteresis loops in all investigated cases, as well as the Bauschniger effect, which became weaker for higher strain ranges. The non-Masing behavior of the SS304 material was very strong, and this can be modeled better with superposition of the kinematic hardening and isotropic hardening, as in the proposed constitutive model. For the incremental test, the prediction was very good, especially in the fifth loading block. The overprediction of the peak stresses in the first block of loading can be reduced, for example, by introducing memory surface contraction, as was proposed in the original model of Jiang and Sehitoglu [13].

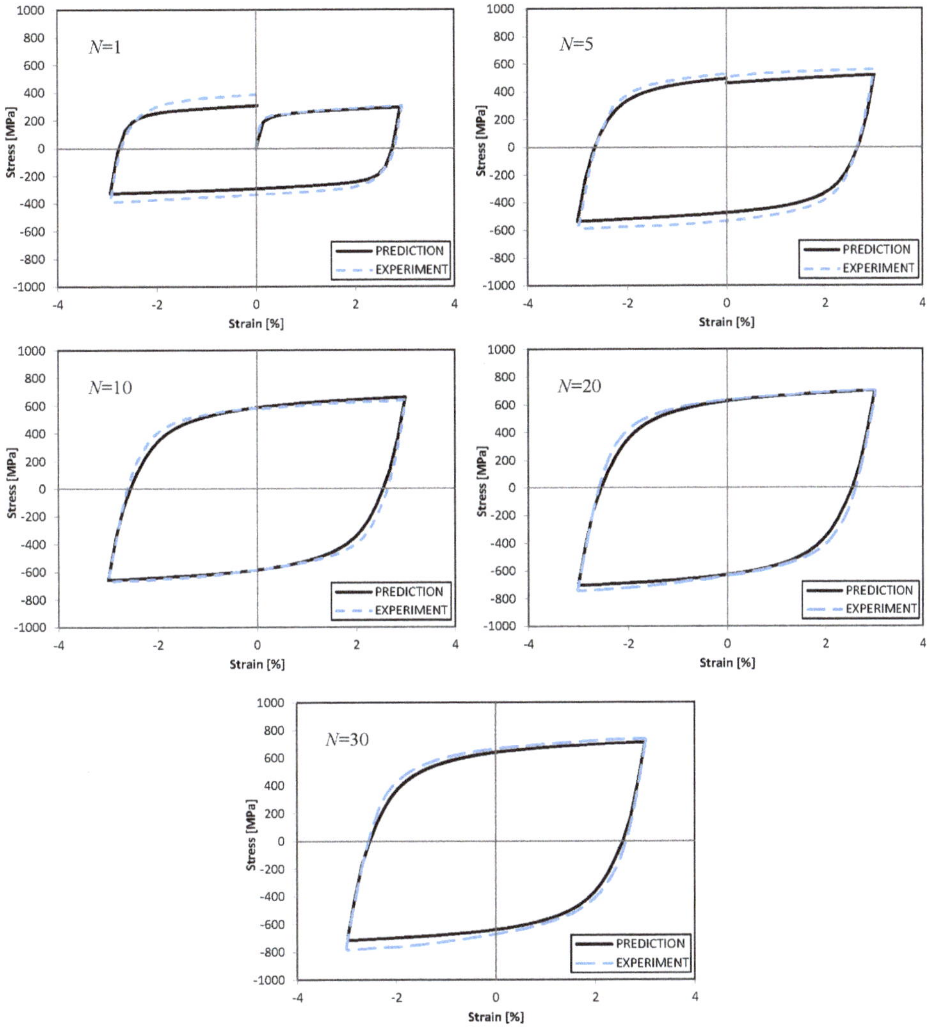

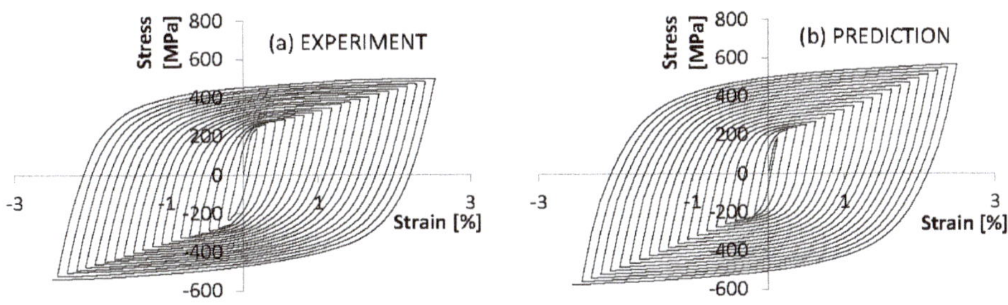

Figure 6. Prediction of the uniaxial test with $\Delta\epsilon = 6\%$ (experimental data were taken from [15]).

Figure 7. Prediction of the incremental test with the linearly-increasing amplitude of the axial stress: (**a**) experiment (experimental data were taken from [15]); (**b**) prediction.

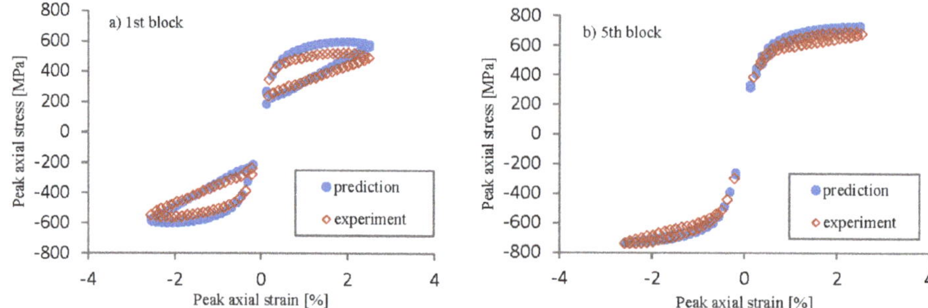

Figure 8. Comparison between a prediction and an experiment in the form of the variation in tensile peak stress for the incremental test (the experiment was taken from [15]): (**a**) first block, (**b**) fifth block.

4. Application to Uniaxial Cyclic Tests of 08Ch18N10T Stainless Steel

4.1. Identification of the Material Parameters for 08Ch18N10T Stainless Steel

The model was also calibrated for original experimental data on austenitic steel 08Ch18N10T [25]. A total of 12 uniaxial specimens were used for the material parameters' identification process (marked by the abbreviation IDF in the following text, each specimen representing a different level of loading). According to the ASTM standard [26], the classic uniform-gauge geometry of the specimen is limited up to the amplitude of the total strain $\epsilon_a = 0.5\%$. For higher strain levels, an hour-glass type geometry is required. According to this standard, the IDF specimens were compiled from uniform-gauge geometry (specimens IDF1–IDF5) and hourglass geometry (specimens IDF6–IDF12); see Figure 9.

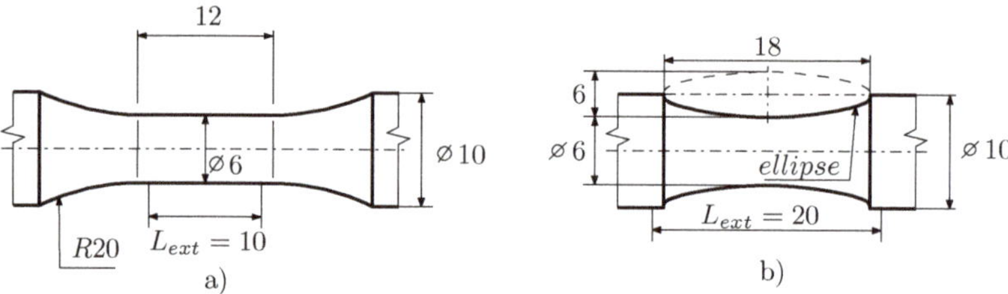

Figure 9. Specimen geometries: (**a**) uniform-gauge; (**b**) hourglass.

Another 17 uniaxial specimens (all with hourglass geometry) were used to verify the prediction ability of the model.

The loading force F applied to the IDF specimen was known, as was the strain field of the surface of the specimen. The strain field was measured by the extensometer in the case of uniform-gauge geometry, or by the digital image correlation method in the case of hourglass geometry. Considering the uniaxial stress field in the cross-section of a specimen, the stress can be determined as $\sigma = F/A$, where A is the cross-section surface of the specimen. This allowed the use of a different calibration process, based on knowledge of the shape of the stress–strain hysteresis loops in all cycles during the experiment to failure.

Let us select one hysteresis stress–strain loop of a point on the specimen representing one loading cycle. This can be optimally simulated by a set of material parameters C_1, γ_1, C_2, γ_2, C_3, γ_3, and σ_y. However, in the next cycle, the optimal set of these parameters can be slightly different, as can the set of parameters of a specimen with different loading conditions. This material model uses the memory surface concept by setting these material parameters as functions of R_M and making these coefficients dependent on the loading history and the loading level conditions.

The material model did not include a simulation of the material damage process, so only experimental data up to damage were used for the calibration. The number of cycles used was N_d, and this number corresponded to the drop in the loading force during the experiment by 2%, due to crack initiation and propagation, leading to failure.

First, the fatigue life was divided into about 10 evenly-spaced parts by selecting hysteresis loops (SHLs), and the cycle number of each selected hysteresis loop (SHL) is given as $n \simeq N_d/k$, where $k = 1, 2, \dots, 10$. The Young modulus E, the Poisson ratio v, and the yield strength σ_y were determined from the tensile test according to the ISO standard [27].

σ_y can be interpreted as the point where the linear part of the tensile curve turns into the non-linear part (see Figure 10). The root mean squared error method ($RMSE$) can be applied to find the point. In the tensile test (or in the first cycle of the cyclic test), the yield strength σ_y corresponded to $RMSE \approx 8$. Applying $RMSE = 8$ to each SHL, the actual yield stress Y was found. This shows the development of the actual yield stress Y during the fatigue life; see Figure 11.

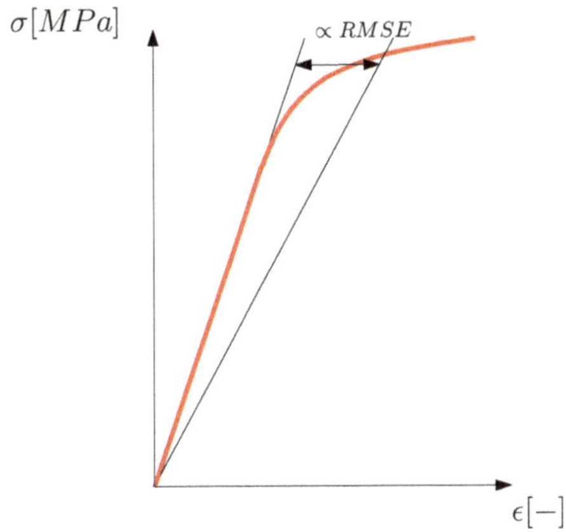

Figure 10. Actual yield stress determination.

Two SHLs were chosen, the bigger one and the smaller one, each with cycle number $n = N_d$ (the last cycle). The Chaboche material model parameters C_1, γ_1, C_2, γ_2, C_3, and γ_3 were found using an optimization process. The target function was set to the optimal shape match between the simulation and experiment of the two SHLs. The result is shown in Figure 12.

Knowing the Chaboche material parameters, a first guess of the memory surface size for each specimen was determined, using Equations (5)–(9). The formulation of R_M and the constant amplitude of the loading conditions resulted in fast saturation of the R_M value for each specimen (after the first cycle), which made the calibration process easier.

The yield stress was now fitted as a function of R_M, using Equations (17)–(19), by finding material parameters A_R, B_R, and C_R. Using the tensile test experimental data and performing a simulation of this test, parameter ϕ_0 was found using Equation (11) as an optimal value of ϕ for the tensile test simulation. The value of function ϕ from Equation (11) was found for SHLs, using a similar optimization process as for determining the Chaboche material parameters. ϕ_∞ was the value of ϕ for $n = N_d$, and from Equation (14), ϕ_∞ was then set as a function of R_M by finding material parameters A_∞, B_∞, C_∞, D_∞, and F_∞. Function ω determined the transition of function ϕ between its border values ϕ_0 and ϕ_∞. Knowing the course of function ϕ during the fatigue life, ω was determined as a function of R_M by finding material parameters A_ω, B_ω, and C_ω from Equation (15). This result was not necessarily optimal, so one more optimization was performed to find better ϕ_∞ and ω material parameters. The target function was set to the best possible match of the amplitude stress response between simulation and experiment during the whole fatigue life (not only SHLs).

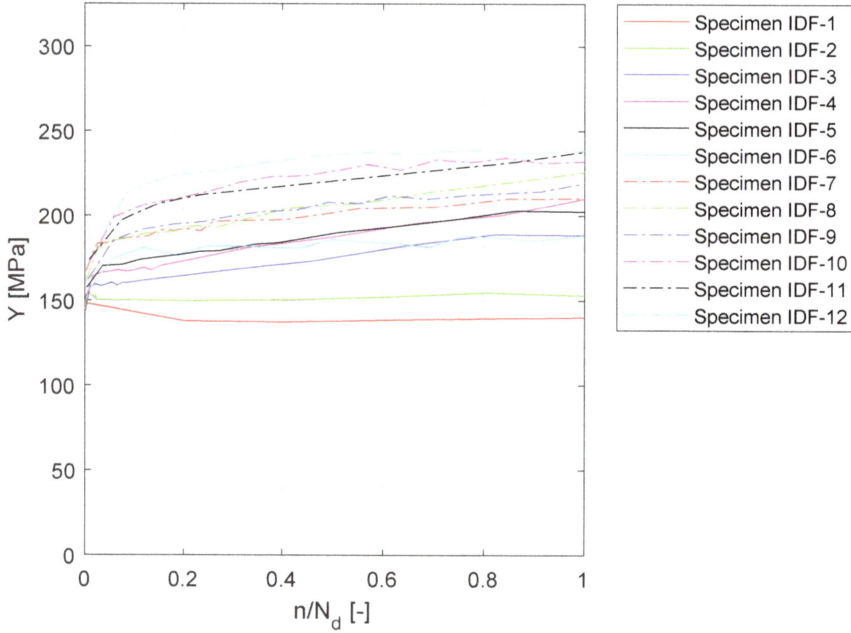

Figure 11. Actual yield stress development during fatigue life evaluated for 08Ch18N10T.

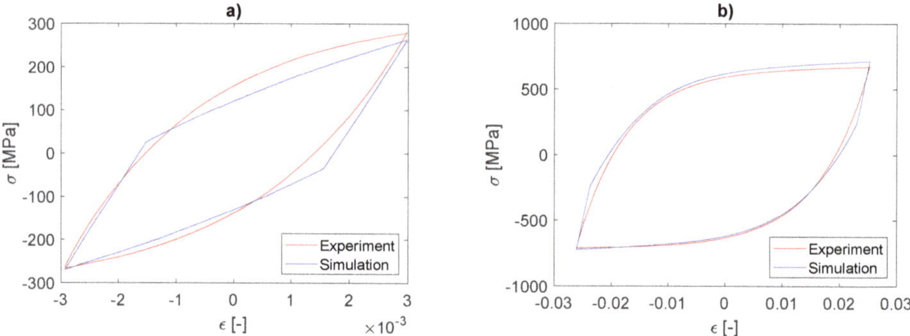

Figure 12. Chaboche coefficients fitting: (**a**) small loop; (**b**) large loop.

The R_M value for each specimen was determined only as a first guess, so a number of iterations of the whole calibration process had to be carried out to find the final and optimal set of material parameters. The optimal material parameters are presented in Table 2.

Table 2. Material parameters of the proposed model for 08Ch18N10T.

E [MPa]	ν	σ_y [MPa]	C_1 [MPa]	γ_1	C_2 [MPa]
210,000	0.3	150	63,400	148.6	10,000
γ_2	C_3 [MPa]	γ_3	A_∞	B_∞	C_∞
911.4	2000	0	-1.441×10^{-9}	1.911×10^{-6}	-8.951×10^{-4}
D_∞	F_∞	A_R [MPa^{-1}]	B_R	C_R [MPa]	R_{M0} [MPa]
1.688×10^{-1}	-10.6	1.264×10^{-4}	-4.709×10^{-2}	3.801	225.4
A_ω	B_ω	C_ω	$R_{M\omega}$ [MPa]	ϕ_0	
0	3.456×10^{-11}	-4.197	130.5	2.318	

4.2. Uniaxial Prediction for 08Ch18N10T Stainless Steel

The proposed model was used for an FE simulation of the uniaxial experimental program. The error of the amplitude of the force between experiment and simulation is formulated as:

$$Error = \frac{F_{a\ exp} - F_{a\ sim}}{F_{a\ exp}} \cdot 100[\%] \tag{26}$$

The mean error over the specimen is defined as:

$$Mean\ Error = \frac{1}{N_d} \sum_{n=1}^{N} Error_n \tag{27}$$

where $Error_n$ is the error in cycle n and N is the overall number of cycles in the simulation. The total error over all specimens is calculated as:

$$Total\ Error = \frac{1}{S} \sum_{s=1}^{S} Mean\ Error_s = 4.83\% \tag{28}$$

where $Mean\ Error_s$ is the mean error of specimen number s and $S = 17$ is the total number of specimens. The results of the experiments and the FEA simulations of all 17 uniaxial specimens are shown in Figures 13–29.

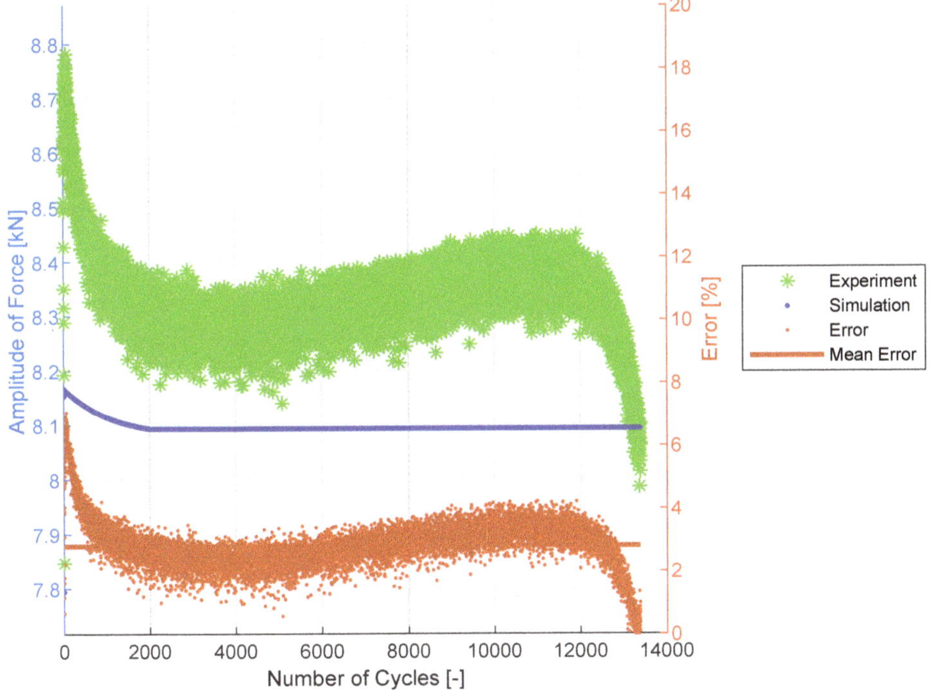

Figure 13. Specimen E9-1.

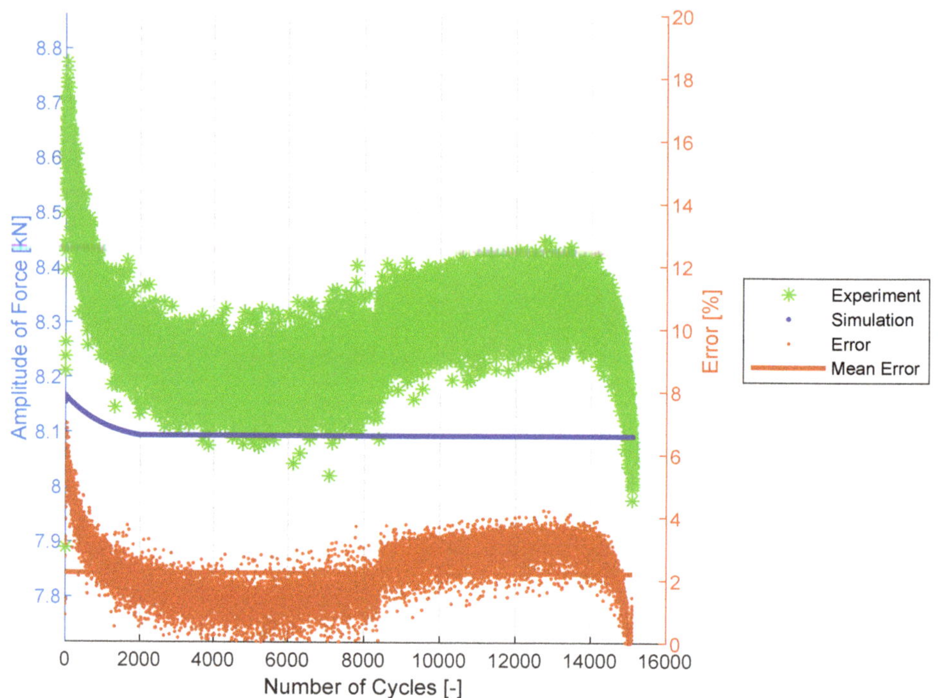

Figure 14. Specimen E9-2.

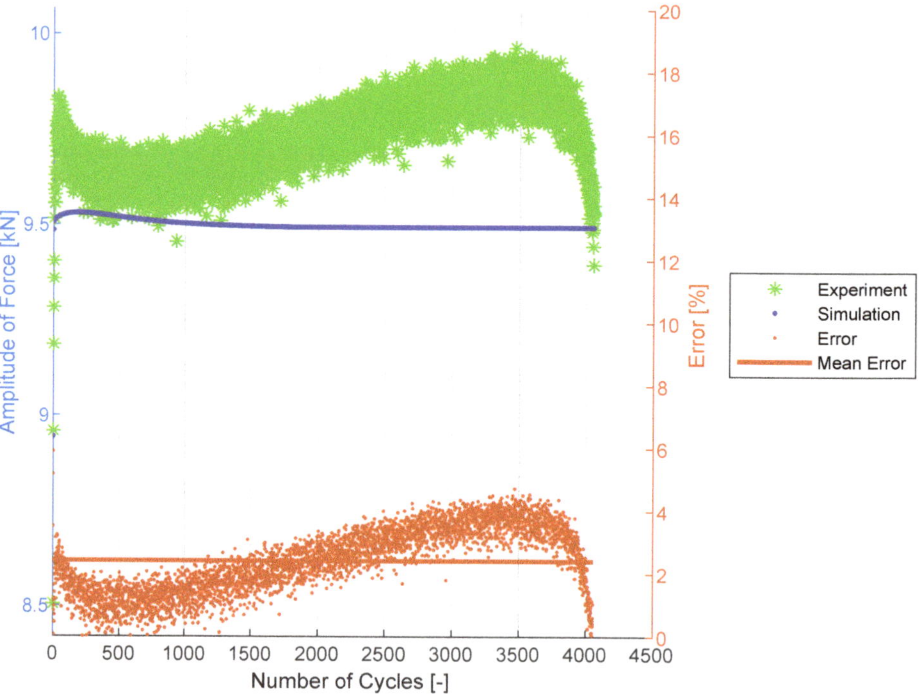

Figure 15. Specimen E9-3.

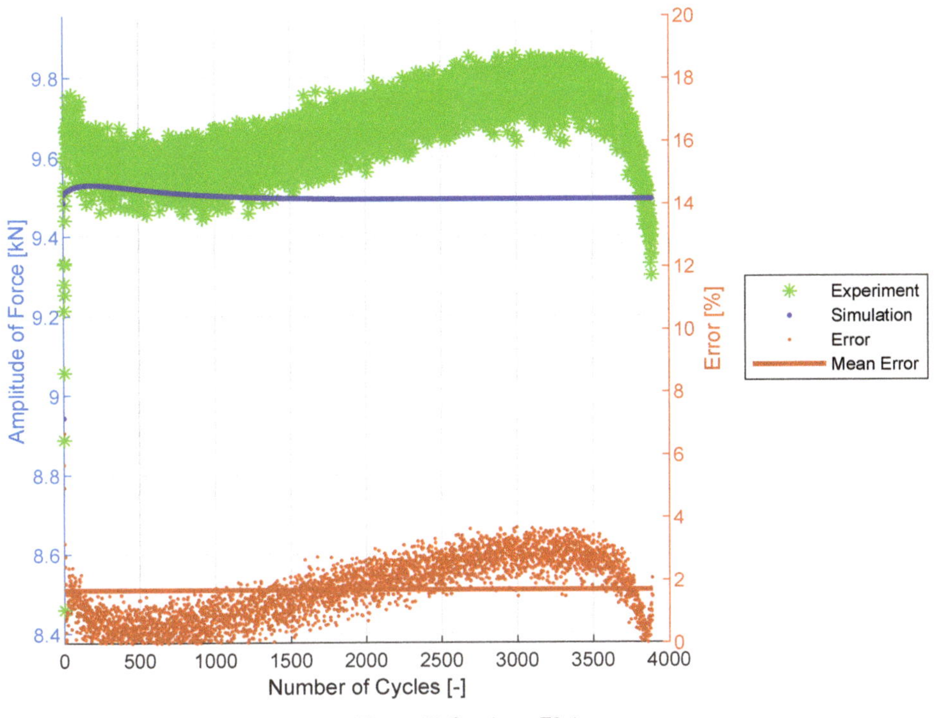

Figure 16. Specimen E9-4.

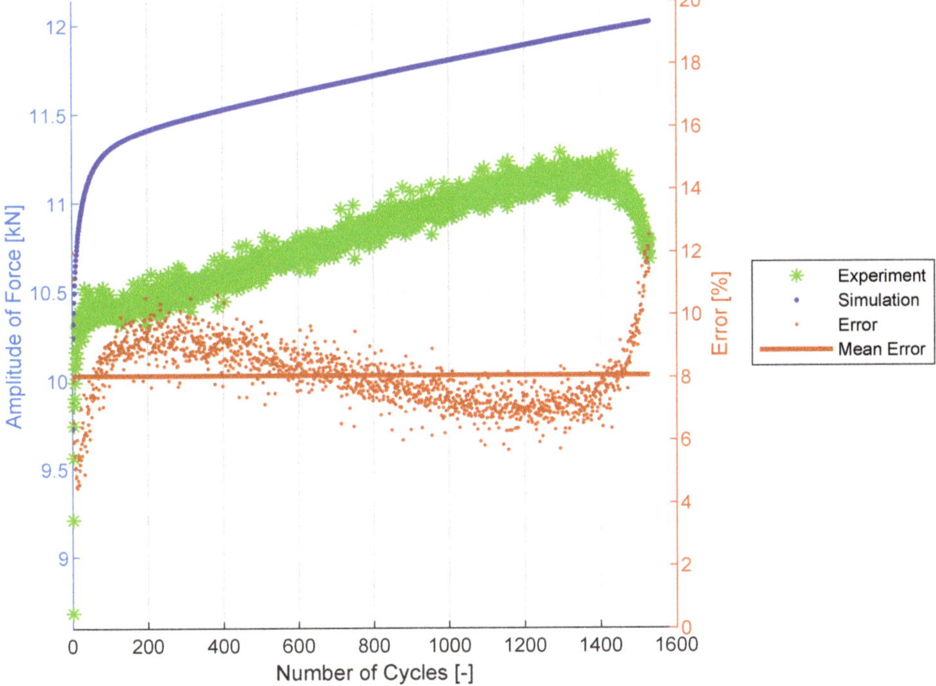

Figure 17. Specimen E9-5.

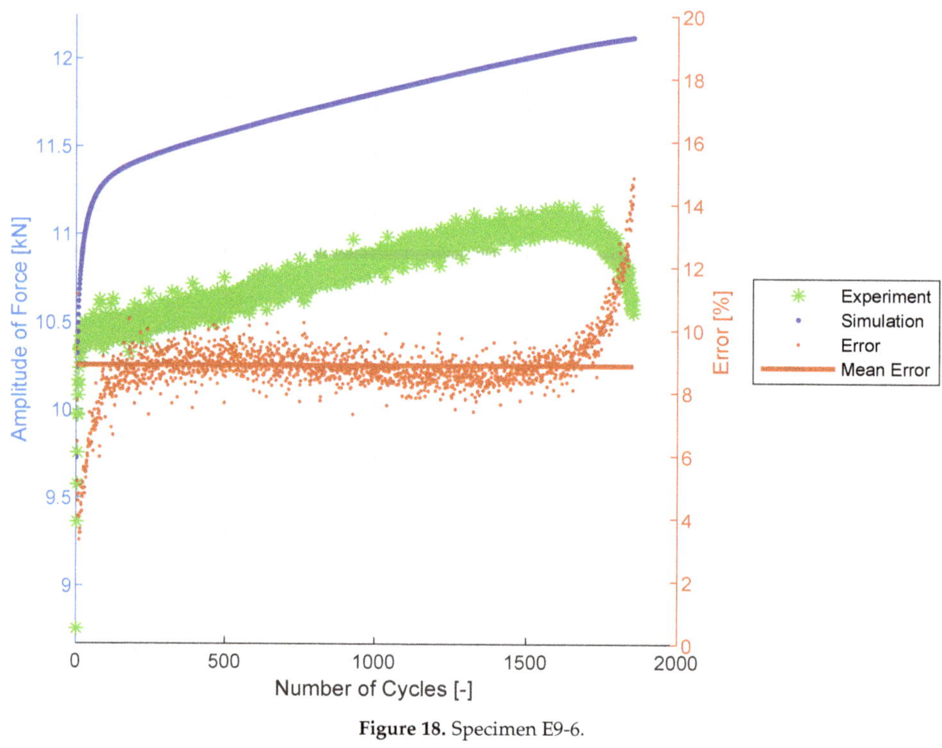

Figure 18. Specimen E9-6.

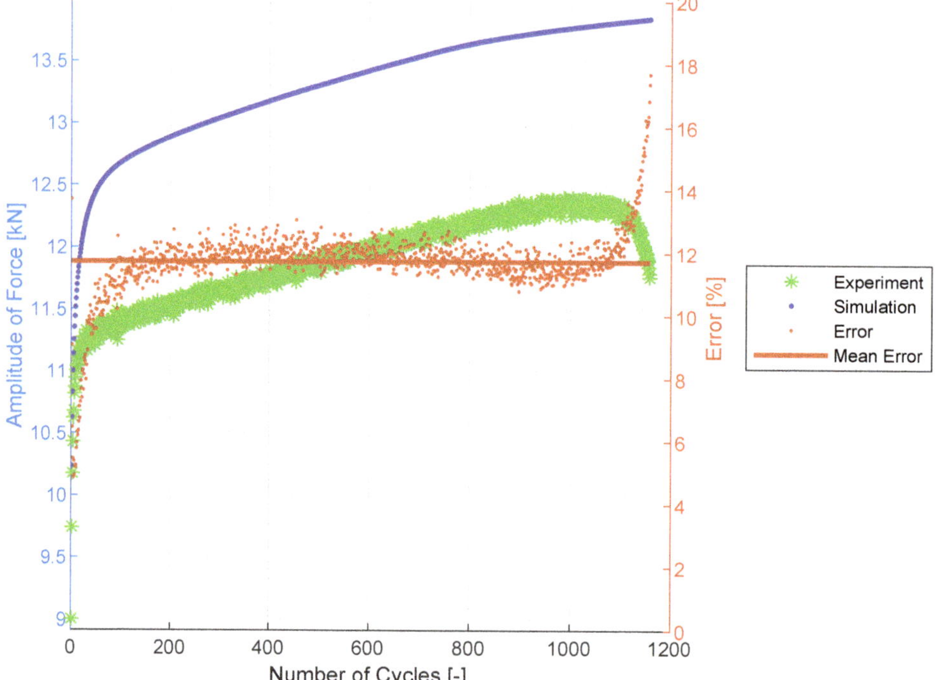

Figure 19. Specimen E9-7.

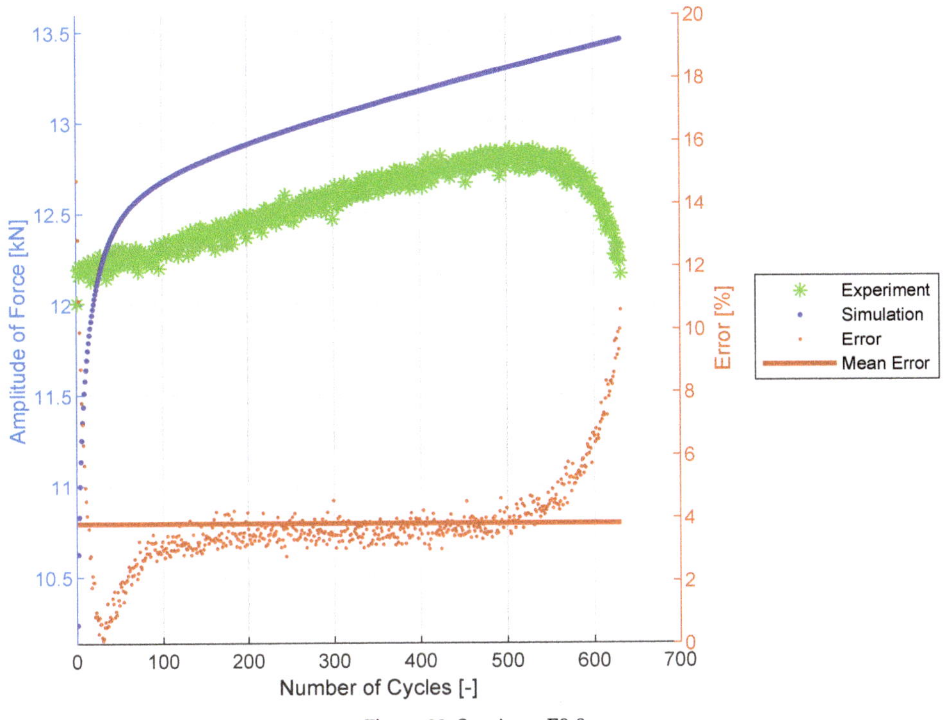

Figure 20. Specimen E9-8.

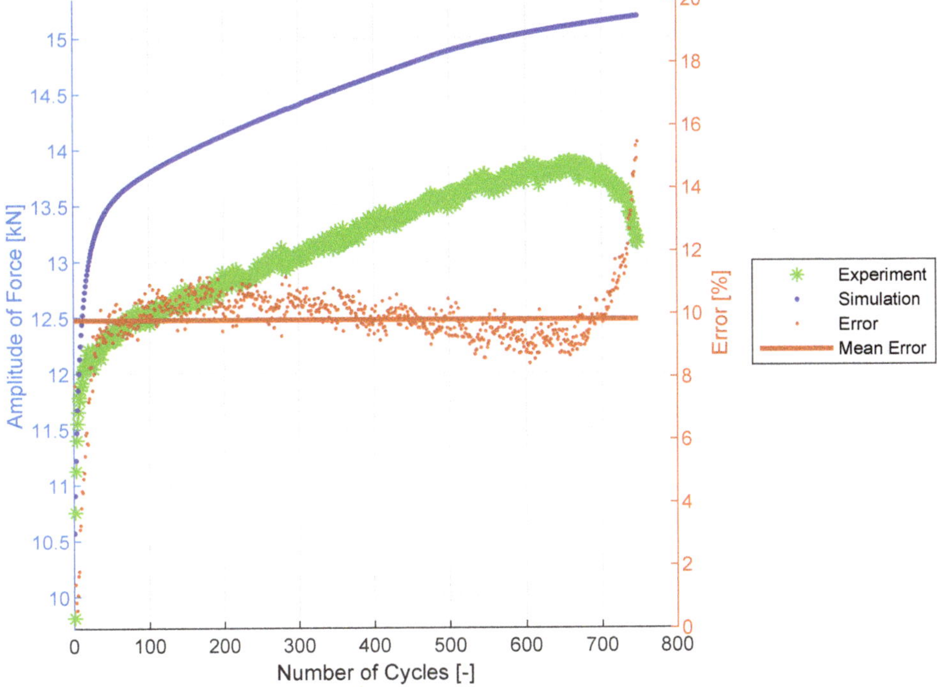

Figure 21. Specimen E9-9.

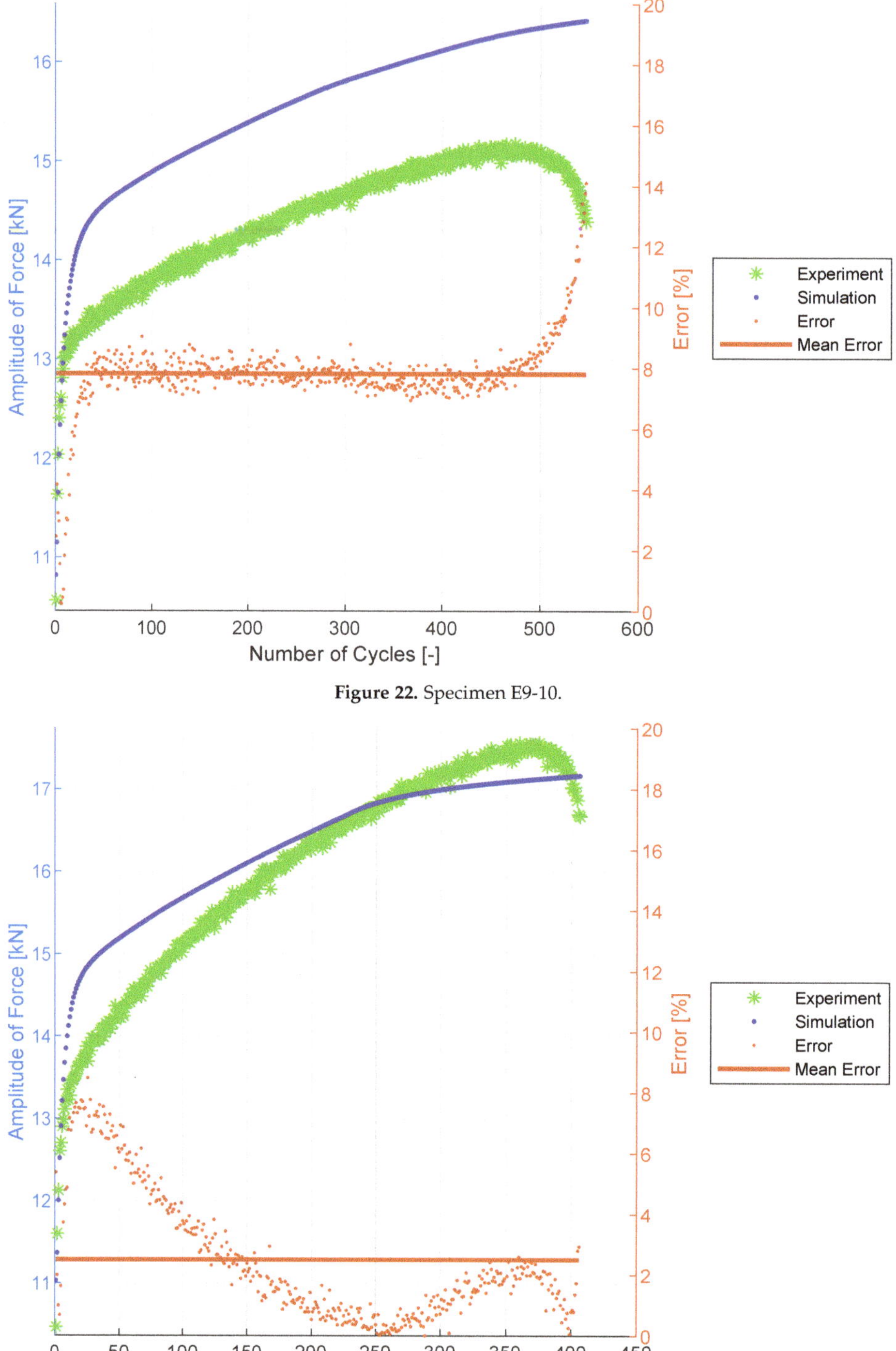

Figure 22. Specimen E9-10.

Figure 23. Specimen E9-11.

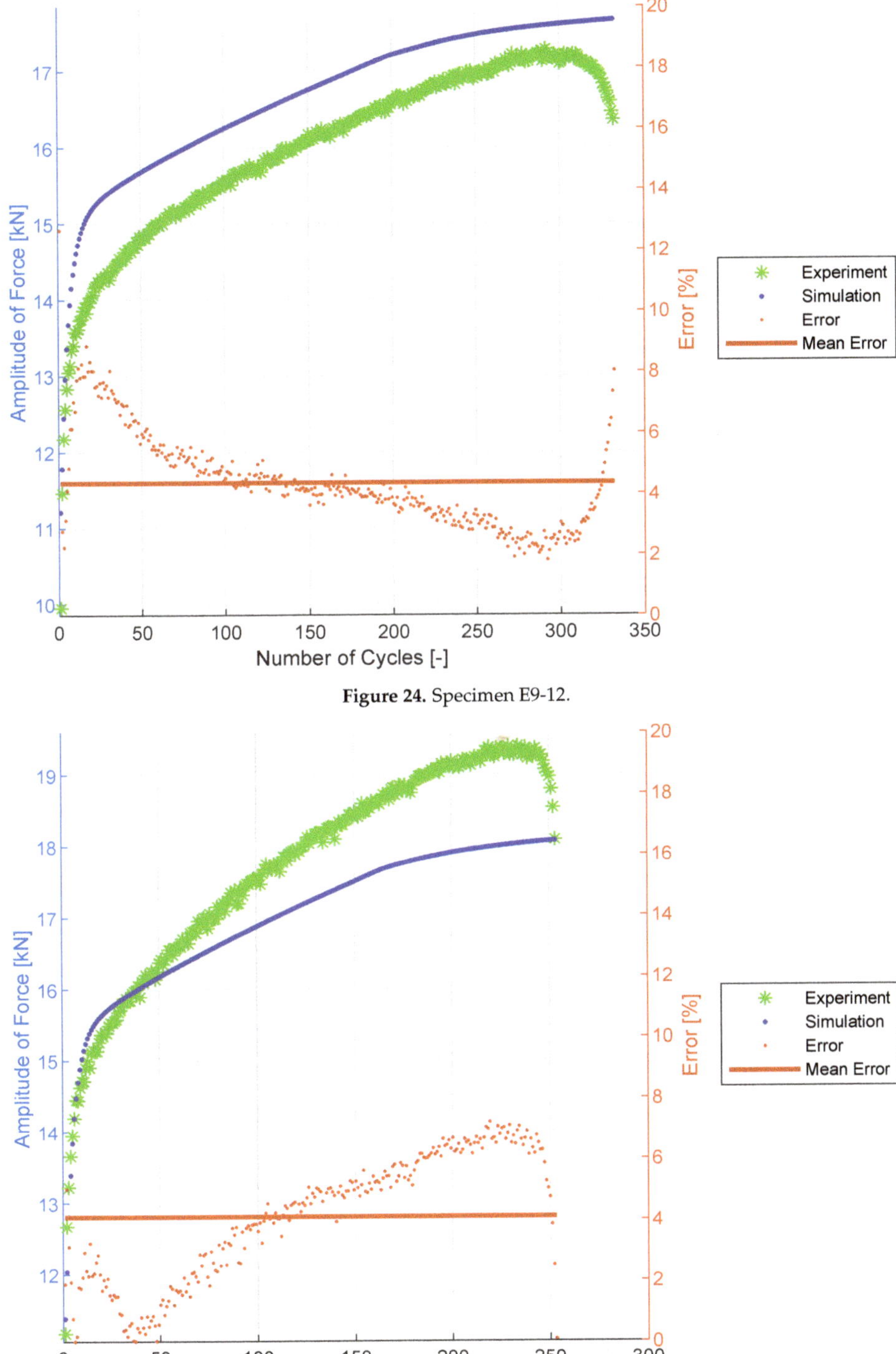

Figure 24. Specimen E9-12.

Figure 25. Specimen E9-13.

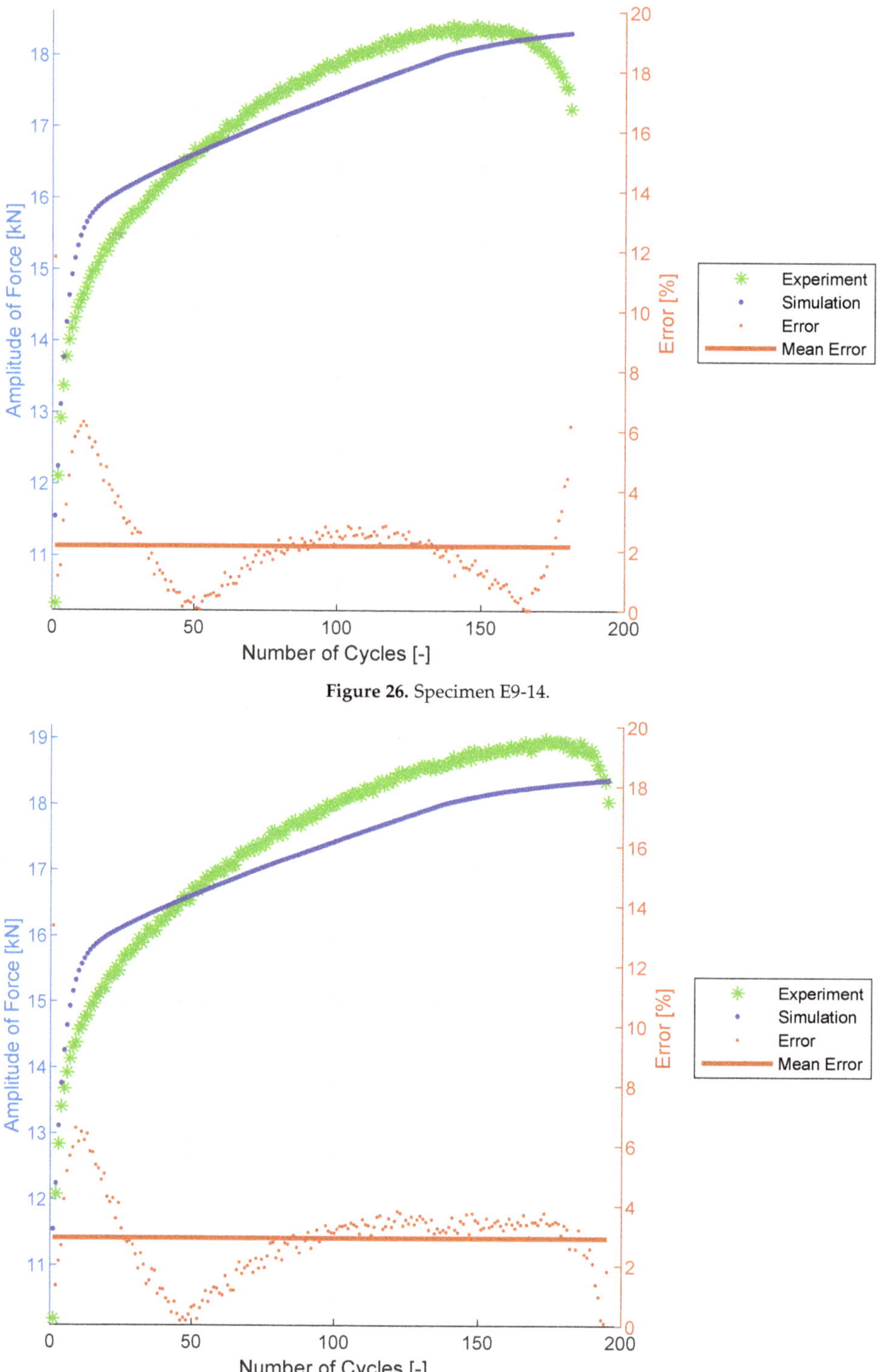

Figure 26. Specimen E9-14.

Figure 27. Specimen E9-15.

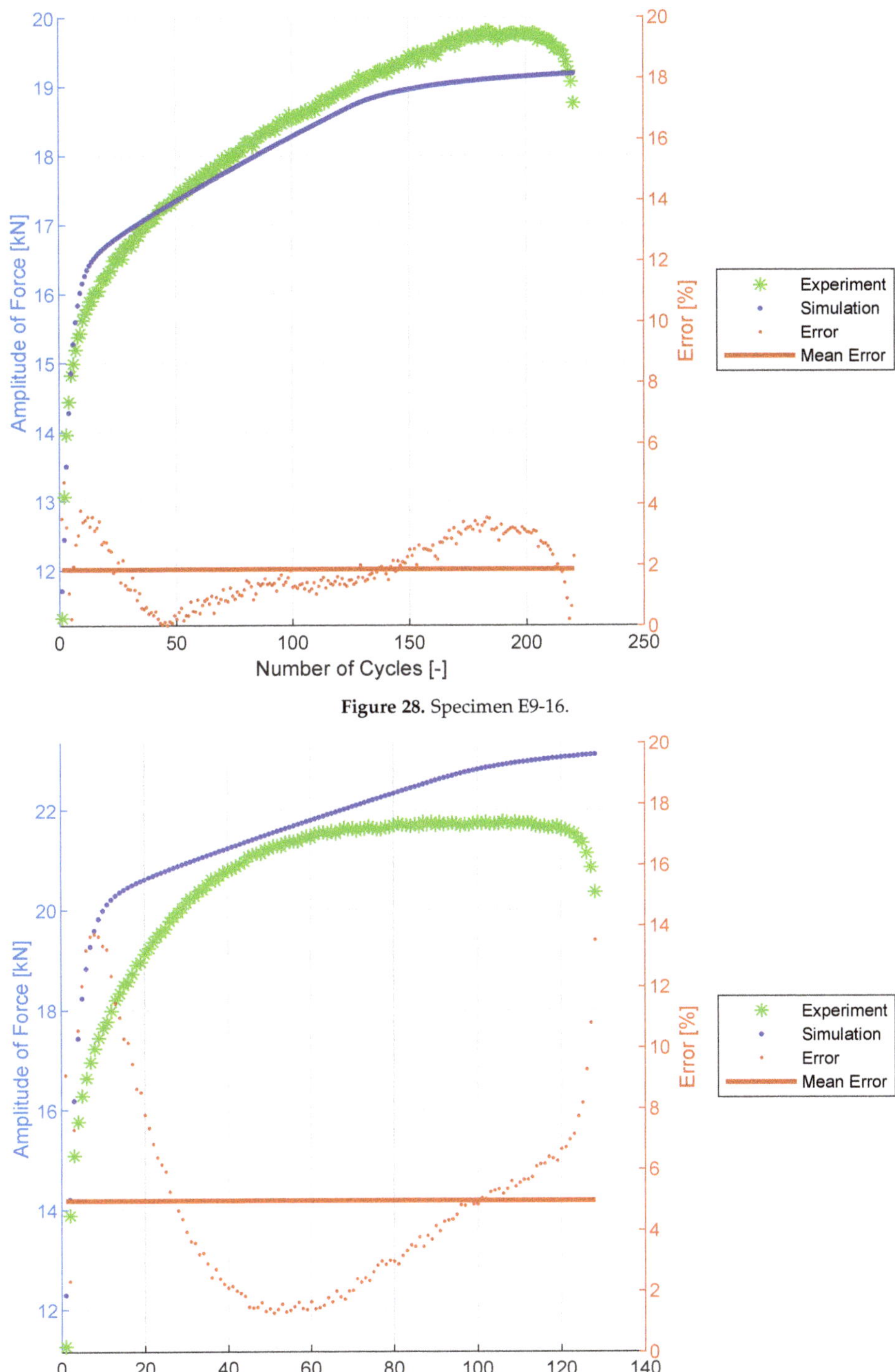

Figure 28. Specimen E9-16.

Figure 29. Specimen E9-17.

5. Discussion

As was shown in the previous section, the proposed model was able to capture the static curve and the cyclic stress–strain curve of SS304 very well. It also simulated well the shapes of the stress–strain hysteresis loops in all investigated cases, as well as the Bauschniger effect. The non-Masing behavior of the SS304 material was very strong, and this can be modeled better by superposing the kinematic and isotropic hardening, as proposed in the new constitutive model.

For the incremental test, the prediction was very good, especially in the fifth loading block, where it outperformed Kang's model [15]. Overprediction of the peak stresses in the first block of loading can be reduced, e.g., by introducing the memory surface contraction, as was proposed in the original model of Jiang and Sehitoglu [13].

The proposed model also provided a good description of the uniaxial tests of austenitic steel 08Ch18N10T. It captured the strain range-dependent cyclic hardening of this material, with an average simulation error of 4.83%. The proposed model slightly overestimated the initial phase of hardening. Furthermore, the model was unable to describe the softening at the end of the fatigue life caused by the fatigue crack growth. This phenomenon was not included in the model.

Figure 14 (Specimen E9-2) shows the attentive reader what seems to be a jump on the error axis at about 8200 cycles. Zooming on the data shows that the amplitude of the force predicted by the simulation was constant, while there was a relatively small gradual increase in the amplitude of the force in the experiment that took place over dozens of cycles. This is a common phenomenon in cyclic testing. Along with the relatively small value of the error between experimental prediction and simulation in that area, it optically intensified the jump effect in the graph.

6. Conclusions

In this paper, a new model of cyclic plasticity was proposed for describing the cyclic hardening of a material, when there is an influence of strain amplitude, based on the Jiang–Sehitoglu memory surface stated in the stress space. The introduction of a new internal variable in the form of virtual back-stress, which characterizes the behavior of the material in the case of a large strain amplitude, significantly reduced the number of material constants. Moreover, these parameters were now relatively easy to identify. Particular effects of cyclic plasticity could be described thanks to the introduction of dependency between selected parameters of the Chaboche kinematic hardening rule and the nonlinear isotropic hardening rule and the radius of memory surface R_M. The model contained 23 parameters in total, two of which were considered as zero for the SS304 material used here. The number of required parameters was less than one-third of the number required for the Kang model [15], while maintaining the accuracy of the description of the stress–strain behavior. Acceptable results were also obtained by the new cyclic plasticity model in simulations of our own experimental data for austenitic steel 08Ch18N10T. The idea of a stress-based memory surface applied to a virtual back-stress can also be used with other nonlinear kinematic hardening rules. The authors will focus on incorporating the modified Abdel-Karim–Ohno hardening rule [8] into the proposed model to get a better prediction of the ratcheting and stress relaxation of stainless steels in future works. Some interesting results of the Abdel-Karim–Ohno model enhanced by a memory surface can be found in [28].

Author Contributions: R.H.: proposed the cyclic plasticity theory; J.F.: application to 08Ch18N10T stainless steel and implementation into ABAQUS; P.G.: consistent tangent modulus determination for efficient implementation into FE codes; T.K.: calibration of the model with memory to SS304 experimental data; A.M.: implementation in ANSYS.

Funding: The paper has been done in connection with project Innovative and additive manufacturing technology—new technological solutions for 3D printing of metals and composite materials, reg. no. CZ.02.1.01/0.0/17_049/0008407 financed by Structural Founds of Europe Union and with the project No. GA19-03282S financed by the Grant Agency of the Czech Republic (GACR). This work was also supported by the ESIF, EU Operational Programme Research, Development and Education, and from the Center of Advanced Aerospace Technology (CZ.02.1.01/0.0/0.0/16_019/0000826), Faculty of Mechanical Engineering, Czech Technical University in Prague.

Conflicts of Interest: The authors declare no conflict of interest.

Abbreviations

The following abbreviations are used in this manuscript:

IDF	identification specimen series
NPP	nuclear power plant
SHL	selected hysteresis loop
SHLs	selected hysteresis loops

References

1. Halama, R.; Sedlák, J.; Šofer, M. Phenomenological Modelling of Cyclic Plasticity. In *Numerical Modelling*; Miidla, P., Ed.; IntechOpen: London, UK, 2012; pp. 329–354. Available online: http://www.intechopen.com/books/numerical-modelling/phenomenological-modelling-of-cyclic-plasticity (accessed on 23 March 2012).
2. Kopas, P.; Saga, M.; Baniari, V.; Vasko, M.; Handrik, M. A plastic strain and stress analysis of bending and torsion fatigue specimens in the low-cycle fatigue region using the finite element methods. *Procedia Eng.* **2017**, *177*, 526–531. [CrossRef]
3. Chen, X.; Chen, X.; Yu, D.; Gao, B. Recent progresses in experimental investigation and finite element analysis of ratcheting in pressurized piping. *Int. J. Press. Vessel. Pip.* **2013**, *101*, 113–142. [CrossRef]
4. Armstrong, P.J.; Frederick, C.O. A Mathematical Representation of the Multiaxial Bauschinger Effect. *Mater. High Temp.* **2007**, *24*, 1–26.
5. Chaboche, J.L.; Van Dang, K.; Cordier, G. Modelization of the Strain Memory Effect on the Cyclic Hardening of 316 Stainless Steel. In Proceedings of the 5th International Conference on Structural Mechanics in Reactor Technology, Berlin, Germany, 13–17 August 1979; pp. 1–10.
6. Ohno, N.; Wang, J.D. Kinematic Hardening Rules with Critical State of Dynamic Recovery. Part I: Formulation and Basic Features for Ratchetting Behavior. *Int. J. Plast.* **1993**, *9*, 375–390. [CrossRef]
7. Abdel-Karim, M.; Ohno, N. Kinematic hardening model suitable for ratchetting with steady-state. *Int. J. Plast.* **2000**, *16*, 225–240. [CrossRef]
8. Halama, R.; Fusek, M.; Šofer, M.; Poruba, Z.; Matušek, P.; Fajkoš, R. Ratcheting Behavior of Class C Wheel Steel and Its Prediction by Modified AbdelKarim-Ohno Model. In Proceedings of the 10th International Conference on Contact Mechanics CM2015, Colorado Springs, CO, USA, 30 August–3 September 2015.
9. Feigenbaum, H.P.; Dafalias, Y.F. Directional distortional hardening in metal plasticity within thermodynamics. *Int. J. Solids Struct.* **2007**, *44*, 7526–7542. [CrossRef]
10. Parma, S.; Plešek, J.; Marek, R.; Hrubý, Z.; Feigenbaum, H.P.; Dafalias, Y.F. Calibration of a simple directional distortional hardening model for metal plasticity. *Int. J. Solids Struct.* **2018**, *143*, 113–124. [CrossRef]
11. Sung, S.J.; Liu, L.W.; Hong, H.K.; Wu, H.C. Evolution of yield surface in the 2D and 3D stress spaces. *Int. J. Solids Struct.* **2011**, *48*, 1054–1069. [CrossRef]
12. Ohno, N. A constitutive model of cyclic plasticity with a nonhardening strain region. *J. Appl. Mech.* **1982**, *49*, 721–727. [CrossRef]
13. Jiang, Y.; Sehitoglu, H. Modeling of cyclic ratchetting plasticity, part I: Development of constitutive relations. *J. Appl. Mech.* **1996**, *63*, 720–725. [CrossRef]
14. Halama, R.; Fojtík, F.; Markopoulos, A. Memorization and Other Transient Effects of ST52 Steel and Its FE Description. *Appl. Mech. Mater.* **2013**, *486*, 48–53. [CrossRef]
15. Kang, G.Z.; Ohno, N.; Nebu, A. Constitutive modeling of strain-range dependent cyclic hardening. *Int. J. Plast.* **2003**, *19*, 1801–1819. [CrossRef]
16. Schäfer , B.J.; Song, X.; Sonnweber-Ribic, P.; Hassan, H.; Hartmaier, A. Micromechanical Modelling of the Cyclic Deformation Behavior of Martensitic SAE 4150—A Comparison of Different Kinematic Hardening Models. *Metals* **2019**, *9*, 368. [CrossRef]
17. Moeini, G.; Ramazani, A.; Myslicki, S.; Sundararaghavan, V.; Könke, C. Low Cycle Fatigue Behaviour of DP Steels: Micromechanical Modelling vs. Validation. *Metals* **2017**, *7*, 265. [CrossRef]
18. Msolli, S. Thermoelastoviscoplastic modeling of RAFM steel JLF-1 using tensile and low cycle fatigue experiments. *J. Nucl. Mater.* **2014**, *451*, 336–345. [CrossRef]

19. Ji, S.; Liu, C.; Li, Y.; Shi, S.; Chen, X. Effect of torsional pre-strain on low cycle fatigue performance of 304 stainless steel. *Mater. Sci. Eng.* **2019**, *746*, 50–57. [CrossRef]

20. Kang, G.; Li, Y.; Gao, Q. Non-proportionally multiaxial ratcheting of cyclic hardening materials at elevated temperatures: Experiments and simulations. *Mech. Mater.* **2005**, *37*, 1101–1118. [CrossRef]

21. Kang, G.; Gao, Q.; Yang, X. A visco–plastic constitutive model incorporated with cyclic hardening for uniaxial/multiaxial ratcheting of SS304 stainless steel at room temperature. *Mech. Mater.* **2002**, *34*, 521–531. [CrossRef]

22. Halama, R.; Markopoulos, A.; Jančo, R.; Bartecký, M. Implementation of MAKOC Cyclic Plasticity Model with Memory. *Adv. Eng. Softw.* **2018**, *113*, 34–46. [CrossRef]

23. Chaboche, J.L.; Lemaitre, J. *Mechanics of Solid Materials*; Cambridge University Press: Cambridge, UK, 1990.

24. Bari, S.; Hassan, T. Anatomy of Coupled Constitutive Models for Ratcheting Simulations. *Int. J. Plast.* **2000**, *16*, 381–409. [CrossRef]

25. Fumfera, J.; Halama, R.; Kuželka, J.; Španiel, M. Strain-Range Dependent Cyclic Plasticity Material Model Calibration for the 08Ch18N10T Steel. In Proceedings of the 33rd conference with international participation on Computational Mechanics 2017, Špičák, Czech Republic, 6–8 November 2017.

26. *ASTM Standard E606-92, 1998, Standard Practise for Strain-Controlled Fatigue Testing*; ASTM International: West Conshohocken, PA, USA, 1998.

27. *Metallic Materials—Tensile Testing—Part 1: Method of Test at Room Temperature*; ISO 6892-1:2016; International Organization for Standardization: Geneva, Switzerland, 2016.

28. Halama, R.; Bartecka, J.; Gal, P. FE Prediction and Extrapolation of Multiaxial Ratcheting for R7T Steel. *Key Eng. Mater.* **2019**, *810*, 76–81. [CrossRef]

Article

Ductile Compressive Behavior of Biomedical Alloys

Christian Affolter [1,*], **Götz Thorwarth** [2], **Ariyan Arabi-Hashemi** [3], **Ulrich Müller** [4] and **Bernhard Weisse** [1]

1   Lab for Mechanical Systems Engineering, Empa, Swiss Federal Laboratories for Materials Science and Technology, Ueberlandstrasse 129, CH-8600 Dübendorf, Switzerland; Bernhard.Weisse@empa.ch
2   IMT Masken und Teilungen AG, Im Langacher 46, CH-8606 Greifensee, Switzerland; gthorwarth@imtag.ch
3   Lab for Advanced Materials Processing, Empa, Swiss Federal Laboratories for Materials Science and Technology, Ueberlandstrasse 129, CH-8600 Dübendorf, Switzerland; ariyan.arabi-hashemi@empa.ch
4   Lab for Nanoscale Materials Science, Empa, Swiss Federal Laboratories for Materials Science and Technology, Ueberlandstrasse 129, CH-8600 Dübendorf, Switzerland; Ulrich.Mueller@empa.ch
*   Correspondence: Christian.Affolter@empa.ch

Received: 16 November 2019; Accepted: 20 December 2019; Published: 29 December 2019

Abstract: The mechanical properties of ductile metals are generally assessed by means of tensile testing. Compression testing of metal alloys is usually only applied for brittle materials, or if the available specimen size is limited (e.g., in micro indentation). In the present study a previously developed test procedure for compressive testing was applied to determine the elastic properties and the yield curves of different biomedical alloys, such as 316L (two different batches), Ti-6Al-7Nb, and Co-28Cr-6Mo. The results were compared and validated against data from tensile testing. The converted flow curves for true stress vs. logarithmic strain of the compressive samples coincided well up to the yield strength of the tensile samples. The developed compression test method was shown to be reliable and valid, and it can be applied in cases where only small material batches are available, e.g., from additive manufacturing. Nevertheless, a certain yield asymmetry was observed with one of the tested 316L stainless steel alloys and the Co-28Cr-6Mo. Possible hypotheses and explanations for this yield asymmetry are given in the discussion section.

Keywords: biomedical alloys; yield asymmetry; compressive testing; flow curve

---

1. Introduction

The origins of the theory of plasticity go back more than 100 years. A complete work can be found in Hill [1], originally published in 1950. The ductile properties of alloys including their flow curve during yielding are commonly determined with tensile testing, which is a well-standardized procedure. Compressive testing may be the appropriate material test for characterizing a ductile alloy, if the data are used to dimension a structure dominantly loaded under compression, or if the sample size and the physical constraints do not allow standard tensile testing. Kulagin et al. [2] applied compressive testing to analyze the effect of severe plastic deformation (SPD) on the microstructure and mechanical properties of a pure titanium. With even further reduced sample size, testing with micro or nano-indentation may be justified, where yield strength and tensile strength can be deduced recursively by fitting with numerical models.

In an earlier project for the development of hard coatings on biomedical alloys, a test procedure was developed by the present authors [3] in accordance to two standards:

- DIN 50106: originally from 1960, revised in 1978 and recently in 2016. It was mainly written for the determination of the compressive strength of brittle (cast) alloys, and it says very little about the allowable type of strain or displacement measurement. The standard is not particularly designed for the measurement of Young's modulus or yield strength [4].

- ASTM E9-09 describes in particular the adequate test setup with respect to axial alignment and parallelism of two hardened bearing blocks, and it addresses the problems related with buckling of slender test samples and 'barrelling' (the non-uniform deformation in the sample's end region due to friction; [5]).

The new test protocol allowed the determination of Young's modulus, yield strength and the entire flow curve (stress–strain curve in the plastic domain), and it was validated in a Round Robin on a 316L (DIN 1.4441 according to [6]) material against standard tensile testing [3]. The compressive test has an additional benefit, because tensile samples tend to undergo necking after a few percent of plastic strain, which will dictate the ultimate tensile strength (when necking has started, the true local stress may still increase, but the evaluated nominal stress drops). The yield curve in compression may evolve to plastic strains far above the strain A_t (ultimate strain at rupture) of a tensile test.

The hard coatings made of diamond-like carbons (*DLC*) developed in a previous project were investigated by micro indentation [7]: A conical imprint in the coating was produced, superimposing plastic strains and high stresses onto the already existing residual stresses from the coating process. The simultaneously created damage in the DLC-coating lead to a high delamination rate at the interface DLC—Substrate and allowed investigation of different types of coating agents and interlayers. The numerical simulation of the micro indentation tests required the actual flow curve of the different substrate materials, in order to predict the residual stresses, the plastic strain, and finally the energy release rates during delamination. As the available material batches were small and the results were used to simulate an indentation experiment (with dominantly compressive and hydrostatic stresses), the tests were performed on relatively small samples in compression, considering the relevant standards.

When tensile or compressive test data are produced, they are generally evaluated as 'nominal stress' vs. 'nominal strain'. These curves have to be converted to true stress vs. logarithmic strain for many state-of-the-art Finite Element (FE) codes such as *Abaqus* from Simulia/3DS [8]. The true stress can be derived as follows:

$$\sigma_{true} = \sigma_{nom} \, (1 + \varepsilon_{nom}), \tag{1}$$

and the logarithmic strain measure is defined as

$$\varepsilon_{ln} = \ln(1 + \varepsilon_{nom}) \tag{2}$$

The conversion of tensile and compression test curves of identical material batches should result in coinciding flow curves at least up to the start of necking in the tensile experiment (at strain A_g). Differences may indicate inappropriate test setups or even wrong boundary conditions.

In the present project, different biomedical alloys were tested in compression, and where the size of the material batch allowed it, tensile tests were performed for comparison. The elastic properties could be determined as well as the entire flow curve under tension/compression, and the conversion of the flow curves to true stress vs. logarithmic strain allowed further interpretation and assessment of the materials. Although the applied methodology in this study is not fully new, the yield curves presented and the verification against tensile data are difficult to find in literature and may be of interest to researchers employing numerical simulation. The observed yield asymmetry in two of the studied alloys was further examined and partly explained.

In the test setup used for this study, the influence of friction could be reduced to a minimum, as shown in [3]. Robinson et al. showed in a previous study how the friction coefficient affects the outcome of compression tests on ring samples [9]: The deformation pattern and the change of the inner ring diameter depend strongly on friction, and the results can be used to recursively determine the friction coefficient or material parameters by means of the Finite Element Method (FEM).

2. Materials and Methods

2.1. Test Method and Mechanical Setup

The method for the compressive tests was in accordance with ASTM E9-09 and DIN 50106, with some specific considerations concerning strain measurement, friction, and evaluation [3]. A setup was chosen, where the two hardened and polished bearing block surfaces can be set to parallel under preload by means of a spherical calotte, but it was an important requirement to fix the block's surface orientation after reaching the preload such that the calotte cannot further rotate under increased load. Two hardened cones were used to increase the working space between the parallel bearing block surfaces such that the clip-on gauge for strain measurement could be placed on the specimen after the cylindrical sample had been aligned in the machine axis and set under preload (cone material: DIN 1.3351; hardened to a Rockwell hardness ≥65 HRC). The chosen setup is shown schematically in Figure 1. The strain measurement was performed symmetrically on two sides of the cylindrical samples with a gauge type "Mini MFA-2" (from *MF Mess- & Feinwerktechnik GmbH*, Velbert, Germany). At a total strain of approx. 4% the transducer was removed and the following deformation was measured via the crossbar displacement to calculate the plastic strain. As described in [3] the load does not change substantially anymore, hence the distortion of the surrounding test setup can be assumed to be almost constant for increased plastic deformation, and the resulting flow curve will not be significantly influenced.

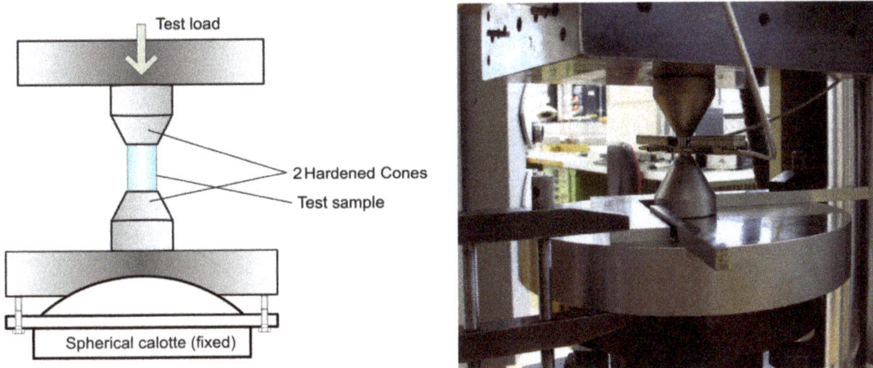

Figure 1. Schematic view of the mechanical setup (**left**) and installation with prepared sample in the test machine (**right**), with large plates adjusted and fixed in parallel.

The tensile tests were performed according to DIN EN ISO 6892-1:2009, and the strains were measured with a transducer type *multiXtens* from Zwick (Ulm, Germany) of class 0.5 according to EN ISO 9513. The strain rates for both test procedures were adjusted carefully, such that the strain rate in the elastic domain for tension was similar to the rate in the elastic domain for compression. The strain rates during plastic deformation were set higher by an order of magnitude, but again were comparable for tension and compression

2.2. Materials

The materials involved in this experimental study were two types of stainless steel 316L, a titanium alloy and a cobalt chromium molybdenum (CoCrMo), defined by the following specifications:

- A grade AISI 316L stainless steel (DIN 1.4441, implant quality), with further specifications as follows: X2CrNiMo 18-15-3 (ISO 5832-1 UNS S31673, ASTM F138), tensile strength between 930 and 1100 MPa. Polished rod with circular cross section (diameter $\varnothing$ = 10 mm; tolerance *h6*). The material will simply be called '1.4441'.

- A second batch of medical grade 316L stainless steel (Ø = 18 mm), where the supplier and more detailed specifications are not known, hereafter called '316L'.
- A titanium alloy type Ti-6Al-7Nb, hereafter called 'TAN'.
- A cobalt based Co-28Cr-6Mo, hereafter called 'CCM'.

The material first used for the development of the test procedure and the validation against tensile tests was the 1.4441 stainless steel in implant quality. All tensile and compressive samples had been produced from the same batch of a polished round rod (diameter Ø = 10.00 mm). Out of this rod, and also with the other materials, the following test samples were manufactured:

- Tensile test: according to DIN 50125, type "F 10 × 50" with rod lengths between 330 and 500 mm, or type "B 6 × 30" for shorter samples, where the outer diameter is given by the *M10*-thread. The 316L samples were of type "B 10 × 50" as the raw material was available with larger diameter.
- Compressive test: on the basis of DIN 50,106 and ASTM E9-09 with cylindrical shape, $d_0 = 10$ mm, and $h = 15$ mm which resulted in a h/d_0 ratio of 1.5.

All results were evaluated as *nominal stress* vs. *nominal strain*. From each set of tensile and compressive tests, the mean curves were taken and converted to *true stress* vs. *logarithmic strain* (cf. [8]). The conversion of the measured nominal strain provides a shift of the data points in the horizontal direction, in addition to the vertical shift for stress.

3. Results

All tests were performed carefully and the data digitally collected for conversion and comparison. Figure 2 shows an overview of all flow curves for plastic yielding, each graph providing the comparison between tensile and compressive tests. For some materials only a limited number of samples was available which does not allow a statistics for the relevant material parameters. For TAN no tensile samples were available at all, hence the test data of the compression tests had to be compared to literature data from Polyakova et al. ('Initial', coarse-grained in [10]).

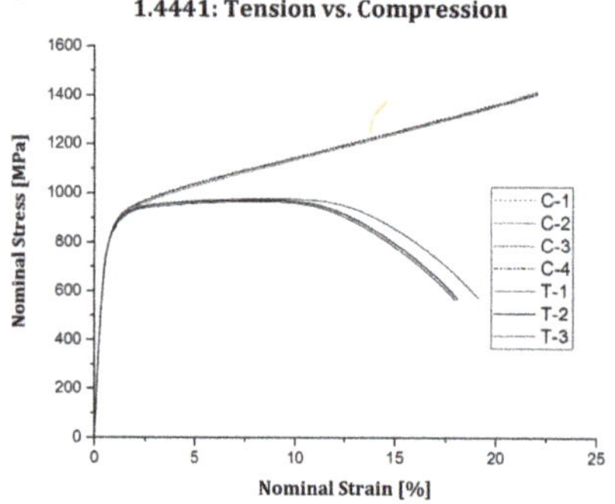

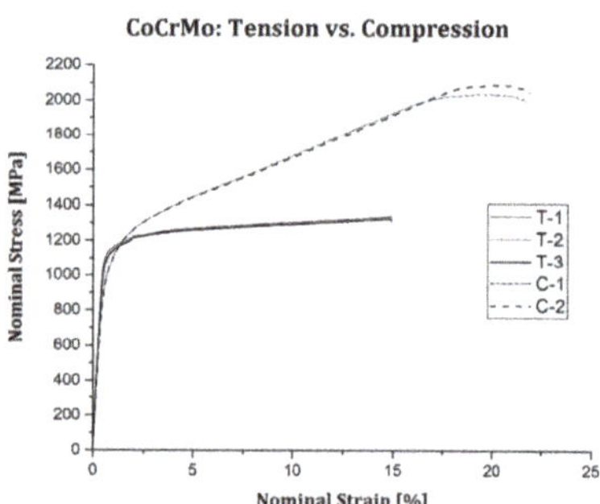

Figure 2. *Cont.*

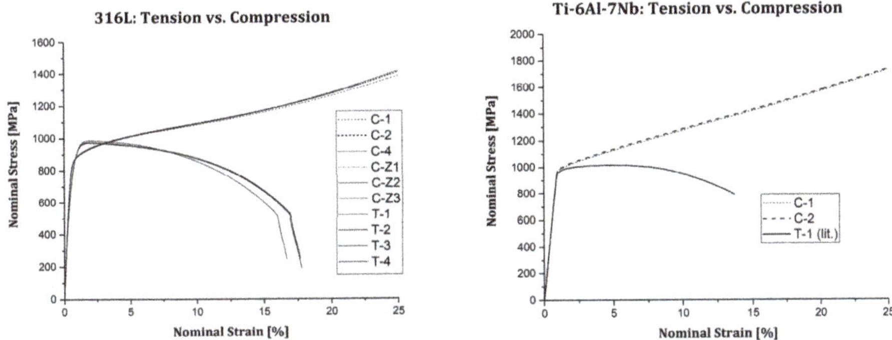

Figure 2. All test data, presented as nominal stress vs. nominal strain.

The most relevant material parameters are provided in Table 1. The results for Young's modulus E and yield strength $R_{p0,2}$ are in all cases comparable between tension and compression. Only the batches of 316L and CCM show a certain discrepancy in the yield strength, which is already indicated in the test curves. TAN was the only material showing a distinct yield limit. For all other materials, the yield strength is determined at the point of 0.2% permanent plastic deformation ($R_{p0,2}$).

Table 1. Overview of material parameters with statistics and comparison of the test results.

| Material | 1.4441 | | 316L | | CCM | | TAN | |
Type of Testing	Tension	Compr.	Tension	Compr.	Tension	Compr.	Tension	Compr.
Young's Modulus E [MPa]	178,714	174,370	179,847	179,085	228,292	220,398	105,000 (lit.)	101,618
$R_{p0,2}$ [MPa]	778.8	777.2	760.4	842.0	1102.3	973.9	975	990.3
R_m [MPa]	970.7	n.a.	977.6	n.a.	1323.7	n.a.	1020	n.a.
$\varepsilon_{Ult.}$ [%]	18.4	n.a.	17.5	n.a	13.4	n.a	13	n.a
No. of samples	3	4	4	6	3	2	(1, lit. [1])	2

[1] from literature [10].

For a further analysis a mean curve was taken from each set of resulting test curves, and these curves were then converted to *true stress* vs. *logarithmic strain* (according to Equations (1) and (2), see also [8]) which is a common and established procedure for the numerical simulation of ductile (plastic) materials. Figure 3 provides for each material a set of converted curves (continuous) versus the original curves (dashed).

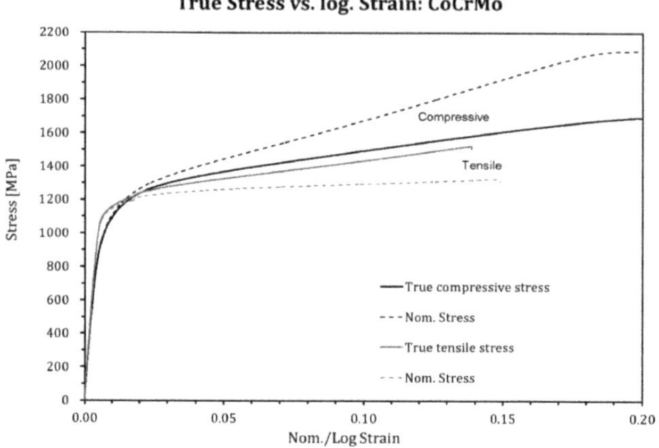

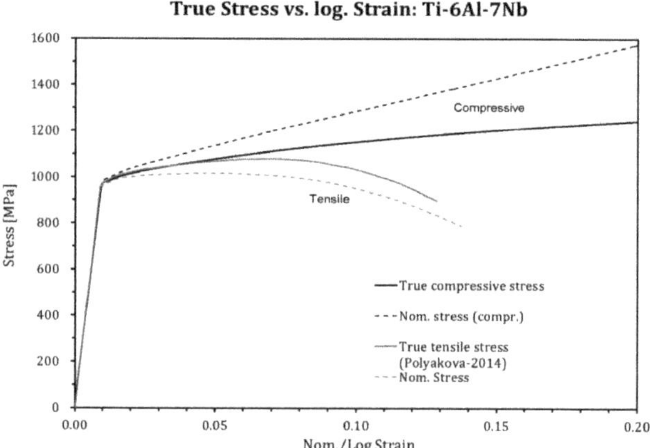

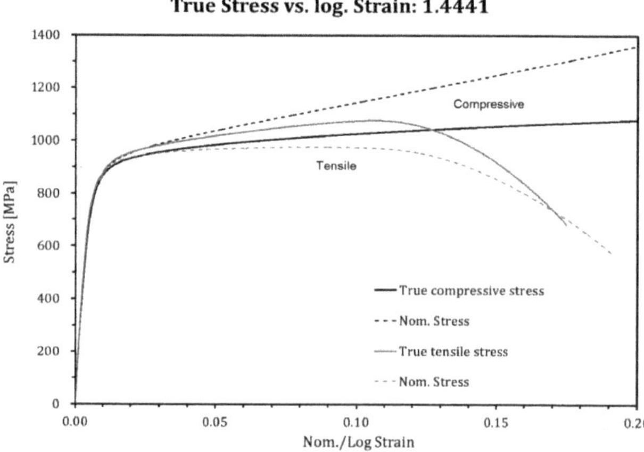

Figure 3. *Cont.*

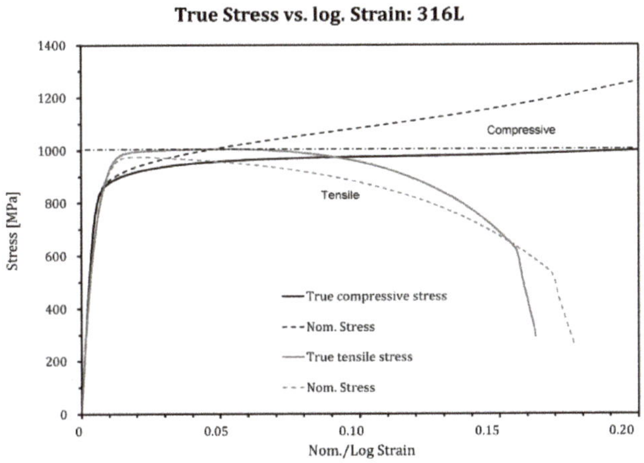

Figure 3. Test data displayed as *True Stress* vs. *Logarithmic Strain*.

The only material showing a fast fracture without antecedent necking of the tensile sample is the Co-28Cr-6Mo (CCM), all other materials show necking of the tensile sample during plastic deformation, which leads to the drop of the tensile curve. This drop is also observed in the converted 'true stress' curve due to the purely mathematical conversion process (as the true local stress in the necking area actually could not be measured).

4. Discussion

When looking at the converted flow curves for tension and compression in the range between yield strength and tensile strength (true stress vs. log strain), both curves usually run parallel up to the tensile strength R_m, where necking of the tensile sample begins (at strain A_g). This is especially true for the two materials 1.4441 and CCM; the stainless steel 1.4441 shows slightly higher strength in tension, CCM on the other hand has a higher strength in compression. TAN shows a similar behavior, but both flow curves start rather tangentially from the yield strength and progress with different curvature.

316L showed another behavior than the three previous materials, as necking of the tensile samples seemed to start shortly after reaching the material's yield strength. Both converted flow curves rather had a horizontal envelope in common at approx. 1000 MPa (cf. Figure 3). For this 316L batch even the yield strength was very different in tension and compression, respectively, see also Table 1. This 'yield asymmetry' could also be observed with CCM, though $R_{p0,2}$ of 316L was higher in compression, but for CCM higher in tension. A detailed look at the flow curves of these two materials is shown in Figure 4 in the transition between elastic and plastic behavior (up to 6% total strain).

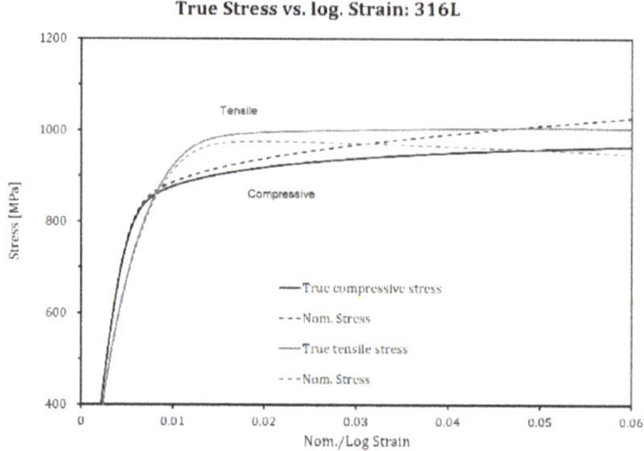

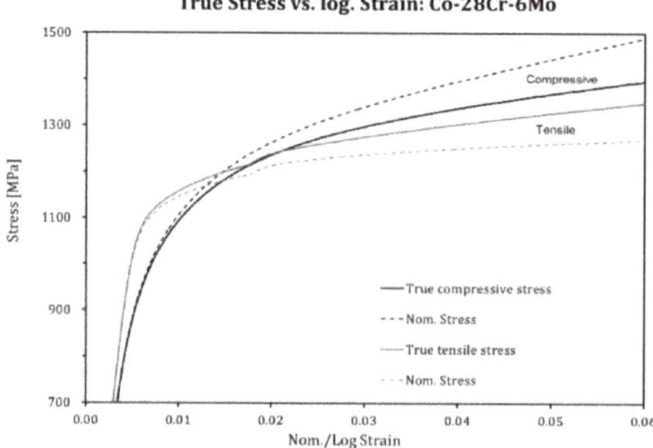

Figure 4. Asymmetry in steel 316L (**top**) and Co-28Cr-6Mo (**bottom**).

There are different possible explanations for the yield asymmetry in the 316L stainless steel. Implant quality 316L has preferred a pure austenitic microstructure. Cold working (or machining and mechanical treatment with consequent heating/thermal treatment) may lead to a martensite phase transformation [11]. Martensite is undesirable for implants as it is a highly magnetic phase [12]. Another mechanism which may lead to yield asymmetry is the creation of 'twin' crystals during plastic deformation. The creation of twins is different under tension and compression, which may also lead to an asymmetric behavior during yielding. Phase transformations then occur simultaneously to grain boundary sliding. Magnetism could not be observed in the tested samples, hence martensite phase transformation seems unlikely in the present case.

In order to further reveal the origin of the yield strength asymmetry of 316L, its microstructure was compared with the microstructure of 1.4441 which does not show a yield strength asymmetry. 316L and 1.4441 have the same chemical composition, and both materials show a high degree of deformation twinning in the as-received state, as shown in Figure 5. Alloy 1.4441 has a larger grain size than 316L. The high density of deformation twins in the as-received state is attributed to large plastic deformations of both materials. Due to the difference of grain size and the high amount of deformation twinning it can be expected that both materials have undergone different thermo-mechanical treatments. The reduced grain size in 316L might be the reason for the yield asymmetry favoring yielding under tension.

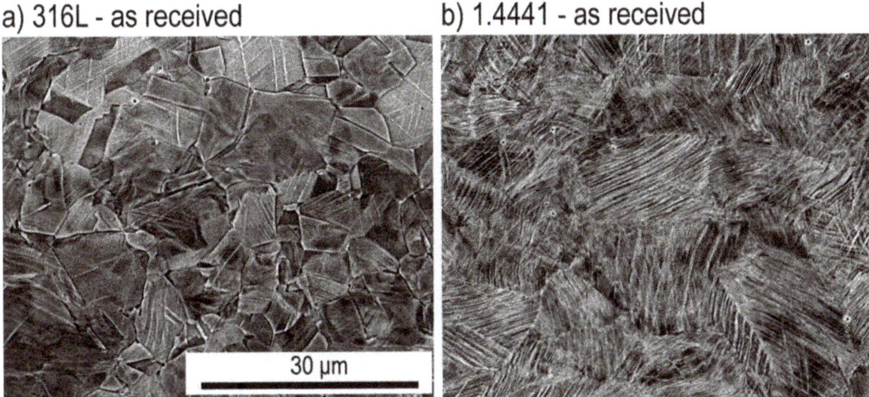

Figure 5. Microscopy using backscattered electrons reveals the microstructures in the as-received state of (**a**) 316L and (**b**) 1.4441.

The different thermo-mechanical treatments can cause textures which can both be responsible for the yield asymmetry in 316L. Texture might have an impact on the yield strength asymmetry. In the present case 316L shows a reduced yield strength for tension. XRD and EBSD results show that 316L has a texture where preferentially grains with a <111> orientation are aligned parallel to the loading direction (Figure 6, left). Twinning for <111> orientated grains is easier for tension than for compression [13,14]. Thus the observed texture is in accordance with the observed yield strength asymmetry. However, since both materials 316L and 1.4441 show a <111> texture, texture does not fully explain the observed asymmetry, as it is not present in the case of 1.4441 (Figure 6, right).

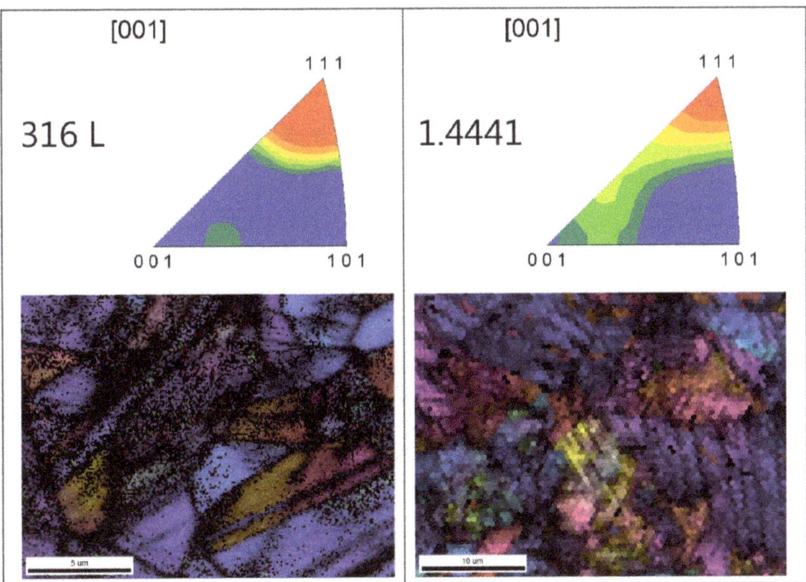

Figure 6. Plot (IPF) and EBSD measurement of the two stainless steel materials.

To obtain insights into the microstructural changes of 316L during compression and tension, XRD measurements were done. The measurements were done at a cross-section of the round specimens which were unstrained and strained to +1.5% and −1.5%. The planes of diffraction of the measurements shown in Figure 7 had their normal axis parallel to the loading direction. In a texture-free sample the (200) peak would have around 50% of the intensity of the (111) peak. Thus in the present case, due to the strong (111) peak, a (111) texture is present as confirmed by the pole figure measurements. The XRD measurements show that the intensity of the (111) peak decreases for tension and increases for compression. This indicates that during tension new twins are formed or existing twins grow. On the other hand during compression it seems that detwinning takes place. That means that existing twins shrink and thus the measured (111) intensity increases.

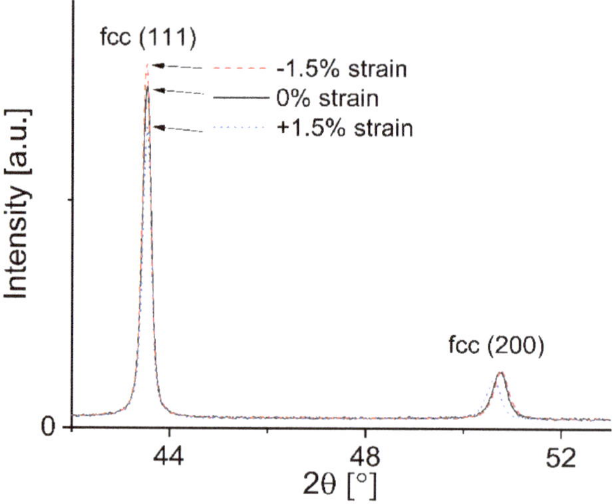

Figure 7. XRD measurements of 316L taken at 0% strain, +1.5% and −1.5% strain. The normal vector of the diffraction planes is aligned parallel to the loading direction.

A final explanation may be that the raw material was inhomogeneous. The rods from which the 316L test samples were machined had an outer diameter of 18 mm, the ones for the 1.4441 samples 10 mm. If the tensile and the compressive test specimen had a differing diameter, and if there was a changing grain size or phase composition from the core to the outside of the rods, i.e., a radial inhomogeneity, then machining might lead to a different microstructure through the tested cross sections of tensile and compression samples, respectively, or it might affect (release) internal stresses. However, tensile and compressive samples had identical diameters for both material batches.

5. Conclusions

Yield curves were produced for different biomedical alloys by means of tensile and compressive testing, and the curves were cross-validated. The new compression test method ([3]) was shown to be reliable and valid, and it can be applied in cases where only small material batches are available. The occurrence of a yield asymmetry with 316L and CoCrMo could not fully be explained by the microstructural analysis. However, the presented curves are a valuable input for future research and applied R&D (implant development) supported by numerical simulation.

Following recommendations can be given for the implementation of plasticity in finite element modelling: Most of the FE codes will require a flow curve defined as true stress vs. (logarithmic) plastic strain. It is appropriate to implement an averaged curve from tension and compression test data, if both results are available, and if the load case can result in tensile and compressive strains (e.g., in a bending load case). For dominantly compressive load cases it is advantageous to implement a test curve from a compressive test (e.g., simulation of indentation tests). It is particularly inappropriate to implement the declining right branch of the flow curve which results from necking in the tensile samples. It is simpler and finally more accurate to extend the approximately linear part of the (averaged) flow curve before necking.

Author Contributions: C.A. led the study, developed the test procedures and performed the validations also by means of FEM modelling. C.A. also considered the continuum mechanical aspects. G.T. considered practical aspects in the study, and ensured its relevance for practical applications in biomedical engineering and production of hard coatings on metals. G.T. also performed metallographic experiments (EBSD) and helped to analyze the observed yield asymmetries. A.A.-H. performed the microstructural analysis: micrographs, XRD, and EBSD. A.A.-H. investigated the origins of the observed yield asymmetry. U.M. helped to develop the compression test procedure which allows to determine the plastic flow curve, and also supported the numerical modelling in the first phase. B.W. organized the funding and supported the study with theoretical thoughts (continuum mechanics, mechanical testing). All authors have read and agreed to the published version of the manuscript.

Funding: The study was partially funded by the Swiss Commission for Technology and Innovation CTI/KTI with the grant No. 8475.1 LSPP-LS.

Acknowledgments: The authors express their gratitude to Hans Michel for all the valuable ideas and practical work in the testing lab, and the continuous efforts to improve the test procedures.

Conflicts of Interest: The authors declare no conflict of interest.

References

1. Hill, R. *The Mathematical Theory of Plasticity*; Oxford University Press: New York, NY, USA, 1998; Volume 11, p. 355.

2. Kulagin, R.; Latypov, M.I.; Kim, H.S.; Varyukhin, V.; Beygelzimer, Y. Cross Flow During Twist Extrusion: Theory, Experiment, and Application. *Metall. Mater. Trans. A* **2013**, *44*, 3211–3220. [CrossRef]

3. Affolter, C.; Müller, U.; Leinenbach, C.; Weisse, B. Compressive Testing of Ductile High-Strength Alloys. *J. Test. Eval.* **2015**, *43*, 1554–1562. [CrossRef]

4. DIN 50106. *Testing of Metallic Materials—Compression Test at Room Temperature*; DIN Deutsches Institut für Normung: Berlin, Germany, 2016.

5. ASTM E9-09. *Standard Test Methods of Compression Testing of Metallic Materials at Room Temperature*; ASTM International: West Conshohocken, PA, USA, 2000; pp. 98–105.

6. DIN EN 10027-2. *Designation Systems for Steels—Part 2: Numerical System, in Designation Systems for Steels—Part 2: Numerical System*; Beuth Verlag GmbH: Berlin, Germany, 2015.

7. Falub, C.V.; Thorwarth, G.; Affolter, C.; Müller, U.; Voisard, C.; Hauert, R. A quantitative in vitro method to predict the adhesion lifetime of diamond-like carbon thin films on biomedical implants. *Acta Biomater.* **2009**, *5*, 3086–3097. [CrossRef] [PubMed]

8. 3DS/Simulia. *ABAQUS/Standard: Abaqus Analysis User's Guide*; 3DS/Dassault Systems: Vélizy-Villacoublay, France, 2014.

9. Robinson, T.; Ou, H.; Armstrong, C.G. Study on ring compression test using physical modelling and FE simulation. *J. Mater. Process. Technol.* **2004**, *153*, 54–59. [CrossRef]

10. Polyakova, V.V.; Anumalasetty, V.N.; Semenova, I.P.; Valiev, R.Z. Influence of UFG structure formation on mechanical and fatigue properties in Ti-6Al-7Nb alloy. In *IOP Conference Series: Materials Science and Engineering*; IOP Publishing: Bristol, UK, 2014; p. 012162.

11. Saltykova, A. *Verformungsinduzierte Martensitbildung und Schädigungsverhalten im Stahl X5CrNi18.10, in Institut für Metallkunde*; TU Bergakademie Freiberg: Freiberg, Germany, 2005; p. 80.

12. Disegi, J. Implant Materials. In *Ti-6Al-7Nb*; Synthes, I.U., Ed.; Synthes: West Chester, PA, USA, 2008.

13. Yang, P.; Xie, Q.; Meng, L.; Ding, H.; Tang, Z. Dependence of deformation twinning on grain orientation in a high manganese steel. *Scr. Mater.* **2006**, *55*, 629–631. [CrossRef]

14. Yapici, G.G.; Karaman, I.; Luo, Z.P.; Maier, H.J.; Chumlyakov, Y.I. Microstructural refinement and deformation twinning during severe plastic deformation of 316L stainless steel at high temperatures. *J. Mater. Res.* **2011**, *19*, 2268–2278. [CrossRef]

 *metals*

Article

Quasi-Particle Approach to the Autowave Physics of Metal Plasticity

Lev B. Zuev * and Svetlana A. Barannikova

Institute of Strength Physics and Materials Science, Siberian Branch of Russian Academy of Sciences, 634055 Tomsk, Russia; bsa@ispms.tsc.ru
* Correspondence: lbz@ispms.tsc.ru; Tel.: +7-3822-491-360

Received: 28 September 2020; Accepted: 28 October 2020; Published: 29 October 2020

Abstract: This paper is the first attempt to use the quasi-particle representations in plasticity physics. The de Broglie equation is applied to the analysis of autowave processes of localized plastic flow in various metals. The possibilities and perspectives of such approach are discussed. It is found that the localization of plastic deformation can be conveniently addressed by invoking a hypothetical quasi-particle conjugated with the autowave process of flow localization. The mass of the quasi-particle and the area of its localization have been defined. The probable properties of the quasi-particle have been estimated. Taking the quasi-particle approach, the characteristics of the plastic flow localization process are considered herein.

Keywords: metals; plasticity; autowave; crystal lattice; phonons; localization; defect; dislocation; quasi-particle; dispersion; failure

1. Introduction

The autowave model of plastic deformation [1–4] admits its further natural development by the method widely used in quantum mechanics and condensed state physics. The case in point is the introduction of the quasiparticle exhibiting wave properties—that is, application of corpuscular-wave dualism [5]. The current state of such approach was proved and illustrated in [5].

Few attempts of application of quantum-mechanical representations are known in plasticity physics. They were focused, for example, on direct quantum-mechanical treatment of specific details of plastic flow mechanisms unexplained by traditional approaches. Thus, Bell [6] paid attention to possible quantization of elastic modules of materials, and Steverding [7] introduced the representation about quantization of elastic waves accompanying a destruction process. Later on, this problem in expanded interpretation was considered by Maugin [8] with application to solitons in elastic media. Gilman [9] and then Oku and Galligan [10] tried to use the tunnel effect to explain the dislocation breaking from obstacles at low temperatures when the thermal activation ceased to operate and estimated the probability of tunneling under these conditions. Petukhov and Pokrovskii [11] presented the well-founded and exact analysis of this phenomenon for dislocations moving in the Peierls potential relief. Recently, these problems have been repeatedly addressed in [12].

The other group of studies was apparently initiated by Morozov, Polak, and Fridman in their crucial work [13]. These authors analyzed the kinetics of growth of a fragile crack in a deforming medium and postulated the existence of a quasiparticle which they associated with the end of the growing fragile crack and called crackon. This idea was further used and developed to study mechanisms of strain-induced microdefects. Thus, Zhurkov [14] introduced the concept of elementary crystal excitation—dilaton—representing a negative density fluctuation. Olemskoi and Katsnelson [15] discussed the occurrence of the shear or destruction nuclei they called frustron.

These works were the first attempts to introduce the quantum representations into plasticity physics. They were crucial, because the existing plastic flow models in actual crystals with defects were completely based on principles of classical mechanics. At the same time, the crystal lattice theory is, generally speaking, quantum-mechanical in character. Such different approaches, obviously, make it difficult to construct a complete and physically well-founded plastic flow pattern.

The idea of the existence of the quasi-particle corresponding to the localized plastic flow autowave arises naturally against this background. Some aspects of this promising problem are studied in the present work. The approach developed here supplements the autowave mechanics of plasticity [1] and in our opinion, can be fruitful for an explanation of the main features of the plastic flow phenomenon.

The present study initiates the use of a quasi-particle model for the explanation of the localized plastic flow development in solids. A similar step matches the main directions in physics of solids [5].

2. Materials and Methods

The studies of localized plastic flow were performed for nineteen metals from the 3rd, 4th, 5th, and 6th periods of the Periodic Table of Elements (see the Table 1). The autowave characteristics of the localized plastic strain required for an analysis were obtained from visual pattern of their localized plasticity recorded by the speckle-photographic method (SPM) of analysis of deformation fields during elongation of flat samples described in detail in [1–4]. SPM has a field of vision of about 100 mm and resolution of about 1 μm comparable to that of optical microscopy. The method was realized using an Automatic Laser Measuring Complex (ALMEC), which enables one to reconstruct the displacement vector fields $r(x, y)$ for the deforming flat sample and calculate all the plastic distortion tensor components—i.e.,

$$\beta_{ij} = \nabla r(x, y) = \begin{pmatrix} \varepsilon_{xx} & \varepsilon_{xy} \\ \varepsilon_{yx} & \varepsilon_{yy} \end{pmatrix} + \omega_z,$$

where $\begin{pmatrix} \varepsilon_{xx} & \varepsilon_{xy} \\ \varepsilon_{yx} & \varepsilon_{yy} \end{pmatrix}$ is the plastic deformation tensor for plane specimen and ω_z is the rotation about the axis z. The plastic distortion tensor components are: longitudinal extension $\varepsilon_{xx} = \partial u / \partial x$, lateral contraction $\varepsilon_{yy} = \partial v / \partial y$, shear $\varepsilon_{xy} = \varepsilon_{yx} = 1/2(\partial v / \partial x + \partial u / \partial y)$, and rotation $\omega_z = 1/2(\partial v / \partial x - \partial u / \partial y)$. Here, $u = r \cos \phi$ and $v = r \sin \phi$ are, respectively, the longitudinal and the transverse components of the displacement vector r; ϕ is the angle the vector r makes with the sample extension axis, x. A computer program was written for creating data files $\varepsilon_{xx}(x, y)$, $\varepsilon_{yy}(x, y)$, $\varepsilon_{xy}(x, y)$ and $\omega_z(x, y)$. The results of calculations can be presented as components of the plastic distortion tensor distributed over the sample (see example in Figure 1).

Values of the parameters the wavelengths λ and the velocities V_{aw} were evaluated from the $X - t$ diagrams (X is the coordinate of the localized strain center in the laboratory system of coordinates and t is time), as described in [1–4] (see example in Figure 2), and the interplanar distance χ and the transverse ultrasonic wave rate in crystal V_t values were borrowed from the handbooks.

Table 1. Effective mass values (amu), calculated by Equation (1).

Mass	Metals										
$m_{ef}^{(emp)}$	Cu	Zn	Al	Zr	Ti	V	Nb	α-Fe	γ-Fe	Ni	Co
	1.8	1.1	0.5	2.0	1.1	1.4	2.3	1.8	1.8	1.9	1.3
$m_{ef}^{(emp)}$	Sn	Mg	Cd	In	Pb	Ta	Mo	Hf	-	-	-
	1.3	4.0	4.2	1.6	1.3	0.6	0.3	4.0	-	-	-

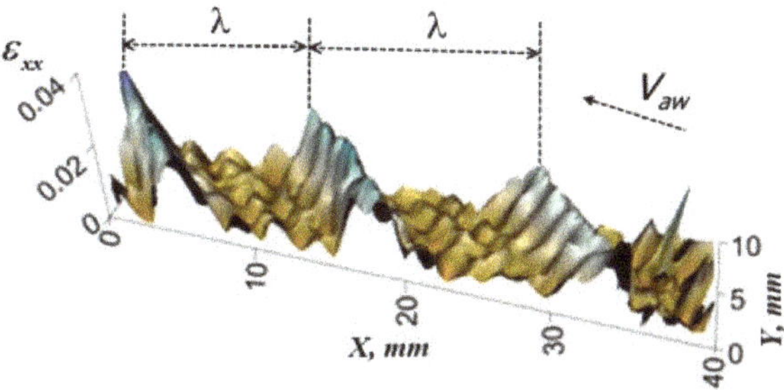

Figure 1. Spatial distribution of local elongations ε_{xx} over the test sample Al.

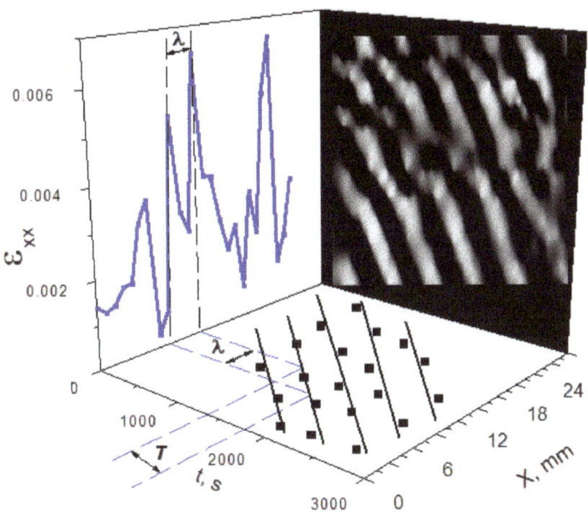

Figure 2. Localized plastic flow pattern: distribution of local elongations ε_{xx} observed for the sample midline; pattern observed for different times (dark and light bands correspond to active and passive material volumes, respectively); illustration of the method $X - t$ diagrams for measuring λ and T values.

3. Results

The results presented below and their interpretations are based on the experimental data obtained previously in the study of the autowave character of plastic flow development in materials of various kinds [1–4].

3.1. On the Possibility of Introduction of the Quasi-Particle

3.1.1. Mass Associated with the Localized Plasticity Autowave

The first well-known attempt of direct application of quantum-mechanical models for interpretation of localized plastic flow autowaves was undertaken by Billingsley [16]. He applied the de Broglie equation $m = h/\lambda V$, written for the mass, to the characteristics of the autowaves described in our works [17–19] and demonstrated that the calculated value correlates with the atomic mass of the investigated metal. In the above equation, h is the Planck constant and λ and V are the autowave wavelength and velocity, respectively.

This idea was further developed by us in [20,21], including an increase in the number of investigated materials and a more correct interpretation of the autowave characteristics of the deformation process. In [16], we wrote the de Broglie equation for the mass in the form

$$m_{ef}^{(emp)} = \frac{h}{\lambda V_{aw}}$$

(1)

where the wavelengths λ and the velocities V_{aw} of autowaves observed at the stages of linear work hardening of metals investigated by the present time we took from our investigations of the development of localized plasticity in nineteen metals. As follows from Table 1, the empirical effective masses calculated from Equation (1) lie within comparatively narrow limits $0.3 \le m_{ef}^{(emp)} \le 4.2$ amu (1 amu $= 1.66 \times 10^{-27}$ kg is the atomic mass unit).

The average effective mass calculated in this way was $\left\langle m_{ef}^{(emp)} \right\rangle = 1.7 \pm 0.3 \approx 2$ amu. The comparatively small variation of this value demonstrates that the masses calculated from Equation (1) are not occasional in character and are determined by the development of the localized plastic flow.

3.1.2. Nature of the Effective Mass

To elucidate the physical meaning of m_{ef}, we took into account that separate regions of a deforming material are involved in the process of plastic flow in a particular, serial order, one by one, during propagation of the phase localized plasticity autowave. Hence, the plastic deformation carriers–dislocations move with acceleration. In this case, the parameter m_{ef} can be interpreted as the attached mass caused by accelerated motion of dislocations in a viscous medium [22]. Such a medium in the case of plastic deformation can be the phonon and electronic gases of a metal crystal the viscosity of which was studied by Al'shits and Indenbom [23]. The dislocation motion dynamics during plastic flow is controlled by the viscous drag force (per unit length) $F_{vis} \sim BV_{disl}$ determined by the viscosity B of the phonon and electronic gases in a crystal [23]. This is correct, if $V_{disl} = const$, but when $V_{disl} \ne const$, the inertial term F_{in} proportional to the dislocation acceleration $\dot{V}_{disl} \ne 0$ is added to the force F_{vis} [22]. In this case,

$$F_{\Sigma} = F_{vis} + F_{in} \sim BV_{disl} + \frac{B}{v} \cdot \dot{V}_{disl} \sim B\left(V_{disl} + \frac{\dot{V}_{disl}}{v} \right)$$

(2)

In metals, the contributions of the phonon and electronic gases to the viscous drag coefficient of dislocations are additive—that is, $B = B_{ph} + B_e$ [23]. In this case, from Equation (2), it follows that

$$F_{\Sigma} \sim (B_{ph}V_{disl} + B_e V_{disl} + \frac{B_{ph}}{v}\dot{V}_{disl} + \frac{B_e}{v}\dot{V}_{disl}).$$

(3)

If v is the frequency of deformation acts, the parameter B/v can be interpreted as the attached mass per unit dislocation length. This coincides with the result obtained by Eshelby [24] according to which the dislocation motion is described by the same equation, as the motion of the Newton particle. Thus, the third and fourth terms in the right-hand side of Equation (3) are the attached mass—that is,

$$F_{in} \sim \frac{B_{ph} + B_e}{v}\dot{V}_{disl} \sim \left(\frac{B_{ph}}{v}\dot{V}_{disl} + \frac{B_e}{v}\dot{V}_{disl} \right).$$

(4)

The factors by the acceleration $\dot{V}_{disl}$ are the attached mass including, obviously, the contributions of the phonon, $m_{ph} \sim \left(B_{ph}/v \right)$, and electron gases, $m_e \sim (B_e/v)$, respectively. The phonon contribution depends weakly on the metal nature, because at $T > T_D$, where T_D is Debye's temperature [25], the metal properties are almost independent of the special features of its phonon spectrum. Thus, for example, the lattice (phonon) heat capacity at these temperatures is independent of the material and temperature [25]. Based on the foregoing, we can set $F_{in}^{(ph)} \sim \left(B_{ph}/v\right)\dot{V}_{disl} \approx const$ and $m_{ph} \sim \left(B_{ph}/v\right) \approx const$. The contribution of electronic gas drag differs from the one described. According to [23], $B_e \approx B_0 + \kappa n$, where n is the electronic gas density (the number of conduction electrons per unit cell), and B_0 and κ are constants. Then, Equation (4) can be written in the form

$$F_{in}^{(e)} \sim \frac{B_e}{v} \cdot \dot{V}_{disl} \sim \frac{B_0 + \kappa n}{v} \cdot \dot{V}_{disl}. \tag{5}$$

In this case, the effective mass is $(B_0 + \kappa n)v^{-1} \sim n$, which is confirmed by the experimental data presented in [20] and shown in Figure 3. In this figure, by analogy with [20], the dimensionless mass $s = m_{ef}^{(emp)}/A$ is used instead of $m_{ef}^{(emp)}$, where A is the atomic mass of the corresponding element. From Figure 3, it is evident that the proportionality $s \sim n$ is satisfied for two periods of the Periodic Table of Elements. This result emphasizes a relationship of the macroscopic characteristics λ and V_{aw} of localized plastic deformation with the parameters of the electronic structure of a metal. The existence of the above-considered relationship of the kinetics of moving dislocations with the properties of the electronic gas is indirectly confirmed, for example, by the data presented in [26–28]. In these works, the authors have related the electric potential on the surface of a deforming metal sample with the entrainment of conduction electrons by moving dislocations during jump-like plastic deformation.

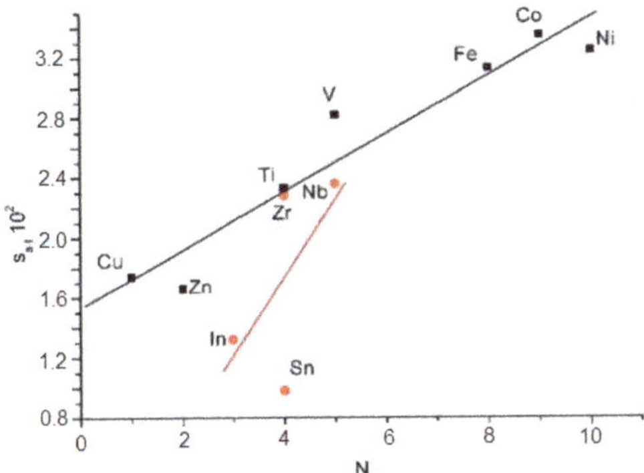

Figure 3. Dimensionless mass of autolocalizon as a function of the of electron number per unit cell: ■ for metals from 4th period; ● for metals from 5th period of the Periodic Table of Elements.

3.1.3. Nature of the Effective Mass

Let us consider instead of masses $m_{ef}^{(emp)}$ found from Equation (1), masses $m_{ef}^{(cal)} = \rho r_{ion}^3$ close in values but calculated independently from reference data. Here, ρ is the density of the material, r_{ion} is the ionic radius, and r_{ion}^3 is the volume per ion in the crystal lattice of a metal. In that case, Equation (1) assumes the form

$$\lambda V_{aw} m_{ef}^{(cal)} = (\lambda V_{aw}) \cdot (\rho r_{ion}^3) = \varsigma, \tag{6}$$

relating the characteristics of the autowave processes of the localized plastic flow λ and V_{aw}, found experimentally [29], and the material characteristics ρ and r_{ion} of the deformable medium, determined independently by other researchers.

Results of calculations of the parameter ς of Equation (6) for the investigated metals are given in Table 2. The average parameter $\langle \varsigma \rangle = \sum_{i=1}^{19} \varsigma_i/19 = (6.9 \pm 0.45) \times 10^{-34}$ J·s appears unexpectedly close to the known Planck constant $h = 6.63 \times 10^{-34}$ J·s [30]; moreover, the ratio $\langle \varsigma \rangle/h = 1.04 \pm 0.06 \approx 1$.

Table 2. The values ς, calculated by Equation (6).

Parameter	Metals										
$\varsigma \cdot 10^{34}$	Cu	Zn	Al	Zr	Ti	V	Nb	α-Fe	γ-Fe	Ni	Co
	11.9	9.3	2.8	6.1	4.9	3.5	4.9	4.6	4.6	6.1	7.1
$\varsigma \cdot 10^{34}$	Sn	Mg	Cd	In	Pb	Ta	Mo	Hf	-	-	-
	8.9	4.9	7.4	9.9	18.4	5.5	3.0	7.3	-	-	-

The result according to which $\langle\varsigma\rangle$ and h are close to each other needs statistical testing. The standard statistical procedure used for this purpose consists in calculation of the Student's t-criterion [31]

$$\hat{t} = \frac{\langle\varsigma\rangle - h}{\sqrt{\hat{\sigma}^2}} \cdot \sqrt{\frac{n_1 \cdot n_2}{n_1 + n_2}} \qquad (7)$$

and its subsequent comparison with the chosen standard criterion $\hat{t}_{0.05} =$ for a 0.95 confidence level. In Equation (7), $n_1 = 19$ is the number of ς_i values obtained from Equation (6), and $n_2 = 1$. The last condition means that the Planck constant was determined with high accuracy—that is, with zero variance [30]. Under such circumstances, the estimated total variance of $\langle\varsigma\rangle$ and h is calculated as

$$\hat{\sigma}^2 = \frac{\sum\limits_{i=1}^{n_1}(\varsigma_i - \langle\varsigma\rangle)^2 + \sum\limits_{1}^{1}(h - h)^2}{n_1 + n_2 - 2} = \frac{\sum\limits_{i=1}^{n_1}(\xi_i - \langle\varsigma\rangle)^2}{n_1 + n_2 - 2}. \qquad (8)$$

From Equations (7) and (8), it follows that $\hat{t} = 0.046 << \hat{t}_{0.05} = 2.13$, where $\hat{t}_{0.05}$ is the value of the Student's criterion for a 0.95 confidence level, and the number of degrees of freedom is $n_1 + n_2 - 2 = 18$. Hence, the difference between $\langle\varsigma\rangle$ and the Planck constant h is statistically insignificant, and the corresponding samples belong to one general population. Thus, $\langle\varsigma\rangle$ calculated from Equation (6) is the Planck constant, and the plastic deformation phenomenon acquires formal relationship with the quantum-mechanical phenomena. The possibility itself of determining the Planck constant from rather rough measurements of the length and velocity of the localized strain-induced autowave seems surprising. However, this fact suggests that quantum-mechanical laws play a more important role in plasticity physics than it is customary to consider.

Nowadays, it is natural to postulate the existence of the quasi-particle associated with the autowave process [5] considering that the effective mass defined by Equation (1) is indeed its mass. This quasi-particle is called autolocalizon [32,33]. Here, it must be taken into account that the developing reasons can be applied only to the phase autowave of the localized plastic flow for which it is possible to measure sufficiently reliably the parameters λ and V_{aw} [29]. The autolocalization characteristics, estimated from their relationships with the localized plastic flow autowaves, are given in Table 3.

Table 3. The principal characteristics of autolocalizon.

Characteristic	Formula	Value
Dispersion law	$\omega(k) \sim 1 + k^2$	-
Mass (amu)	$m_{a-l} \equiv m_{ef} = h/\lambda V_{aw}$	1.7 ± 0.2
Velocity (m/s)	$V_{a-l} \equiv V_{aw}$	$10{-}5 \ldots 10{-}4$
Momentum (J·s/m)	$p = \hbar k = h/\lambda$	$(6 \ldots 7)\cdot10{-}32$

3.2. Autolocalizon and Localized Plastic Flow

Now it is necessary to demonstrate that the introduction of the autolocalizon simplifies the explanation of the special features of plastic flow kinetics, in particular, associated with stress or strain jumps. Here, we use the quasi-particle approach to reach some of these goals.

3.2.1. Some Characteristics of Autowave Plastic Flow

First, it is necessary to discuss some details of the Portevin–Le Chatelier effect manifested through jump-like stress increments and decays of the deforming stress caused by them. This effect is often observed during deformation of binary Al-Cu or Al-Mg alloys and has been studied in ample detail [6,28,34–38]. During jump-like deformation, the sample lengthens discretely, and the amplitude of each jump is determined by the properties of the deforming material.

Analyzing this phenomenon, we use the natural assumption that an integer number $m = 1, 2, 3, \ldots$ of autowaves with length λ fits into the sample length. In this case (see Figure 4), Equation (1) is written in the form

$$\lambda = \frac{h}{\rho r_{ion}^3 V_{aw}} \tag{9}$$

so that with strain growth, the sample length increases by

$$\delta L \approx \lambda m \approx \frac{h}{\rho r_i^3 V_{aw}} \cdot m \sim m. \tag{10}$$

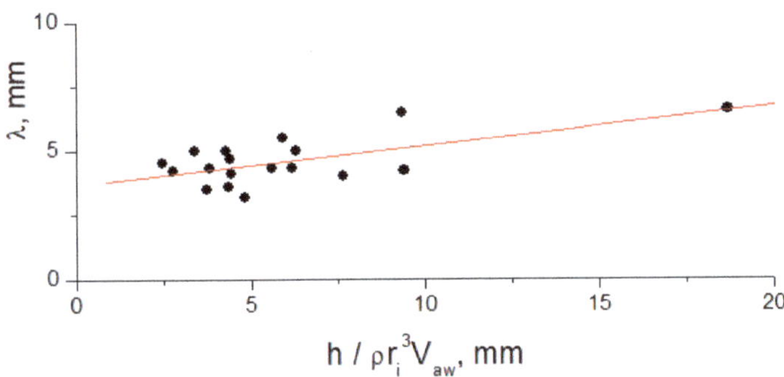

Figure 4. Verification of Equation (9). Values of the wavelengths λ and velocities V_{aw} were evaluated from the $X - t$ diagrams (see Figure 2), and values of ion radius r_i and density ρ were borrowed from the handbooks.

Equation (10) can be used to estimate the jump-like sample elongation δL if m = 1, $V_{aw} \approx 3 \times 10^{-3}$ m/s, and $\rho \cdot r_i^3 \approx 1.7$ amu $\approx 3 \times 10^{-27}$ kg. Under these conditions, $\delta L \approx 10^{-4}$ m. This is in agreement with the experimentally observed jump length caused by the Portevin–Le Chatelier effect [28].

It follows from Equation (10) that the sample length changes discretely. This means that the jump-like deformation can be considered as a process of fine tuning of the sample length to the existing autowave pattern of plastic strain localization, and Equation (10) becomes obligatory in character. In addition, from Equation (10) it follows that $\delta L \sim V_{aw}^{-1}$. The data presented by us in [29] demonstrate that the autowave velocity is proportional to the velocity of the traverse of a test machine—that is, $V_{aw} \sim V_{mach}$ In this case, a decrease in the jump amplitude should be expected with increasing stretching rate. There are convincing experimental evidences of such change—for example, in [39] for deformation of aluminum at the temperature of 1.4 K.

3.2.2. Autowave Length as the Free Path Length of the Autolocalizon

It is well-known that all the quasi-particles are unobservable in principle. Their existence can be verified by the calculations of the quantities founded experimentally. The example of similar evaluation is considered below for the length of localized plastic flow autowave.

In the context of the quasi-particle approach, a deforming medium can be considered as a binary mixture of quasi-particles of two kinds: phonons and autolocalizons. In this case, quantitative laws of plastic flow can be explained by the Brownian motion of autolocalizons in the viscous phonon gas. Such characteristic, as the autowave length $\lambda \approx 10^{-2}$ m, important for autowave processes of plastic flow localization, can be identified with the length of the resultant displacement $R = \left(\sum\limits_{i=1}^{i=n} r_i^2 \right)^{1/2}$ [40] in the process of the Brownian random walk of autolocalizons in the phonon gas during the time $\tau = 2\pi/\omega \approx 10^3$ s. Thus, according to [40],

$$R \approx \sqrt{\frac{k_B T}{\pi B r_{a-l}} \cdot \tau} \approx \sqrt{\frac{2 k_B T}{\omega B r_i}} \tag{11}$$

where k_B is the Boltzmann constant. If we accept that the autolocalizon radius is $r_{a-l} \approx r_{ion} \approx 10^{-10}$ m and that the dynamic viscosity of phonon gas is $B \approx 10^{-4}$ Pa·s [23], then the displacement at T = 300 K will be $R \approx 10^{-2}$ m—that is, it will agree with the length λ of the localized deformation autowave.

3.2.3. Elastic-Plastic Invariant of Deformation and Autolocalizon

From Equation (11) it follows that the quantity

$$\frac{R^2 \omega}{2\pi} \approx \frac{k_B T}{\pi B r_{a-l}} \approx 1.3 \cdot 10^{-7} m^2/s \tag{12}$$

can be identified with the product λV_{aw} with which it coincides by dimensionality and value [29]. In that case, the basic equation of autowave mechanics of plasticity—the elastic-plastic invariant of deformation (see Figure 5)—can be re-written as

$$\hat{Z}_{calc} \equiv \frac{\lambda V_{aw}}{\chi V_t} = \frac{k_B T}{\pi B r_{ion} \chi V_t} \approx \frac{\hbar \omega_D}{\pi B r_{ion} \chi V_t}. \tag{13}$$

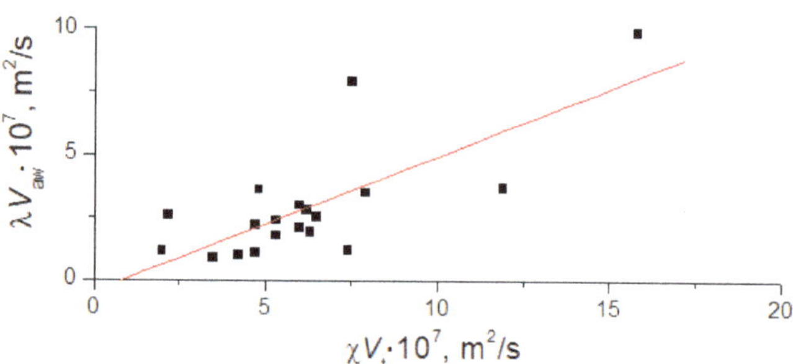

Figure 5. Verification of the elastic-plastic invariant of deformation (13).

Here, χ is the interplanar distance and V_t is the transverse ultrasonic wave rate in a crystal [1]. Equation (13) is valid by $T > T_D$, when all thermal vibrations have been excited in the crystal, and it is possible to set $k_B T_D \approx \hbar \omega_D$.

Evidently, the value $\hat{Z}_{calc}$ calculated with the help of Equation (13) is defined only by the material (lattice) constants. Results of its calculations are given in Table 4.

The average calculated value is $\langle \hat{Z} \rangle_{calc} = 0.48 \pm 0.1$, and the ratio $\langle \hat{Z} \rangle_{exp} / \langle \hat{Z} \rangle_{calc} = 0.97 \approx 1$—that is, it almost coincides with the experimentally observed value.

This conclusion is important, because it demonstrates that the elastic-plastic strain invariant is a versatile characteristic of the deforming medium. Equation (13) can be used to calculate this parameter, knowing in fact only one characteristic of the plastic deformation process, namely, the phonon gas viscosity B. Thus, by introduction of the autolocalizon, some problems of solid state plasticity have been solved. This proves the correctness of the idea being developed.

Table 4. Calculated and experimentally founded values $\hat{Z}$.

Value	Metals										
	Cu	Zn	Al	Zr	Ti	V	Nb	α-Fe	γ-Fe	Ni	Co
$\hat{Z}_{calc}$	0.75	0.31	1.05	0.55	0.32	0.45	0.33	0.54	0.34	0.35	0.5
$\hat{Z}_{exp}$	0.42	0.5	4.47	0.33	0.42	0.85	0.44	0.46	0.48	0.35	0.38
	Sn	Mg	Cd	In	Pb	Ta	Mo	Hf			
$\hat{Z}_{calc}$	0.65	0.63	0.27	1.18	1.4	2.7	0.4	0.33			
$\hat{Z}_{exp}$	0.48	0.98	0.24	0.78	0.5	1.1	0.2	0.24			

4. Discussion

4.1. Elementary Excitation Spectrum of a Plastically Deforming Medium

In general, the deformation processes involve interrelated elastic waves and plastic autowaves whose interaction provides the basis for the construction of the two-component localized plastic flow model proposed in [29]. The well-known dispersion relations characterize both components. The close relationship between the elastic and plastic deformation components allows us to potentially obtain a common dispersion law for the elastic and plastic strains.

4.1.1. Hybridized Spectrum of an Elastic-Plastic Medium

The model relating the acoustic and deformation processes in solids envisages the possibility of hybridization of the elementary excitation spectra of the elastic (phonon) and deformation (autolocalizon) subsystems. The hybridized spectrum can be obtained by overlaying the linear dispersion relation for transverse phonons $\omega \approx V_t k$ far away from the Brillouin zone boundary—i.e., at $\omega \to \omega_D$ [25]—and the parabolic curve of the dispersion relation for localized plasticity autowaves (autolocalizons) [29]. The result of such overlay for polycrystalline aluminum is shown in Figure 6.

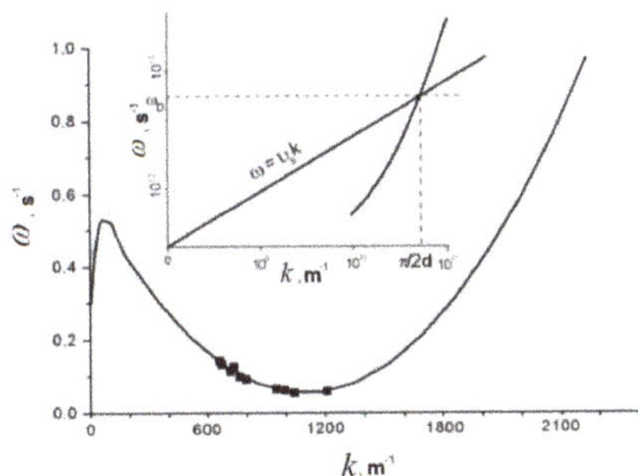

Figure 6. Generalized dispersion law for localized plastic flow autowaves ($\omega \sim 1 + k^2$) and elastic waves ($\omega \sim k$); in the insert, the high-frequency range of spectrum is represented.

The estimate shows that the coordinates of the intersection point of the plots for high-frequency range are physically justified: $\hat{\omega} \approx \omega_D \approx 10^{13}$ Hz, and the wave number $\hat{k} = 2\pi/\hat{\lambda}$ corresponds to the minimal possible wavelength $\lambda_{min} = \hat{\lambda} \approx 2\chi$. This confirms the applicability of the plastic flow description by the interaction of the phonon gas and the quasi-particles (autolocalizons).

The situation near the point of intersection of the linear and parabolic dependences in the low-frequency region is more interesting at small values of the wave number where the hybridized dependence $\omega(k)$ has a local maximum. Its existence can be due to dislocations induced by elastic nonlinear strain of initially defect-free crystal according to the mechanism of condensation of long-wavelength phonons proposed in [41]. Individual dislocations appearing because of this process are unstable, but their energy decreases with decreasing or increasing wave number k. We can assume that the first case corresponds to the formation and development of the low-energy dislocation structures [42] (in the limit of the localized plastic strain autowaves), and the second case leads to the origin of fragile microcracks because of interaction of a small number of dislocations.

According to the dispersion relation for autowaves of localized plastic flow, in the vibrational spectrum $\omega(k)$ there is a gap at $0 \leq \omega_0 \approx 10^{-2}$ Hz. Since at any temperature $\hbar\omega_0 << k_B T$, the spontaneous plastic strain localization should be excited at temperatures as low as is wished. Indeed, the jump-like deformation associated with flow localization was repeatedly observed at $T \leq 1$ K [39]. The hybridized dependence $\omega(k)$ shown in Figure 6 demonstrates its striking similarity to the dispersion curve $E(k)$ of superfluid helium [5]. The minimum in the parabolic branch of this curve corresponds to the creation of rotons–elementary excitations in the medium with a quadratic dispersion law. This suggests that the autolocalizon is an analog of the roton created by the localized plastic strain and determines the kinetics of this process.

4.1.2. Autolocalizon Dispersion and Effective Mass

Using the dispersion relation for the localized plasticity autowaves [1,29], it is possible to calculate by the standard method [5] the effective autolocalizon mass corresponding to the localized plastic flow of the autowave:

$$m_{a-l} = \hbar \frac{\partial^2}{\partial k^2}[\omega(k)], \tag{14}$$

equal to 0.1 amu for Al and 0.6 amu for Fe. The effective masses calculated from Equation (1) $m_{ef} = h/\lambda V_{aw}$ and given in Table 1 are 0.5 and 1.8 amu for aluminum and iron, respectively—that is, they have the same orders of magnitude. These estimates confirm the correctness of the assumption that quasi-particles (autolocalizons) with an average mass of ~1.7 amu are associated with the aotowave processes of plastic flow localization. According to [5], the effective roton mass is $m_{rot} = 0.64$ amu. The close values of the effective masses of the autolocalizon and the roton also testify in favor of the above-discussed hypothesis.

4.1.3. Condensation of Quasi-Particles during Plastic Flow

There is interesting possibility to refine the nature of the elastic-plastic strain invariant. Equation (13) for the elastic-plastic invariant written in the form (see Figure 7)

$$\frac{h}{\lambda V_{aw}} = 2\frac{h}{\chi V_t} \tag{15}$$

makes sense of the equality of masses, where $h/\lambda V_{aw} = m_{a-l}$ is the de Broglie autolocalization mass and $h/\chi V_t = m_{ph}$ is the phonon mass.

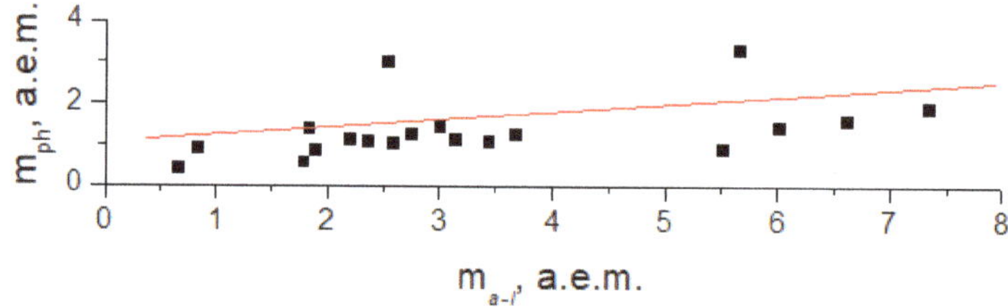

Figure 7. Correlation dependence of phonon and autolocalizon masses for Equation (15).

In this case, Equation (15) is reduced to the equality $m_{a-l} \approx 2m_{ph}$ that can be considered as the condition of creation of the autolocalizon with the momentum

$$p_{a-l} = (m_{a-l}V_{aw}) = \left[2\left(m_{ph}V_t\right)^2(1 + \cos\alpha)\right]^{1/2} = \sqrt{2}m_{ph}V_t(1 + \cos\alpha)^{1/2}, \tag{16}$$

as shown in Figure 8. Here, 2 is the angle between the momenta of phonons.

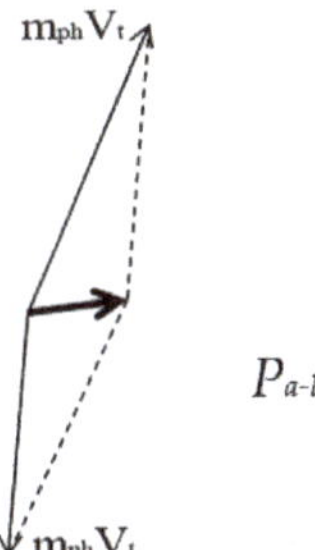

Figure 8. A scheme of autolocalizon generation in a two-phonon process. P_{a-l} is a momentum of autolocalization.

The probability of this process was first calculated in [43] to explain the creation of rotons in superfluid helium. Reissland [44] considered the process of combining two phonons into one in detail. The process discussed in the present work is close to the quantum mechanism of defect formation in crystals considered in [41]. It was demonstrated in this work that the creation of dislocations and dislocation ensembles (dislocation walls, grain boundaries, and other defects) in an ideal crystal is the result of Bose's condensation of long-wavelength phonons under the influence of deformation or temperature.

The foregoing provides the basis for the development of new representations on the stages of the localized plastic flow of solids. According to these representations, quasi-particles of definite types correspond to each deformation stage. The collapse of the localized plastic flow autowave observed at the stage of prefailure of materials described in [29] is a typical example of this condensation of quasi-particles in a plastic flow. This process can be considered as the production of the crackon [13]—that is, as the beginning of the failure process.

Experimental data allow us to describe qualitatively the process of generation of structural defects. At the stage of elastic deformation, only phonons, which redistribute the elastic stresses over the crystal, exist in the crystal. In the process of phonon condensation, according to the mechanism proposed by us in [41], the plastic shear quanta (dislocations) are formed in the crystal, thereby giving rise to the plastic flow. The intermediate stages at which the defects of the above-mentioned types, such as dilatons, frustrons, or the so-called dislocation precursors [45–47], can precede the stage of dislocation formation. Finally, the collapse of the localized plastic flow autowaves can be considered as the process of condensation of the autolocalizons with formation of the destruction quantum–crackon and its further motion during cracking.

4.1.4. General Meaning of Autolocalizon Introduction

The amazingly close values of the parameter $\langle\varsigma\rangle$ calculated from Equation (6) considered above, which contains the macroscopic parameters λ and V_{aw} that are determined experimentally, to the reference value of the Planck constant h indicate a direct relationship of the macroscopic localization phenomena of plastic flow with the properties of the lattice. With application to the process of localized plastic flow, this means that the spatiotemporal microstructural inhomogeneity of the deforming medium (the dislocation submicrostructure) defines much less the measurable macromechanical characteristics, such as the flow stress, yield point, work hardening coefficient, and plastic flow localization pattern than it is supposed in traditional dislocation models.

For these reasons, we can determine that the mechanical properties are substantially controlled by the lattice characteristics of the elastic medium and hence, by the laws of quantum mechanics. From this point of view, the obvious importance of Equation (6) relating the quantities that are traditionally considered as diverse is additionally confirmed. The existence of such relationship has been conventionally neglected in different plastic flow models; however, Equation (6) explicitly indicates wrongfulness of this opinion.

4.1.5. Plastic Flow as a Macroscopic Quantum Phenomenon

The consistent development of representations stated above suggests the opportunity to refer the localized plastic flow to a series of macroscopic quantum phenomena, such as superfluidity, superconductivity [48], and quantum Hall's effect [49]. The quantization of the corresponding characteristics of the medium in these phenomena is manifested on the macroscopic scale and can be observed directly (Table 5); the macroscopic equations describing these characteristics involve the Planck constant.

Table 5. Macroscopic quantum phenomena.

Phenomenon	Quantum Characteristic	
	Value	**Formula**
Superconductivity [5]	Magnetic flux	$\Phi = \frac{\pi \hbar c}{e} \cdot m$
Superfluidity [5]	Rotational velocity of vortices in superfluidity HeII	$v = \frac{\hbar}{A_{He}} \cdot \frac{1}{r} \cdot m$
Quantum Hall effect [49]	The Hall resistance	$R_H = \frac{h}{e^2} \cdot \frac{1}{m}$
Serrated plastic deformation	Elongation during deformation jump	$\delta L = \frac{h}{\rho r_{ion}^3 V_{aw}} \cdot m$

Notation: e is the electron charge, c is the speed of light, r is the radius of the vortex, $m = 1, 2, 3 \dots$.

Formally, the possible quantum character of the plastic flow is indicated by the close analogy between the forms of dispersion curves for the plastic flow autowaves and for superfluid helium [5] shown in Figure 6. Moreover, it can be noted that the quadratic dispersion curve of the localized plasticity autowaves is similar to the spectrum of elementary excitations in a superconductor [5].

The similarity of physical natures of the macroscopic phenomena is because the effects underlying them do not admit a description in the context of models based largely on the additive properties of individual strain carriers. For the plastic flow, the dislocations, each of which describes "the shear quantum", can be considered as such carriers [50].

There are some additional reasons in favor of the above-considered formal similarity between the plastic flow and the superfluidity. Indenbom [51] first paid attention to the analogy between the dislocations and quantum vortices in superfluid helium or quantized flows in the second-order superconductors. This analogy can be expanded if we consider that in a deforming material, as well as in superfluid helium, two regimes of motion co-exist with different velocities. First, this is slow motion of separate volumes accompanying the change of the body shape as a whole, and second, this is motion of dislocations with high velocities providing this change of the body shape. The high viscosity of the crystal material $\eta \approx 10^6$ Pa·s [6] corresponds to the first of them, whereas the low viscosity of the phonon gas $B \approx 10^{-4}$ Pa·s [23] corresponds to the second of them. It is well known that the velocity of viscous motion is proportional to the applied force and inversely proportional to the viscosity [23]. Hence we can write down that $V = F_1/B$ or $V = F_1/\eta$, where $F_1 = \sigma L$ is the Peach-Kehler force per unit length of a dislocation [52]. For the phase autowave at the stage of linear work hardening, $\sigma >> 100$ MPa, $L \approx \lambda \approx 10^{-2}$ m, and $\eta \approx 10^6$ Pa·s. The estimated velocity is in agreement with $V_{aw} \approx$

10^{-5} m/s observed in [28]. At low viscosity $B \approx 10^{-4}$ Pa·s, we obtain $V_{disl} \approx 10^2$ m/s, which is in agreement with the velocity of dislocation motion over the barrier [23].

A similar approach based on the quantum-mechanical representations seems surprising for experts in plasticity physics, since the spatial scale of the macroscopic plasticity phenomena considerably exceeds the scales for which the quantum approach is traditionally used. Thus, the ratio of the autowave ($\lambda \approx 10^{-2}$ m) and dislocation ($b \approx 10^{-10}$ m) scales $\lambda/b \approx 10^8$. However, already on the dislocation level, the strain quantization seems natural, at least because of the stepwise behavior of the crystal lattice, and the Burgers vector b can be interpreted as a shear strain quantum, as pointed out by many authors [24,41,50]. Thus, the use of the wave-quasi-particle dualism for a description of strain localization processes is quite justified.

5. Conclusions

The close relationship of the plasticity effects with the crystal lattice characteristics during macroscopic plastic deformation is a possible reason for the existence of quantum effects in the macroscopic strain processes. In other words, generation of the localized plastic flow autowaves—self-organization of the deforming medium—is realized in the elastic medium whose behavior is governed by laws of the quantum mechanics. Therefore, the energy redistribution in the process of forming the localized plasticity autowaves in the system capable of self-organization is also subordinate to the quantum nature of crystals.

Experimental results and their interpretation indicate the importance of the consideration of the close interrelation of defect ensembles with the phonon subsystem in crystals. Such interrelation can be described correctly both by the mechanism of interaction of waves and autowaves and by the mechanism of interaction of quasi-particles, phonons and autolocalizons. The two-component model of the plastic flow developed on this basis correctly explains the correlations of the macroscopic scale during plastic flow localization in deforming metals and alloys. Thus, the quasi-particle approach completes the traditional analysis of plastic flow phenomenon. This suggests that auto-localizon can be used for the development of the physical theory of plasticity.

Author Contributions: Both authors contributed to the manuscript. Conceptualization, L.B.Z.; Investigation, S.A.B. The manuscript was prepared by L.B.Z. and S.A.B. All authors have read and agreed to the published version of the manuscript.

Funding: The research was founded by Russian Scientific Foundation, grant No. 16-19-10025 P.

Acknowledgments: We thank our colleagues in the Strength Physics Laboratory of ISPMS SB RAS for helpful discussion of this study.

Conflicts of Interest: The authors declare no conflict of interest.

References

1. Zuev, L.B.; Barannikova, S.A. Autowave physics of material plasticity. *Crystals* **2019**, *9*, 458. [CrossRef]
2. Zuev, L.B. Autowave mechanics of plastic flow in solids. *Phys. Wave Phenom.* **2012**, *20*, 166–173. [CrossRef]
3. Zuev, L.B. Wave phenomena in low-rate plastic flow of solids. *Ann. Phys.* **2001**, *10*, 965–984. [CrossRef]
4. Zuev, L.B. On the waves of plastic flow localization in pure metals and alloys. *Ann. Phys.* **2007**, *16*, 287–310. [CrossRef]
5. Brandt, N.B.; Kulbachinskii, V.A. *Quasi-Particles in Condensed State Physics*; Fizmatlit: Moscow, Russia, 2007; 631p.
6. Bell, J.F. *Mechanics of Solids*; Springer: Berlin/Heidelberg, Germany, 1973; 596p.
7. Steverding, B. Quantization of stress waves and fracture. *Mater. Sci. Eng.* **1972**, *9*, 185–189. [CrossRef]
8. Maugin, G.A. Solitons in elastic solid. *Mech. Res. Commun.* **2011**, *38*, 341–349. [CrossRef]
9. Gilman, J.J. Escape of dislocations from bound states by tunneling. *J. Appl. Phys.* **1968**, *39*, 6086–6090. [CrossRef]
10. Oku, T.; Galligan, J.M. Quantum mechanical tunneling of dislocations. *Phys. Rev. Lett.* **1969**, *22*, 596–597. [CrossRef]

11. Petukhov, B.V.; Pokrovskii, V.L. Quantum and classic motion of dislocations in the potential Peierls relief. *J. Exp. Theor. Phys.* **1972**, *63*, 634–647.

12. Iqbal, S.; Sarwar, F.; Rasa, S.V. Quantum mechanics tunneling of dislocations: Quantization and depinning from Peierls barrier. *World J. Cond. Mater. Phys.* **2016**, *6*, 103–108.

13. Morozov, E.M.; Polak, L.S.; Fridman, Y.B. On variation principles of crack development in solids. *Sov. Phys. Dokl.* **1964**, *156*, 537–540.

14. Zhurkov, S.N. Dilaton mechanism of the strength of solids. *Solid Stat. Phys.* **1983**, *25*, 1797–1800.

15. Olemskoi, A.I.; Katsnelson, A.A. *A Synergetic of Condensed Medium*; URSS: Moscow, Russia, 2003; 335p.

16. Billingsley, J.P. The possible influence of the de Broglie momentum-wavelength relation on plastic strain "autowave" phenomena in "active materials". *Int. J. Solids Struct.* **2001**, *38*, 4221–4234. [CrossRef]

17. Danilov, V.I.; Barannikova, S.A.; Zuev, L.B. Localized Strain Autowaves at the Initial Stage of Plastic Flow in Single Crystals. *Tech. Phys.* **2003**, *48*, 1429–1435. [CrossRef]

18. Zuev, L.B.; Danilov, V.I. Plastic deformation viewed as evolution of an active medium. *Int. J. Solids Struct.* **1997**, *34*, 3795–3805. [CrossRef]

19. Zuev, L.B.; Danilov, V.I. A self-excited wave model of plastic deformation in solids. *Philos. Mag. A* **1999**, *79*, 43–57. [CrossRef]

20. Zuev, L.B. The linear work hardening stage and de Broglie equation for autowaves of localized plasticity. *Int. J. Solids Struct.* **2005**, *42*, 943–949. [CrossRef]

21. Zuev, L.B.; Danilov, V.I.; Barannikova, S.A.; Zykov, I.Y. Plastic flow localization as a new kind of wave processes in solids. *Mater. Sci. Eng. A* **2001**, *319–321*, 160–163. [CrossRef]

22. Landau, L.D.; Lifshits, E.M. *Course of Theoretical Physics. Fluid Mechanics*; Pergamon Press: Oxford, UK, 1987; Volume 6, 539p.

23. Alshits, V.I.; Indenbom, V.L. Mechanisms of dislocation drag. In *Dislocations in Crystals*; Nabarro, F.R.N., Ed.; North-Holland: Amsterdam, The Netherlands, 1986; Volume 7, pp. 43–111.

24. Eshelby, J.D. Continual theory of defects. *Solid Stat. Phys.* **1956**, *3*, 79–173.

25. Newnham, R.E. *Properties of Materials*; University Press: Oxford, UK, 2005; p. 378.

26. Bobrov, V.S.; Zaitsev, S.I.; Lebyodkin, M.A. Statistics of dynamic processes at low temperature serrated deformation of metals. *Phys. Solid Stat.* **1990**, *32*, 3060–3065.

27. Bobrov, V.S.; Lebyodkin, M.A. The role of dynamic processes at low temperature serrated deformation of aluminium. *Phys. Solid Stat.* **1993**, *35*, 1881–1889.

28. Lebyodkin, M.; Bougherira, Y.; Lebedkina, T.; Entemeyer, D. Scaling in the local strain-rate field during jerky flow in an Al-3%Mg alloy. *Metals* **2020**, *10*, 134. [CrossRef]

29. Zuev, L.B. Autowave Plasticity. In *Localization and Collective Modes*; Fizmatlit: Moscow, Russia, 2018; 208p.

30. Atkins, P.W. *Quanta. A Handbook of Conceptions*; Clarendon Press: Oxford, UK, 1974; 319p.

31. Hudson, D.J. *Statistics*; CERN: Geneva, Switzerland, 1964; 242p.

32. Zuev, L.B.; Barannikova, S.A.; Maslova, O.A. The features of localized plasticity autowaves in solids. *Mater. Res.* **2019**, *22*. [CrossRef]

33. Barannikova, S.A.; Danilov, V.I.; Zuev, L.B. Plastic strain localization in Fe-3%Si single crystals and polycrystals under tension. *Tech. Phys.* **2004**, *49*, 1296–1300. [CrossRef]

34. Lebyodkin, M.A.; Zhemchuzhnikova, D.A.; Lebedkina, T.A.; Aifantis, E.C. Kinematics of formation and cessation of type B deformation bands during the Portevin-Le Chatelier effect in an Al-Mg alloy. *Results Phys.* **2019**, *12*, 867–869. [CrossRef]

35. Kubin, L.P.; Chihab, K.; Estrin, Y.Z. The rate dependence of the Portevin—Le Chatelier effect. *Acta Metall.* **1988**, *36*, 2707–2718. [CrossRef]

36. Kubin, L.P.; Estrin, Y.Z. The critical condition for jerky flow. *Phys. Stat. Solid* **1992**, *172*, 173–185. [CrossRef]

37. Rizzi, E.; Hähner, P. On the Portevin—Le Chtelier effect: Theoretical modeling and numerical results. *Int. J. Plast.* **2004**, *29*, 121–165. [CrossRef]

38. Zaiser, M.; Aifantis, E.C. Randomness and slip avalanches in gradient plasticity. *Int. J. Plast.* **2006**, *22*, 1432–1455. [CrossRef]

39. Pustovalov, V.V. Serrated deformation of metals and alloys at low temperatures. *Phys. Low Temp.* **2008**, *34*, 871–913. [CrossRef]

40. Rumer, Y.B.; Ryvkin, M.S. *Thermodynamics, Statistical Physics and Kinetics*; Mir Publ.: Moscow, Russia, 1980; 600p.

41. Umezava, H.; Matsumoto, H. *Thermo Field Dynamics and Condensed States*; North-Holland Publ. Comp.: Amsterdam, The Netherlands, 1982; 504p.

42. Kuhlmann-Wilsdorf, D. The low energetic structures theory of solid plasticity. In *Dislocations in Solids*, Nabarro, F.R.N., Duesbery, M.S., Eds.; Elsevier: Amsterdam, The Netherlands, 2002; pp. 213–338.

43. Landau, L.D.; Khalatnikov, I.M. The theory of viscosity HeII. *J. Exp. Theor. Phys.* **1949**, *19*, 637–650.

44. Reissland, J.A. *The Physics of Phonons*; John Wiley and Sons LTL: London, UK, 1973; 365p.

45. Psakhie, S.G.; Zolnikov, K.P.; Kryzhevich, D.S. Elementary atomistic mechanism of crystal plasticity. *Phys. Lett. A* **2007**, *367*, 250–253. [CrossRef]

46. Psakhie, S.G.; Shilko, E.V.; Popov, M.V.; Popov, V.L. The key role of elastic vortices in the initiation of shear cracks. *Phys. Rev. E* **2015**, *91*, 63302.

47. Dmitriev, A.I.; Nikonov, A.Y.; Filippov, A.E.; Psakhie, S.G. Molecular Dynamics Study of the Evolution of Rotational Atomic Displacements in a Crystal Subjected to Shear Deformation. *Phys. Mesomech.* **2019**, *22*, 375–381. [CrossRef]

48. Tilley, D.R.; Tilley, J. *Superfluidity and Superconductivity*; IOP Publ.: Bristol, UK, 1990; 268p.

49. Imry, Y. *Introduction to Mesoscopic Physics*; University Press: Oxford, UK, 2002; 236p.

50. Katanaev, M.O. Geometric theory of defects. *Phys. Uspekhi* **2005**, *175*, 705–733. [CrossRef]

51. Indenbom, V.L. The structure of real crystals. In *The Modern Crystallography*; Vainstein, B.K., Fridkin, V.M., Indenbom, V.L., Eds.; Nauka Publ.: Moscow, Russia, 1979; pp. 297–341.

52. Hull, D.; Bacon, D.J. *Introduction in Dislocations*; Elsevier: Oxford, UK, 2011; 272p.

Publisher's Note: MDPI stays neutral with regard to jurisdictional claims in published maps and institutional affiliations.

Article

Scaling in the Local Strain-Rate Field during Jerky Flow in an Al-3%Mg Alloy

Mikhail Lebyodkin [1,*], Youcef Bougherira [2], Tatiana Lebedkina [1,3] and Denis Entemeyer [1]

[1] Laboratoire d'Etude des Microstructures et de Mécanique des Matériaux (LEM3), CNRS, Université de Lorraine, Arts & Métiers ParisTech, 7 rue Félix Savart, 57070 Metz, France; tatiana.lebedkin@univ-lorraine.fr (T.L.); denis.entemeyer@univ-lorraine.fr (D.E.)

[2] Unité de recherche en optique et photonique (UROP), CDTA, Université de Sétif1, El Bez, Sétif 1900, Algeria; ybougherira@cdta.dz

[3] Center of Excellence "LabEx DAMAS", Université de Lorraine, 7 rue Félix Savart, 57070 Metz, France

* Correspondence: mikhail.lebedkin@univ-lorraine.fr; Tel.: +33-(0)3-72-74-77-71

Received: 25 November 2019; Accepted: 13 January 2020; Published: 16 January 2020

Abstract: Jerky flow in alloys, or the Portevin-Le Chatelier effect, presents an outstanding example of self-organization phenomena in plasticity. Recent acoustic emission investigations revealed that its microscopic dynamics is governed by scale invariance manifested as power-law statistics of intermittent events. As the macroscopic stress serrations show both scale invariance and characteristic scales, the micro-macro transition is an intricate question requiring an assessment of intermediate behaviors. The first attempt of such an investigation is undertaken in the present paper by virtue of a one-dimensional (1D) local extensometry technique and statistical analysis of time series. The data obtained complete the missing link and bear evidence to a coexistence of characteristic large events and power laws for smaller events. The scale separation is interpreted in terms of the phenomena of self-organized criticality and synchronization in complex systems. Furthermore, it is found that both the stress serrations and local strain-rate bursts agree with the so-called fluctuation scaling related to general mathematical laws and unifying various specific mechanisms proposed to explain scale invariance in diverse systems. Prospects of further investigations including the duality manifested by a wavy spatial organization of the local bursts of plastic deformation are discussed.

Keywords: Portevin-Le Chatelier effect; aluminum alloys; local extensometry; avalanches; self-organized criticality; synchronization; fluctuation scaling

---

1. Introduction

Plastic flow of alloys is prone to instability caused by the interaction of mobile dislocations with solute atoms diffusing in their elastic fields, or dynamic strain aging (DSA) [1]. In the mathematical sense, the instability stems from the DSA transforming the monotonous dependence of the flow stress on the plastic strain rate, $\sigma(\dot{\varepsilon})$, into an N-shaped function with an interval of negative strain-rate sensitivity (SRS). In tensile tests with a constant applied strain rate, $\dot{\varepsilon}_a$, this nonlinearity must give rise to stress serrations caused by repetitive strain-rate jumps within localized deformation bands [2–13]. Such jerky flow is observed in various alloys and is well-known as the Portevin-Le Chatelier, or PLC, effect [14]. It is generally agreed that the PLC effect is governed by the DSA mechanism. Were the motion of all dislocations identical, it would result in periodic relaxation oscillations [2,3]. The intrinsic strain heterogeneity leads to complex behaviors of real materials.

The application of various methods of time-series analysis to serrated deformation curves showed that jerky flow cannot be described as random fluctuations about the ideal case of periodic oscillations [15–21]. Moreover, distinct kinds of complex dynamics arise when experimental conditions are varied. In particular, whereas power-law distributions of stress drop amplitudes and durations, tantamount to scale-free dynamics, are found close to the high-strain rate boundary of the instability domain, peaked histograms reflecting intrinsic scales of deformation processes are reported for lower strain rates [15,17]. The latter behavior is neither random but is characterized by various scaling properties. For example, dynamic chaos was detected for strain rates in the middle of the instability domain using such model objects as CuAl single crystals and AlMg polycrystals [16,17]. Multifractal features of stress serration series were identified at all strain rates for AlMg alloys [17,21].

In view of the persisting scaling behaviors, the further step to the comprehension of the plastic flow in the instability conditions implies similar analyses with resolutions higher than that of the mechanical tests. At the same time, high-resolution measurements carried out on materials not subject to macroscopic instability, e.g., by measuring strain bursts during the deformation of microsamples or recording acoustic emission (AE) in the case of bulk samples, revealed ubiquitous scale-free statistics testifying to inherently intermittent or, more exactly, avalanche-like deformation processes in a microscopic scale range [22–27]. Application of the AE technique to jerky flow showed that AE events occur in the same range of amplitudes both at the instants of stress serrations and during stable flow, either before the critical strain ε_{cr} for the onset of the PLC effect or during reloading between stress drops [28,29]. Moreover, in contrast to the diverse statistics of stress serrations, the AE amplitudes obey power-law distributions at all strain rates, in agreement with the avalanche dynamics conjectured to be general for various materials [30–32]. This apparent contradiction was solved due to the observation of a correlation between the occurrence of macroscopic stress serrations and the clustering (chaining) of AE events [29]. Accordingly, taking into account a relationship between the relaxation oscillations and the phenomenon of synchronization in lattice models [33,34], the transition from power-law distributions of AE to distinct kinds of statistics of stress serrations was ascribed to the synchronization of dislocation avalanches, depending on the strain rate [35,36].

An important consequence of this complexity, almost unexplored so far, is that the apparent behavior may depend on the scale of observation. This statement is also confirmed by another body of studies, based on optical methods of recording the local displacement field on the specimen surface, e.g., digital image correlation [37] or speckle interferometry [38], and developed independently of the research into intermittence. The spatial and/or temporal resolutions of these methods occupy a somewhat intermediate position with regard to the AE and mechanical tests. It occurs that they reveal another ubiquitous feature of the macroscopically smooth flow, namely, wavy patterns in the time evolution of the spatial repartition of plastic strain [39–43]. Moreover, a few recent works reported on a duality in the sense that the strain-rate bursts obey scaling distributions and, at the same time, are organized in wavy structures in the spatiotemporal maps [44–46].

As far as the PLC effect is concerned, although the macroscopic instability is caused by exceptional strain heterogeneities, the intermittence reflected in the evolution of the local strains has not been examined so far. The studies applying optical (and also thermal) methods of surveying local strains were mainly focused on the kinematics and evolution of the average parameters of PLC bands [4–12]. In particular, it was established for various alloys that the fast straining is characterized by a quasi-continuous propagation of deformation bands along the tensile axis, and this pattern is progressively replaced by a correlated and finally random occurrence of short-term localized bands when $\dot{\varepsilon}_a$ is decreased, although a quasi-continuous band propagation may take place at low $\dot{\varepsilon}_a$ in alloys with complex microstructures. [47,48].

The present work was aimed at a statistical analysis of the intermittence revealed in the local strain-rate maps during jerky flow of a model AlMg alloy. The choice of the object of investigation was not only motivated by the availability of a large amount of literature data on the properties of jerky flow and AE for such materials, but also by the robustness of these properties regarding the Mg content in a range from 2% to 5%. To provide the best basis for the comparison of behaviors on different scales, experiments were done on the same Al-3%Mg alloy as in [28,29]. The interest to such an analysis is twofold. On the one hand, both the stress serrations and AE represent a global temporal response of the sample to deformation processes taking place in its various parts. It is known that in nonlinear systems, the relationship between the global and local responses may not be trivial [49]. Therefore, to assess spatiotemporal behavior is essential for the understanding of the dislocation dynamics during jerky flow. Besides seizing the spatial aspect, the interest of this method is motivated by its intermediate time and amplitude resolution, enabling filling the gap between the scale ranges pertinent to the AE and deformation curves.

The first analysis reported here was realized in the high strain-rate regime. This type of jerky flow attracts particular interest due to the observation of power-law statistics for both these scale ranges. Therewith, the respective exponents do not take on the same values for the AE and stress serrations but are considerably higher in the former case [50]. Thus, it is meaningful to perform the analysis in an intermediate scale range. Besides, the nature of such scale-free statistics is a matter of current debate. Alternative interpretations in terms of either self-organized criticality (SOC)—a paradigm of avalanche dynamics first suggested to explain earthquakes statistics [51,52]—or turbulent flow were suggested to explain this regime of the PLC effect [15–17,53,54]. Recently, it has been argued that SOC is related to a general phenomenon known as a fluctuation scaling and stating on a power-law relationship between the average and variance of fluctuations in complex systems [55,56]. The present study allowed for a verification of such scaling in the conditions of jerky flow.

2. Materials and Methods

2.1. Experiment

Polycrystalline tensile samples with a dog-bone shape and a $25 \times 6.8 \times 2.5$ mm^3 gage part were cut from a cold-rolled sheet of Al-3% Mg in the rolling direction, solution treated by annealing at 400 °C for two hours and quenched in water. Polycrystalline grains had an approximately equiaxed shape and size between 30 and 100 μm [57]. Mechanical tests were implemented using a Zwick 1476 testing machine (Zwick, Ulm, Germany) with a software package testExper. The specimens were deformed at room temperature at a constant crosshead velocity corresponding to the nominal (i.e., referred to the initial specimen length) applied strain rate $\dot{\varepsilon}_a = 6 \times 10^{-3}$ s^{-1}.

The local extensometry technique was described in detail earlier (e.g., [58]) and is briefly presented below. Using a mask, a sequence of 20 1-mm wide black and white stripes was painted normal to the tensile axis on one side of the specimen. A charge-coupled device (CCD) Line Scan Sensor ZS16D (H.D. Rudolph GMBH, Reinheim, Germany) with a sampling frequency of 10^3 Hz and pixel size of 1.3 μm was set to record the longitudinal displacements of the intersections of black-and-white transitions with the centerline of the specimen, $x_i(t)$, where index i designates such successive survey points, starting from the upper edge of the field of vision. Thus, the so obtained spatiotemporal kinematic field corresponds to the Lagrangian representation [59]. With a 20 mm field of vision of the CCD camera, this setup implements a series of 20 local extensometers. As the total strain before necking was about 25% for all samples, usually about 15 upper extensometers remained in the field of vision during the test.

To avoid any bias caused by a possibly unequal sensitivity of the CCD camera to black-white and white-black transitions, local strains $\varepsilon_{i,i+2}(t)$ were calculated using next-nearest-neighbor transitions corresponding to the same kind of light contrast:

$$\varepsilon_i(t) = \ln \frac{x_{i+2}(t) - x_i(t)}{x_{i+2}(0) - x_i(0)} \tag{1}$$

where the second index in the strain notation is omitted for simplification. The local strain rate field $\dot{\varepsilon}_i(t)$ was calculated by the numerical derivation of the local strains.

As described previously [58], digital noise in $x_i(t)$ records was reduced using the running-average technique. It was found that the optimum window size for denoising at the strain rate explored in the present work corresponds to the average duration of stress drops, about 80 ms. This value allowed for revealing strain heterogeneities without smoothing out the steep fronts of such intense strain localizations as the PLC bands (see the spatiotemporal maps presented below). To avoid arbitrariness, the same running-average window was used for all the tested samples. It was checked that a variation of the window size by one-third does not reduce the scaling interval nor lead to a noticeable biasing of the estimate of the power-law exponent, which would exceed the experimental uncertainty. The reduction of the window down to 20 ms did not bias this estimate either. However, it was impractical because of a considerable increase in the scatter of the calculated distributions.

It is noteworthy that although such measurements are limited to one dimension, their advantage is due to a favorable combination of the time and space resolution. Even more importantly for the present study, the method based on the direct measurement of the markers positions is free from the approximations that are inevitable in the techniques using the local strain field calculation from the displacements of a randomly deposited speckle pattern.

2.2. Statistical and Fluctuation Scaling Analysis

The processing of stress serrations was presented in detail in previous works (e.g., [17,29,50]). The main precaution consists of taking into account a slow average increase in the size $\Delta\sigma$ of stress drops, which may accompany the material work hardening. Various approaches have been tested earlier, e.g., the normalization of $\Delta\sigma$ values by a polynomial fit of their time dependence, $\Delta\sigma(t)$, or, alternatively, normalization of the deformation curve itself by either a running average or a polynomial fit. In the present work, the latter approach with polynomial fitting was applied so that serrations were determined using the curves $\varsigma(t) = \sigma(t)/\overline{\sigma(t)}$.

As will be seen in the next section, the local strain-rate fluctuations did not noticeably evolve in the strain interval of interest (before the onset of necking). Due to this stationarity, the analysis was applied directly to the as-measured $\dot{\varepsilon}_i(t)$ signal. The procedure was the same as in the case of $\varsigma(t)$ signals, i.e., drops on a $\dot{\varepsilon}_i(t)$ curve were taken as a characteristic of its intermittence. It was also checked that the distributions are virtually the same for downward and upward serrations (drops and rises in the signal).

To simplify the notations, the amplitudes of $\dot{\varepsilon}_i(t)$ serrations are further designated as A; τ is used for durations; $\Delta\varsigma$ and Δt respectively denote such characteristics of stress serrations. Hereinafter, histograms of a variable u, where u refers to any of these four quantities, will be traced for the data rescaled by their average value, i.e., $u/<u>$. This reduction makes it possible to compare statistical distributions of different quantities by bringing together their probability density functions (PDF). Besides, it allows avoiding arbitrariness in the choice of the bin size by using a unique dimensionless bin in all cases.

Since power-law statistics mean that large values are rare, various approaches were suggested to correctly handle the high-scale limit [60–62]. To provide a basis for comparison with the earlier statistical studies of stress and AE signals accompanying the PLC effect [29,50], the direct method of calculation of PDF by counting the fraction $n(u)/N$ of events within intervals $u \pm \delta u/2$ was applied. Here, n is the number of events within the corresponding bin, and N is the total number of events in the dataset. The bin size δu was taken as a constant in the intervals rich of events, but it was increased in the deprived regions until gathering a meaningful number of events (five in this work). Accordingly, the probability density was calculated to take into account the bin variation,

$$PDF(u) = \frac{n}{N\delta u}. \tag{2}$$

More exact methods of detecting power-law distributions in empirical data have been developed during the last decade [61,62]. As verified for stress serrations and acoustic emission (e.g., [50]), the direct calculation with the varied-bin correction renders satisfactory estimates of the power-law exponents for the signals characterizing the PLC effect, in the sense that the deviation from the values obtained by the exact methods does not exceed the experimental uncertainty. Accordingly, this method was adapted in the present work.

The fluctuation scaling was tested following the procedure described in [56]. The measured signals, $\varsigma(t)$ or $\dot{\varepsilon}_i(t)$, are represented by times series with equidistant data points. Let us designate the studied series as Y_i, where $i = 1 \ldots K$, K is the number of data points in the analyzed time interval $T = K\delta t$, and δt is the sampling time for the corresponding signal. Fluctuations y_i of the signal about its mean value $\overline{Y}$ are calculated as $y_i = |Y_i - \overline{Y}|$. The scaling analysis is started by covering the interval T with a grid of k non-overlapping bins with an equal size r. The members of the new time series $z_k(r)$ are defined as the sum of y_i-values within the kth bin. The fluctuation scaling is verified if a power-law relationship between the variance $\mathrm{var}[z_k(r)]$ and the mean $E[z_k(r)]$,

$$\mathrm{var}[z_k(r)] \propto E[z_k(r)]^p \tag{3}$$

is satisfied when the bin size r is varied. In the present work, linear and logarithmic bin expanding rules were tested and rendered close results.

3. Results

3.1. Spatiotemporal Maps

An example of mapping of the evolution of local strains and strain rates onto a serrated $\sigma(t)$ curve is presented in Figure 1. Figure 1a displays stress fluctuations that are typical of the high-$\dot{\varepsilon}_a$ behavior of the PLC effect and are often considered as a signature of the propagative nature of deformation bands. Indeed, the propagation is explicitly demonstrated in plots (c), (b), and (d): Figure 1b shows a step-wise character of $\varepsilon_i(t)$ dependences for three local extensometers (data for other extensometers, as well as raw $x_i(t)$ records, are not shown); Figure 1c presents the corresponding $\dot{\varepsilon}_i(t)$ curves. Then, the ensemble of data for all the local extensometers is used to build up a spatiotemporal map $(t, x_i(t), \dot{\varepsilon}_i(t))$ in Figure 1d. The propagative character of deformation bands may already be assumed from the observation that each step in $\varepsilon_i(t)$ and the respective intense peak in $\dot{\varepsilon}_i(t)$, which reflect the passage of a deformation band through the ith extensometer, are systematically shifted in time for successive extensometers. Indeed, the map of Figure 1d visualizes the onward passage of PLC bands through the entire sequence of extensometers. Their progression is seen as inclined bright lines imaging the (quasi)continuous shift of the local region with the highest $\dot{\varepsilon}_i$ value —a PLC band—along the tensile axis. The nucleation of each band is known to require a rise in the stress, followed by a drop back to the general stress level when the band has been formed [2–13], as can be traced, e.g., at $t \approx 32.5$ s, when a band is nucleated within the field of vision of the CCD camera. In other cases, when the band occurs outside the visible part (the nucleation often takes place near the specimen head), the corresponding stress peak slightly precedes the band entering the field of vision. The lower-amplitude stress fluctuations between two subsequent stress rise/drop events accompany the deformation band propagation along the gage length (see [4–12,47,48]).

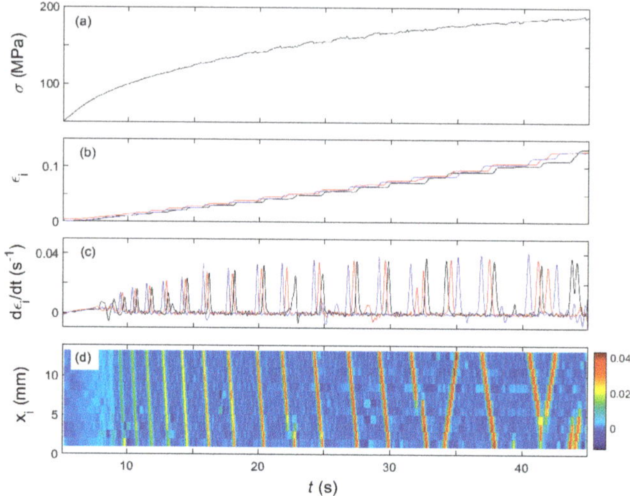

Figure 1. Example of synchronization of stress and strain responses. (**a**) Portion of a stress-time curve $\sigma(t)$ corresponding to serrated deformation. (**b**) Evolution of strains $\varepsilon_i(t)$ measured by three local extensometers selected near the opposite ends and in the middle of the field of vision of the charge-coupled device (CCD) camera. (**c**) The corresponding local strain rates $\dot{\varepsilon}_i(t)$. (**d**) The strain-rate map constructed using 14 local extensometers remaining within the field of vision during the entire test. $x_i(t)$ gives the position of the center of the ith extensometer. The color bars quantify the strain-rate scale (s^{-1}). The peak $\dot{\varepsilon}_i$ value within bright lines exceeds $\dot{\varepsilon}_a$ by a factor of about 7 (cf. [9,47,48]).

The description of high-$\dot{\varepsilon}_a$ behavior in the literature was often reduced to such sequences of stress rises and drops, the so-called type *A* serrations, whereas the intervals of band propagation were considered as smooth plastic flow (e.g., [63]). In other words, smaller and rather irregular stress fluctuations accompanying the band propagation were believed to be caused by the experimental noise and disregarded as a part of the collective dislocation dynamics. However, the statistical analysis proved that the entire set of stress fluctuations obeys power-law distributions and therefore cannot be ascribed to random noise [15,17,29]. In the cited works, these small irregularities were interpreted as being due to fluctuations in the width and velocity of the propagating PLC bands. Figure 1c,d testify that there may be another source of small stress variations. Indeed, besides the PLC bands, these plots reveal heterogeneity at finer scales. Even if the bright lines dominate the overall contrast in Figure 1d, the presence of weaker heterogeneities, which must also contribute to the irregular stress fluctuations, can be discerned outside them. Although the analysis of stress serrations presented in the next section does not allow for separating different contributions because of the global character of the stress response, such discrimination becomes possible by virtue of the local extensometry. The small strain heterogeneities will be analyzed in Section 3.3.

3.2. Statistics of Stress Serrations

A comprehensive statistical analysis of both the amplitude and time characteristics of stress serrations for similar alloys was reported elsewhere (e.g., [15,17,29]). The data succinctly presented below are aimed to verify the conclusions on the scale-free distributions of stress drop size at high $\dot{\varepsilon}_a$ and, notably, provide a basis for comparison with the local strain-rate statistics. Figure 2a shows a typical statistical sample of $\Delta\varsigma$ values for one of the deformed specimens in a time interval corresponding to a stabilized jerky flow. For each specimen, the processed dataset comprised more than 100 data points (up to 200 values). It can be recognized in the figure that the data scatter is denser in the range of small $\Delta\varsigma$ and becomes rarer far from the ordinate origin. This qualitative observation agrees with the results of the statistical analysis illustrated in Figure 2b. Indeed, the examples of PDF bear witness to a power law, PDF($\Delta\varsigma$)$\sim\Delta\varsigma^{-\beta}$, over more than two orders of magnitude of $\Delta\varsigma$, with β values close to 1, in consistence with the β-range between 1 and 1.5 reported in the papers cited above.

Similar to [17,29,54], the power-law nature of the probability law was also verified for the durations, PDF(Δt)$\sim\Delta t^{-\alpha}$, although in a narrower scale range, this lack is common with the previous investigations, perhaps because of a typically worse time resolution provided in mechanical tests in comparison with the force resolution. According to the constraints imposed by such probability laws, it was also verified that the conditional averages of $\Delta\varsigma$ for a given Δt (here designated as [$\Delta\varsigma$]) and the corresponding duration values are also related by a power law, [$\Delta\varsigma$]$\sim\Delta t^h$, with the exponents satisfying the scaling relation [64]:

$$\alpha = h(\beta - 1) + 1. \tag{4}$$

A quantitative example of such a relation will be illustrated below for the case of the local strain-rate serrations.

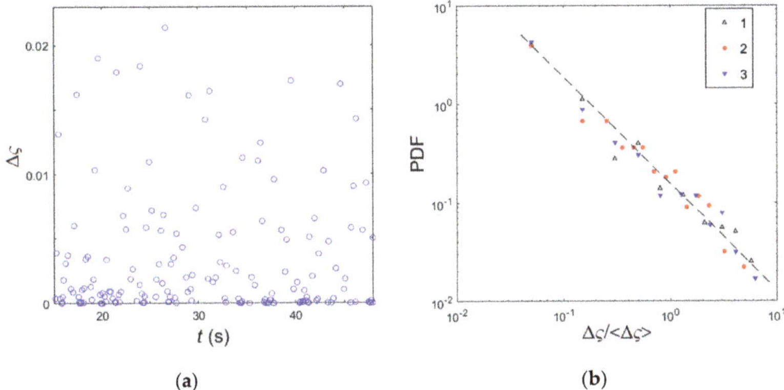

Figure 2. Example of processing of stress serrations. (**a**) Scatter of normalized values of stress serrations, $\Delta\varsigma$, in a time interval corresponding to a stabilized jerky flow for one of the samples. (**b**) Examples of probability density functions (PDF) of $\Delta\varsigma$ serrations for three samples. A dashed line is traced as a guide for the eye to show the slope β of the power-law distributions in log-log coordinates. The least-square evaluation renders the following mean and standard deviation for the estimates of β: 1—1.00 ± 0.07; 2—1.03 ± 0.07; 3—1.04 ± 0.06.

3.3. Statistics of Local Strain-Rate Serrations

Similar to the representation of stress serrations, Figure 3a gives an example of the scatter of amplitudes A of $\dot{\varepsilon}$ oscillations (index i is omitted when its omission is not misleading) gathered by one local extensometer in a strain range corresponding to jerky flow, i.e., beyond ε_{cr}. Figure 3b displays examples of PDF functions for two extensometers in different cross-sections of the same specimen. Although the extensometers gather information from short sections of the deforming material, as contrasted with the global stress response, the $\dot{\varepsilon}$ signals are significantly richer due to the high sensitivity of the CCD camera to the markers' displacements. The analyzed datasets typically comprised 350 to 550 events. The other two charts present examples of verification of the power-law behavior for the probability of durations and the scaling relation between the amplitudes and durations of the events (see Section 3.2). As in the case of stress serrations, the power-law is much less convincing for the durations than for amplitudes. The direct reason for this, namely the worse accuracy of time measurements, was mentioned above in relation with the analysis of stress serrations. However, this problem was also reported for the AE that provides a much higher time resolution [29]. It was suggested that the deviations may also be caused by a possible overlapping of dislocation avalanches, whereas amplitude distributions were shown to be rather robust with regard to this effect [65]. Despite these difficulties, the estimates of various exponents satisfy Equation (4) quite well (see the legend of Figure 3).

The local extensometry also provides an additional view on the relationship between local and global aspects of the plastic flow heterogeneity and intermittence. When strain is calculated using distant markers, i.e., on a larger base covered by more than one local extensometer, the number of the recorded $\dot{\varepsilon}$ bursts gradually decreases with increasing the width of such a "composed" extensometer. The decrease may be caused by two factors. First, the time intervals during which the PLC bands propagate through the extensometer and overshadow the weak heterogeneities become more important. Their fraction with regard to the total examined period is more significant the larger the extensometer. Second, the local $\dot{\varepsilon}$ bursts occurring quasi-simultaneously in its different sites would give rise to a sole event in the resultant signal. As a matter of example, the number of bursts recorded over a 16-mm wide base is typically less than 200. It should be underlined that although it is similar to the number of stress serrations, the information provided even by the largest extensometer is not equivalent to that stemming from the stress signal. In particular, most of the PLC band nucleation events occur outside the field of vision of the CCD camera.

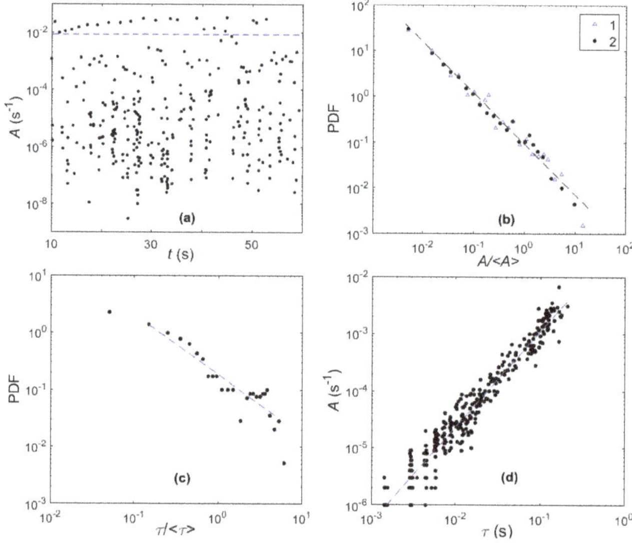

Figure 3. Processing of the local strain-rate responses. (**a**) Scatter of amplitudes of local strain-rate fluctuations, A, gathered by a local extensometer in the middle part of a specimen. The time interval corresponds to jerky flow ($\varepsilon \geq \varepsilon_{cr}$). (**b**) Examples of PDF for two separate extensometers. Similar to Figure 2, a dashed line is traced as a guide for the eye. The least-square evaluation renders the following mean and standard deviation for the estimates of the power-law exponent: 1—1.09 ± 0.04; 2—1.11 ± 0.04. (**c**) PDF of durations τ of serrations recorded by the extensometer 1 of Figure 3b. The respective slope α = 1.05 ± 0.11. (**d**) Amplitude–duration relationship for the same extensometer. The slope calculated for the conditional averages h = 1.65 ± 0.04.

Figure 4 brings together examples of PDF for three kinds of signals, namely, stress (cf. Figure 2b), strain rate from the shortest extensometer (cf. Figure 3b), and from the largest 16-mm wide extensometer. The latter was calculated using a similar procedure as in the case of the local extensometers. Namely, the highest amplitudes of $\dot{\varepsilon}$ bursts occurring during the PLC band propagation and characterized by an intrinsic scale were disregarded. It can be seen that although the three kinds of signals highlight different aspects of the intermittence, the corresponding PDF dependences are quite close. Similar to the case of stress serrations, some tendency to a shallower slope for $\dot{\varepsilon}$ fluctuations measured by large extensometers, as compared with the individual ones, could be conjectured. However, the accuracy of the estimates does not allow for a certain conclusion.

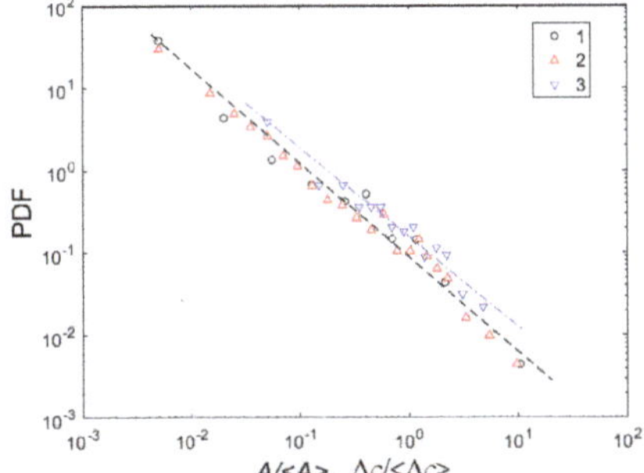

Figure 4. Comparison of PDF for local and global responses: 1—Strain-rate response over a 16-mm base, $\beta = 1.05 \pm 0.06$. 2—One local extensometer, $\beta = 1.11 \pm 0.04$. 3—Stress serrations, $\beta = 1.03 \pm 0.07$. Dashed line is traced for the local extensometer; dash-and-dot line for stress serrations.

Since the examined $\dot{\varepsilon}$ fluctuations occur all over the surveyed gage length and also at any time, it would be enlightening to perform a similar analysis before the onset of the PLC instability. Indeed, similar fluctuations already occurred in the seemingly elastic region, thus corroborating an important role of microplastic flow on the early stages of deformation [66]. However, the number of events recorded by each local extensometer in relatively short intervals before ε_{cr} was statistically insufficient. To provide a significant dataset for a given specimen, the data were collected from all its extensometers. Figure 5 presents the calculation results in two ways. Curves 1 and 2 display examples of PDF obtained for one of the specimens at the macroscopically elastic stage and at the stage of the elastoplastic transition, respectively. Curve 3 shows the PDF obtained using the entirety of these data for both stages together. The data bear evidence that the dependences obtained for $\varepsilon < \varepsilon_{cr}$ also obey a power law with the exponent close to 1.

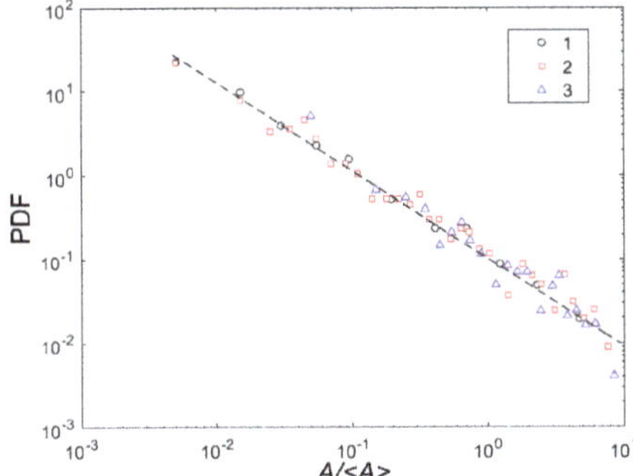

Figure 5. PDF dependences for local strain-rate fluctuations before the onset of the Portevin–Le Chatelier (PLC) instability: 1—Macroscopically elastic part, $\beta = 1.04 \pm 0.05$. 2—Elastoplastic transition, $\beta = 0.98 \pm 0.04$. 3—The totality of data for $\varepsilon < \varepsilon_{cr}$, $\beta = 1.06 \pm 0.04$.

To verify that this result is not biased by the blending of data from various local extensometers, the same procedure was applied to the interval $\varepsilon \geq \varepsilon_{cr}$, and the statistical analysis was repeated for the blended dataset. The results of such processing are presented in Figure 6, which are to be compared with the dependences for individual extensometers in Figure 3b. The PDF shows clear power-law behavior with $\beta = 1.04$ (± 0.05), which agrees with the results illustrated in Figure 3b, thus corroborating the validity of the analysis of blended datasets in the interval $\varepsilon < \varepsilon_{cr}$.

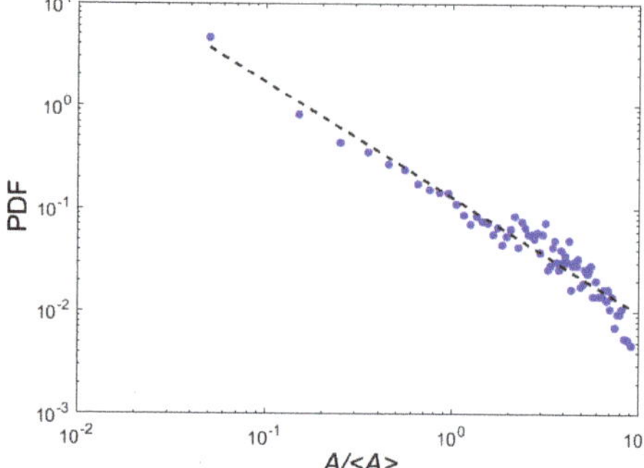

Figure 6. Example of a PDF calculated for one of the specimens using data from all extensometers ($\varepsilon > \varepsilon_{cr}$). The dashed line indicates the slope $\beta = 1.04 \pm 0.05$.

Figure 6 highlights two additional nuances that are vaguely discernible in the dependences obtained for individual extensometers. First, it shows a clear cutoff at large scales, which is typical of empirical power-law statistics and is considered as evidence for the existence of natural limits of scale-invariant behavior in real systems [30,31]. Second, the cutoff is preceded by a bump leading to a distinction between small events obeying scale invariance and large events deviating from the extrapolation of the power-law dependence. Hints to this deviation can also be recognized in Figures 3–5, but it has become clear due to blending of many signals. Such a deviation from SOC-like behavior at large scales was reported for earthquake models and attributed to the triggering of a sequence of avalanches, providing that the triggering avalanche is powerful enough to store sufficient elastic energy [67].

3.4. Fluctuation Scaling

Figure 7 presents examples of verification of the fluctuation scaling for two samples and different kinds of signals. To provide a general view, both linear (Figure 7a) and logarithmic (Figure 7b) bin expanding rules are illustrated. The major result of this section is that the entirety of examples allows for a conclusion that all dependences exhibit fluctuation scaling in a significant scale range.

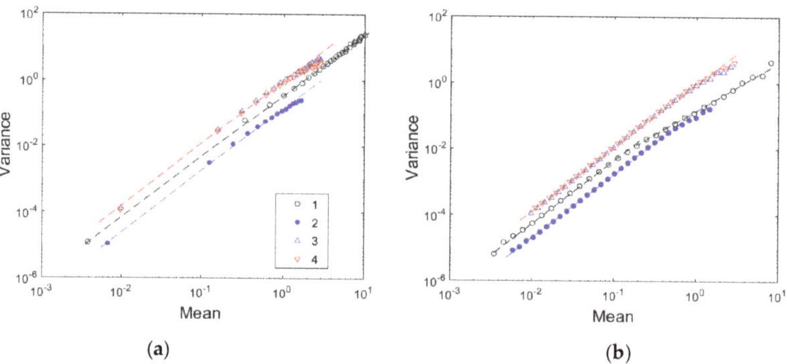

(a)   (b)

Figure 7. Examples of fluctuation scaling for two specimens. (**a**) Linear bin expansion. (**b**) Logarithmic bin expansion. The designations are unique for both charts: 1—Deformation curve; 2—Large extensometer; 3 and 4—Examples for local extensometers. The units are not shown explicitly and simply correspond to the processed signal. All slope values are close to 1.8 (see the body of the text), except for the large-scale behavior of curve 1 in Figure 7b, corresponding to $p = 1.34 \pm 0.05$.

Some specific observations are noteworthy. The analysis of deformation curves leads to a distinction of two qualitatively different cases, one corresponding to a single power-law dependence and another revealing a crossover between two power laws. Curve 1 in Figure 7a illustrates the first case. Indeed, it verifies the power-law behavior for the normalized deformation curve $\varsigma(t)$ over the entire scale range covering three orders of magnitude of the mean value. The index p (see Equation (3)) is approximately 1.86 ± 0.02. A similar slope is also observed at small scales for another sample presented by the example of Figure 7b, while the right-hand part is characterized by a lower p value, $p = 1.34 \pm 0.05$. The $\dot{\varepsilon}$ signals from both narrow and large extensometers show one power-law in each case, with a gradual deviation at large scales. Moreover, the respective slopes agree well with each other and also with the former of the above values obtained for $\varsigma(t)$-curves. More exactly, the totality of estimates varies between approximately 1.73 and 1.9.

4. Discussion and Conclusions

As recalled in the Introduction, the high-$\dot{\varepsilon}_a$ plastic flow of alloys prone to the PLC effect is characterized by power-law statistical distributions on multiple scales. Whereas for materials non-subject to plastic instability, avalanche-like intermittent behavior is only visible on fine scales revealed by the AE [25–27] or the local extensometry technique [44–46]; it engrosses deformation curves in the DSA conditions, giving rise to a macroscopically unstable deformation. At the same time, there is no consensus in the literature on the PLC effect as to which of the observed serrations should be taken into consideration in the analysis of jerky flow. In the early research, only stress rises followed by sharp drops were believed to pertain to plastic instability at high $\dot{\varepsilon}_a$ (type A serrations). Such serrations are caused by the nucleation of new PLC bands which then propagate along the specimen. Taken alone, they would be described by some characteristic scale [63]. However, the deformation curves display a much larger spectrum of stress fluctuations that, taken together, manifest power-law distributions of both amplitudes and durations [15,17,29,50]. Recent studies showed that the AE accompanying the plastic deformation of alloys obeys such critical-type statistics at all strain rates, even though the decrease in $\dot{\varepsilon}_a$ results in a transition to peaked distributions of stress drops [29,36]. Therewith, the AE does not only occur at the instants of stress drops but also during reloading periods corresponding to macroscopically stable deformation. It was conjectured that at high $\dot{\varepsilon}_a$ values, the avalanche mechanism of plastic flow may extend to a wide scale range, including serrated deformation curves. However, the AE and stress fluctuations pertain to two outermost scales, while a large intermediate interval remained unexplored so far. One of the major aims of the present work was to fill this gap by virtue of the local extensometry technique. Moreover, this approach opens a way to assess the spatiotemporal dynamics of dislocations, as compared with the only temporal aspect revealed by the AE or deformation curves. The first results obtained for an AlMg alloy confirm the scale-free statistics of plastic deformation and therefore corroborate the hypothesis of a critical-type dislocation dynamics controlling the plastic flow of dynamically strain-aging alloys in the high-$\dot{\varepsilon}_a$ regime.

At the same time, two aspects of experimental observations need clarification. First, Figure 3 bears evidence that not all events obey scale invariance, but there exists a distinction between small and large $\dot{\varepsilon}$-bursts. Whereas scale-free distributions characterize small events, strong bursts corresponding to the passage of the PLC bands reveal a characteristic scale. Second, whereas the power-law exponents are close to 1 for the bursts of σ and $\dot{\varepsilon}$, the literature data report on β values varying from about 1.5 to almost 3 for the AE energy in the same experimental conditions [29].

Let us first consider the former question. To explain the transition from scale-free to peaked distributions of stress drops with a decrease in $\dot{\varepsilon}_a$, a hypothesis of the coexistence of SOC and synchronization phenomena was advanced in [29]. It can be highlighted as follows. A PLC band generates a strong strain heterogeneity giving rise to incompatibility stresses that may relax via small deformation events. At low $\dot{\varepsilon}_a$, small events effectively smooth out the heterogeneities. As a result, many sections of the specimen reach the instability threshold (the top of the N-shaped SRS-function) quasi-simultaneously, and a new intense PLC band is formed. Its development is stopped by the elastic unloading of the machine-specimen system, this process being the main factor determining the characteristic scale of stress serrations. This reasoning was confirmed by the observation of bimodal distributions of stress drops at low $\dot{\varepsilon}_a$, i.e., a power law with β about 1 at small scales and a peak for deep stress drops. Very similar behavior was found in seismicity models revealing SOC for small earthquakes but also characteristic catastrophic earthquakes [67]. The data of Figure 6 in the present paper testify to a general character of the trend to scale separation. Indeed, the $\dot{\varepsilon}$ signal does not only display a characteristic scale corresponding to the PLC bands, but the smaller events also show a tendency to a deviation from the scale-invariant distribution and the formation of a bump on its right-hand side, which is clearly revealed by summing signals from many local sources.

The application of the AE method to the PLC effect provided an explicit explanation of the scale separation [29,36]. It occurred that despite the coexistence of two scale ranges for stress fluctuations at low $\dot{\varepsilon}_a$, there is virtually no distinction between the amplitudes of AE events occurring at the instants of stress drops and during smooth reloading. Instead, a distinction occurs between their durations because stress serrations are accompanied by clustering of the AE, giving rise to events in a millisecond range for deep stress drops, while individual AE events accompanying the reloading have durations in the microsecond range. Based on these observations, abrupt serrations with a characteristic size were attributed to a brief synchronization of many dislocation avalanches, proceeding through their successive triggering [29,36,50].

It is clear which changes to this picture may occur when $\dot{\varepsilon}_a$ is increased. The faster loading must be accommodated by a more intense plastic deformation and, therefore, lead to a globally stronger clustering and also the superposition of dislocation avalanches, as was confirmed by virtue of the AE technique [29,36]. At the same time, the smoothing of the recurrent heterogeneities becomes less efficient because of shorter reloading periods, so that there constantly exist some parts of the material close to the instability threshold. This has two consequences. On the one hand, the strain gradient sustained at the front of the deformation band leads to its propagation along the specimen. On the other hand, avalanches of any size may occur, and the critical behavior of both amplitudes and the durations of events is virtually observed in the entire scale range spreading from the AE to stress serrations. This conjecture is consistent with the observation of scale-free distributions of $\dot{\varepsilon}$ fluctuations in the present work. However, it should be noted that there is an essential difference between the global σ and AE responses and the local $\dot{\varepsilon}$ response. Namely, while the scale separation disappears from the former ones due to a strong overall plastic activity, it persists in the local response because of the recurrent propagation of the PLC bands.

The supposed effect of $\dot{\varepsilon}_a$ conforms to the observation made in [29] that the exponent β of AE distributions is lesser at higher $\dot{\varepsilon}_a$, which is in agreement with the higher probability of larger (composed) events. Moreover, as pointed out in [50], the same factor may be responsible for lower β values in the case of σ and $\dot{\varepsilon}$ in comparison with the AE, because the time resolution of these signals does not allow resolving microsecond events. Thus, it may be suggested that the β values correctly characterizing dislocation avalanches in the DSA conditions are provided by the AE technique, while the apparent value of 1 rendered by σ and $\dot{\varepsilon}$ signals should be considered with precaution. Thus, it would be challenging to verify this suggestion by drastically increasing the time resolution of the local extensometry.

Another challenge concerns the choice between the specific mechanisms that can be responsible for critical-type statistics. The entirety of the results discussed above, including the literature data, may be explained within the framework combining the SOC and synchronization phenomena. In particular, it explains the paradox noticed in the first studies of statistical distributions of stress serrations. Namely, SOC models require vanishing driving rates [51,52]. Even if the high $\dot{\varepsilon}_a$ values necessary for type A behavior still correspond to quasi-static tests, the fact that peaked distributions were found at slow loading and scale-invariant behavior at a relatively fast loading remained puzzling. The above results bear evidence to SOC manifestations in the entire strain-rate range corresponding to the PLC domain, although the synchronization of dislocations avalanches gives rise to scale separations depending on $\dot{\varepsilon}_a$.

An alternative interpretation of scale-free statistics was proposed in [53,54] on the basis of the analysis of Lyapunov exponents [68] for type *A* serrations in a physically-based model of the PLC effect, considering the reactions and dynamics of distinct dislocation ensembles. It was found that the Lyapunov spectrum, which quantifies the rate of divergence of close trajectories of dynamical systems, is similar to that characterizing turbulent flow [69,70]. However, while this model successfully predicts the transitions from the states with a characteristic scale of stress serrations to scale-invariant behavior when $\dot{\varepsilon}_a$ is varied, it does not account for the scale separation at the same strain rate.

In view of these alternative hypotheses, the question of the specific mechanism of the critical-type behavior of the PLC effect needs further investigation. A special interest deserves a conceptually different approach to the power-law statistics proposed recently [55,56]. Among various issues, it conjectures that the SOC-type behavior may be related to a general feature of complex systems of various nature, which are known as fluctuation scaling or Taylor's law (after investigations in ecology [71]) and are expressed by Equation (3). Moreover, it states that although numerous system-specific dynamical models were proposed with more or less success in each discipline, the statistical power laws emerging in such systems may be explained on a general basis as stemming from the mathematical convergence toward the so-called Tweedie distributions [72]. In the spirit of this concept, fluctuation scaling was verified in Section 3.4 for the experimental data obtained in the present paper. It was found that both σ and $\dot{\varepsilon}$ signals obey approximately the same power law with the slope p about 1.8, which is similar to p-values obtained for some sandpile models [55,56]. From the general point of view, this result attests to the behavior of the PLC effect at high $\dot{\varepsilon}_a$ as belonging to a certain class of complex dynamical systems. Namely, the exponent $1 < p < 2$ corresponds to compound Poisson–gamma distributions, which are particularly used to mimic the process of capturing clusters in ecological data such as biomasses [73]. Therewith, the observation of a crossover in the scaling for some deformation curves (Figure 7b) may be indicative of multifractality [74], which is in consistence with multifractal behaviors detected for the PLC effect [17,21,28]. More specifically, since the sandpile models serve as a paradigm of SOC, the above-mentioned similarity between p-values corroborates the conclusion that the statistical distributions presented here are compatible with the conjecture of SOC-like behavior. On the whole, the first results of such an analysis attest to the PLC effect as a candidate for the investigation of the applicability of these general concepts to collective behaviors in plastic deformation.

In summary, it may be concluded that the local extensometry brings information that is complementary to the AE or mechanical tests. This new angle of view of the problem of plastic heterogeneity requires extending such investigations to a wide strain-rate range, in particular, to verify alternative interpretations of a strikingly rich pattern of various dynamical regimes of plastic flow in model binary alloys. Furthermore, this approach is also promising for investigations of plastic deformation in such novel materials as metallic glasses and high-entropy alloys that are also prone to jerky flow [75–77].

Finally, although the present paper was devoted to a statistical analysis of the intermittence of plastic flow in the conditions of plastic instability, a qualitatively different phenomenon arising from these experiments is worth noting. Similar to several other materials for which similar data have been reported [44–46], the scale of observation provided by the local extensometry reveals both intermittent and wavy patterns of local strain heterogeneities beyond the PLC bands (see Figure 1). Therewith, the intermittence manifests itself as localized bright spots corresponding to $\dot{\varepsilon}_i$ bursts, and the arrangement of such bursts along inclined straight lines reveals the propagation of strain localizations. Such a duality, unstudied so far, presents a great interest for the understanding of correlations between temporal instabilities and spatial heterogeneities in the system of crystal defects.

Author Contributions: Conceptualization, M.L.; methodology, M.L., D.E., and T.L.; software, M.L. and Y.B.; validation, Y.B. and T.L.; formal analysis, Y.B., M.L., D.E., and T.L.; investigation, Y.B., M.L., D.E., and T.L.; resources, M.L.; data curation, Y.B.; writing—original draft preparation, Y.B.; writing—review and editing, M.L.; visualization, Y.B. and M.L.; supervision, M.L.; project administration, M.L.; funding acquisition, M.L. All authors have read and agreed to the published version of the manuscript.

Funding: This research was partly funded by the French State through the program "Investment in the future" operated by the National Research Agency, in the framework of the LabEx DAMAS (ANR-11-LABX-0008-01) and the research program RESEM managed by IRT M2P (Institut de Recherche Technologique en Matériaux, Métallurgie et Procédés).

Conflicts of Interest: The authors declare no conflict of interest. The funders had no role in the design of the study; in the collection, analyses, or interpretation of data; in the writing of the manuscript, or in the decision to publish the results.

References

1. Van den Beukel, A. Theory of the effect of dynamic strain aging on mechanical properties. *Phys. Status Solidi A* **1975**, *30*, 197–206. [CrossRef]
2. Penning, P. Mathematics of the Portevin-Le Chatelier effect. *Acta Metall.* **1972**, *20*, 1169–1175. [CrossRef]
3. Estrin, Y.; Kubin, L.P. Collective Dislocation Behaviour in Dilute Alloys and the Portevin-Le Chatelier Effect. *J. Mech. Behav. Mater.* **1990**, *2*, 255–292. [CrossRef]
4. Chihab, K.; Estrin, Y.; Kubin, L.P.; Vergnol, J. The kinetics of the Portevin-Le Chatelier bands in an Al-5at%Mg alloy. *Scr. Metall.* **1987**, *21*, 203–208. [CrossRef]
5. Ziegenbein, A.; Hähner, P.; Neuhäuser, H. Correlation of temporal instabilities and spatial localization during Portevin-Le Chatelier deformation of Cu-10 at.% Al and Cu-15 at.% Al. *Comput. Mater. Sci.* **2000**, *19*, 27–34. [CrossRef]
6. Chmelík, F.; Ziegenbein, A.; Neuhäuser, H.; Lukáč, P. Investigating the Portevin-Le Chatelier effect by the acoustic emission and laser extensometry techniques. *Mater. Sci. Eng. A* **2002**, *324*, 200–207. [CrossRef]
7. Shabadi, R.; Kumar, S.; Roven, H.J.; Dwarakadasa, E.S. Characterisation of PLC band parameters using laser speckle technique. *Mater. Sci. Eng. A* **2004**, *364*, 140–150. [CrossRef]
8. Louche, H.; Vacher, P.; Arrieux, R. Thermal observations associated with the Portevin–Le Châtelier effect in an Al–Mg alloy. *Mater. Sci. Eng. A* **2005**, *404*, 188–196. [CrossRef]
9. Ait-Amokhtar, H.; Vacher, P.; Boudrahem, S. Kinematics fields and spatial activity of Portevin-Le Chatelier bands using the digital image correlation method. *Acta Mater.* **2006**, *54*, 4365–4371. [CrossRef]
10. Jiang, H.; Zhang, Q.; Chen, X.; Chen, Z.; Jiang, Z.; Wu, X.; Fan, J. Three types of Portevin-Le Chatelier effects: Experiment and modeling. *Acta Mater.* **2007**, *55*, 2219–2228. [CrossRef]
11. Ranc, N.; Wagner, D. Experimental study by pyrometry of Portevin–Le Châtelier plastic instabilities—Type A to type B transition. *Mater. Sci. Eng. A* **2008**, *474*, 188–196. [CrossRef]
12. Lebyodkin, M.A.; Zhemchuzhnikova, D.A.; Lebedkina, T.A.; Aifantis, E.C. Kinematics of formation and cessation of type B deformation bands during the Portevin-Le Chatelier effect in an AlMg alloy. *Results Phys.* **2019**, *12*, 867–869. [CrossRef]
13. Tamimi, S.; Andrade-Campos, A.; Pinho-da-Cruz, J. Modelling of the Portevin-Le Chatelier effects in aluminium alloys: A review. *J. Mech. Behav. Mater.* **2015**, *24*, 67–78. [CrossRef]
14. Portevin, A.; Le Chatelier, F. Sur un phénomène observé lors de l'essai de traction d'alliages en cours de transformation, C.R. *Acad. Sci.* **1923**, *176*, 507–510.
15. Lebyodkin, M.A.; Brechet, Y.; Estrin, Y.; Kubin, L.P. Statistics of the catastrophic slip events in the Portevin-Le Chatelier effect. *Phys. Rev. Lett.* **1995**, *74*, 4758–4761. [CrossRef]
16. Ananthakrishna, G.; Noronha, S.J.; Fressengeas, C.; Kubin, L.P. Crossover from chaotic to self-organized critical dynamics in jerky flow of single crystals. *Phys. Rev. E* **1999**, *60*, 5455–5462. [CrossRef]
17. Bharathi, M.S.; Lebyodkin, M.; Ananthakrishna, G.; Fressengeas, C.; Kubin, L.P. Multifractal Burst in the Spatiotemporal Dynamics of Jerky Flow. *Phys. Rev. Lett.* **2001**, *87*, 165508. [CrossRef]
18. Kugiumtzis, D.; Kehagias, A.; Aifantis, E.C.; Neuhäuser, H. Statistical analysis of the extreme values of stress time series from the Portevin–Le Châtelier effect. *Phys. Rev. E* **2004**, *70*, 036110. [CrossRef]
19. Sarkar, A.; Webber, C.L., Jr.; Barat, P.; Mukherjee, P. Recurrence analysis of the Portevin–Le Chatelier effect. *Phys. Lett. A* **2008**, *372*, 1101–1105. [CrossRef]
20. Iliopoulos, A.C.; Nikolaidis, N.S.; Aifantis, E.C. Portevin Le Chatelier effect and Tsallis nonextensive statistics. *Phys. A* **2015**, *438*, 509–518. [CrossRef]
21. Lebyodkin, M.A.; Lebedkina, T.A. Multifractal analysis of evolving noise associated with unstable plastic flow. *Phys. Rev. E* **2006**, *73*, 036114. [CrossRef] [PubMed]

22. Dimiduk, D.M.; Woodward, C.; LeSar, R.; Uchic, M.D. Scale-free intermittent flow in crystal plasticity. *Science* **2006**, *312*, 1188–1190. [CrossRef] [PubMed]

23. Csikor, F.F.; Motz, C.; Weygand, D.; Zaiser, M.; Zapperi, S. Dislocation Avalanches, Strain Bursts, and the Problem of Plastic Forming at the Micrometer Scale. *Science* **2007**, *318*, 251–254. [CrossRef] [PubMed]

24. Maaß, R.; Wraith, M.; Uhl, J.T.; Greer, J.R.; Dahmen, K.A. Slip statistics of dislocation avalanches under different loading modes. *Phys. Rev. E* **2015**, *91*, 042403. [CrossRef]

25. Weiss, J.; Grasso, J.-R. Acoustic Emission in Single Crystals of Ice. *J. Phys. Chem. B* **1997**, *101*, 6113–6117. [CrossRef]

26. Weiss, J.; Grasso, J.-R.; Miguel, M.-C.; Vespignani, A.; Zapperi, S. Complexity in dislocation dynamics: Experiments. *Mater. Sci. Eng. A* **2001**, *309*, 360–364. [CrossRef]

27. Weiss, J.; Rhouma, W.B.; Richeton, T.; Dechanel, S.; Louchet, F.; Truskinovsky, L. From Mild to Wild Fluctuations in Crystal Plasticity. *Phys. Rev. Lett.* **2015**, *114*, 105504. [CrossRef]

28. Lebyodkin, M.A.; Kobelev, N.P.; Bougherira, Y.; Entemeyer, D.; Fressengeas, C.; Lebedkina, T.A.; Shashkov, I.V. On the similarity of plastic flow processes during smooth and jerky flow in dilute alloys. *Acta Mater.* **2012**, *60*, 844–850. [CrossRef]

29. Lebyodkin, M.A.; Kobelev, N.P.; Bougherira, Y.; Entemeyer, D.; Fressengeas, C.; Gornakov, V.S.; Lebedkina, T.A.; Shashkov, I.V. On the similarity of plastic flow processes during smooth and jerky flow: Statistical analysis. *Acta Mater.* **2012**, *60*, 3729–3740. [CrossRef]

30. Zaiser, M. Scale invariance in plastic flow of crystalline solids. *Adv. Phys.* **2007**, *55*, 185–245. [CrossRef]

31. Papanikolaou, S.; Cui, Y.; Ghoniem, N. Avalanches and plastic flow in crystal plasticity: An overview. *Model. Simul. Mater. Sci. Eng.* **2018**, *26*, 013001. [CrossRef]

32. Maaß, R.; Derlet, P.M. Micro-plasticity and recent insights from intermittent and small-scale. *Acta Mater.* **2018**, *143*, 338–363. [CrossRef]

33. Pérez, C.J.; Corral, Á.; Díaz-Guilera, A.; Christensen, K.; Arenas, A. On self-organized critically and synchronization in lattice models of coupled dynamic systems. *Int. J. Mod. Phys. B* **1996**, *10*, 1111–1151. [CrossRef]

34. Strogatz, S.H. From Kuramoto to Crawford: Exploring the onset of synchronization in populations of coupled oscillators. *Phys. D* **2000**, *143*, 1–20. [CrossRef]

35. Shashkov, I.V.; Lebyodkin, M.A.; Lebedkina, T.A. Multiscale study of acoustic emission during smooth and jerky flow in an AlMg alloy. *Acta Mater.* **2012**, *60*, 6842–6850. [CrossRef]

36. Lebedkina, T.A.; Bougherira, Y.; Entemeyer, D.; Lebyodkin, M.A.; Shashkov, I.V. Crossover in the scale-free statistics of acoustic emission associated with the Portevin-Le Chatelier instability. *Scr. Mater.* **2018**, *148*, 47–50. [CrossRef]

37. Sutton, M.A.; Hild, F. Recent Advances and Perspectives in Digital Image Correlation. *Exp. Mech.* **2015**, *55*, 1–8. [CrossRef]

38. Jacquot, P. Speckle Interferometry: A Review of the Principal Methods in Use for Experimental Mechanics Applications. *Strain* **2008**, *44*, 57–69. [CrossRef]

39. Zuev, L.B.; Danilov, V.I.; Kartashova, N.V.; Barannikova, S.A. The self-excited wave nature of the instability and localization of plastic deformation. *Mater. Sci. Eng. A* **1997**, *234–236*, 699–702. [CrossRef]

40. Sarafanov, G.F. Plastic-strain-softening waves in crystals. *Phys. Solid State* **2001**, *43*, 263–269. [CrossRef]

41. Zuev, L.B. On the way of plastic flow localization in pure metals and alloys. *Ann. Phys.* **2007**, *16*, 286–310. [CrossRef]

42. McDonald, R.J.; Efstathiou, C.; Kurath, P. The wavelike plastic deformation of single crystal copper. *J. Eng. Mater. Technol. Trans. ASME* **2009**, *131*, 031013. [CrossRef]

43. Zuev, L.B.; Barannikova, S.A. Autowave physics of material plasticity. *Crystals* **2019**, *9*, 458. [CrossRef]

44. Fressengeas, C.; Beaudoin, A.J.; Entemeyer, D.; Lebedkina, T.; Lebyodkin, M.; Taupin, V. Dislocation transport and intermittency in the plasticity of crystalline solids. *Phys. Rev. B* **2009**, *79*, 014108. [CrossRef]

45. Mudrock, R.N.; Lebyodkin, M.A.; Kurath, P.; Beaudoin, A.J.; Lebedkina, T.A. Strain-rate fluctuation during macroscopically uniform deformation of a solution-strengthened alloy. *Scr. Mater.* **2011**, *65*, 1093–1096. [CrossRef]

46. Lebyodkin, M.; Amouzou, K.; Lebedkina, T.; Richeton, T.; Roth, A. Complexity and anisotropy of plastic flow of α-Ti probed by acoustic emission and local extensometry. *Materials* **2018**, *11*, 1061. [CrossRef] [PubMed]

47. Zhemchuzhnikova, D.A.; Lebyodkin, M.A.; Lebedkina, T.A.; Kaibyshev, R.O. Unusual behavior of the Portevin-Le Chatelier effect in an AlMg alloy containing precipitates. *Mater. Sci. Eng. A* **2015**, *639*, 37–41. [CrossRef]

48. Zhemchuzhnikova, D.; Lebyodkin, M.; Yuzbekova, D.; Lebedkina, T.; Mogucheva, A.; Kaibyshev, R. Interrelation between the Portevin Le-Chatelier effect and necking in AlMg alloys. *Int. J. Plast.* **2018**, *110*, 95–109. [CrossRef]

49. Laurson, L.; Alava, M.J. Local waiting times in critical systems. *Eur. Phys. J. B* **2004**, *42*, 407–414. [CrossRef]

50. Lebedkina, T.A.; Zhemchuzhnikova, D.A.; Lebyodkin, M.A. Correlation versus randomization of jerky flow in an AlMgScZr alloy using acoustic emission. *Phys. Rev. E* **2018**, *97*, 013001. [CrossRef]

51. Bak, P.; Tang, C.; Wiesenfeld, K. Self-organized criticality. *Phys. Rev. A* **1988**, *38*, 364–374. [CrossRef] [PubMed]

52. Watkins, N.W.; Pruessner, G.; Chapman, S.C.; Crosby, N.B.; Jensen, H.J. 25 Years of Self-organized Criticality: Concepts and Controversies. *Space Sci. Rev.* **2016**, *198*, 3–44. [CrossRef]

53. Bharathi, M.S.; Ananthakrishna, G. Chaotic and power law states in the Portevin-Le Chatelier effect. *Europhys. Lett.* **2002**, *60*, 234–240. [CrossRef]

54. Ananthakrishna, G.; Bharathi, M.S. Dynamical approach to the spatiotemporal aspects of the Portevin–Le Chatelier effect: Chaos, turbulence, and band propagation. *Phys. Rev. E* **2004**, *70*, 026111. [CrossRef]

55. Eisler, Z.; Bartos, I.; Kertész, J. Fluctuation scaling in complex systems: Taylor's law and beyond. *Adv. Phys.* **2008**, *57*, 89–142. [CrossRef]

56. Kendal, W.S. Self-organized criticality attributed to a central limit-like convergence effect. *Phys. A* **2015**, *421*, 141–150. [CrossRef]

57. Shashkov, I.V. Multiscale Study of the Intermittency of Plastic Deformation by Acoustic Emission Method. Ph.D. Thesis, Université de Lorraine, Metz, France, 2012.

58. Roth, A.; Lebedkina, T.A.; Lebyodkin, M.A. On the critical strain for the onset of plastic instability in an austenitic FeMnC steel. *Mater. Sci. Eng. A* **2012**, *539*, 280–284. [CrossRef]

59. Landau, L.D.; Lifshitz, E.M. *Fluid Mechanics*, 2nd ed.; Course of Theoretical Physics; Butterworth-Heinemann: Oxford, UK, 1987; Volume 6.

60. Pickering, G.; Bull, J.M.; Sanderson, D.J. Sampling power-law distributions. *Tectonophysics* **1995**, *248*, 1–20. [CrossRef]

61. Clauset, A.; Shalizi, C.; Newman, M. Power-law distributions in empirical data. *SIAM Rev.* **2009**, *51*, 661–703. [CrossRef]

62. Deluca, A.; Corral, Á. Fitting and goodness-of-fit test of non-truncated and truncated power-law distributions. *Acta Geophys.* **2013**, *61*, 1351–1394. [CrossRef]

63. Pink, E.; Weinhandl, H. The distribution of stress-drop sizes in serrated flow of an aluminum alloy and a mild steel. *Scr. Mater.* **1998**, *39*, 1309–1316. [CrossRef]

64. Kertész, J.; Kiss, L.B. The noise spectrum in the model of self-organized criticality. *J. Phys. A* **1990**, *23*, L433. [CrossRef]

65. Lebyodkin, M.A.; Shashkov, I.V.; Lebedkina, T.A.; Mathis, K.; Dobron, P.; Chmelik, F. Role of superposition of dislocation avalanches in the statistics of acoustic emission during plastic deformation. *Phys. Rev. E* **2013**, *88*, 042402. [CrossRef] [PubMed]

66. Dudarev, E.F.; Deryugin, E.E. Microplastic deformation and yield strength of polycrystals. *Sov. Phys. J.* **1982**, *25*, 510–519. [CrossRef]

67. Carlson, J.M.; Langer, J.S.; Shaw, B.E. Dynamics of earthquake faults. *Rev. Mod. Phys.* **1994**, *66*, 657–670. [CrossRef]

68. Abarbanel, H.D.I.; Brown, R.; Kennel, M.B. Local Lyapunov exponents computed from observed data. *J. Nonlinear Sci.* **1992**, *2*, 343–365. [CrossRef]

69. Heslot, F.; Castaing, B.; Libchaber, A. Transitions to turbulence in helium gas. *Phys. Rev. A* **1987**, *36*, 5870–5873. [CrossRef]

70. Yamada, M.; Ohkitani, K. Lyapunov spectrum of a model of two-dimensional turbulence. *Phys. Rev. Lett.* **1988**, *60*, 983–986. [CrossRef]

71. Taylor, L.R. Aggregation, variance and mean. *Nature* **1961**, *189*, 732–735. [CrossRef]

72. Jørgensen, B.; Martinez, J.R.; Tsao, M. Asymptotic-behavior of the variance function. *Scand. J. Stat.* **1994**, *21*, 223–243.

73. Lecomte, J.-B.; Benoît, H.P.; Ancelet, S.; Etienne, M.-P.; Bel, L.; Parent, E. Compound Poisson-gamma vs. delta-gamma to handle zero-inflated continuous data under a variable sampling volume. *Methods Ecol. Evol.* **2013**, *4*, 1159–1166. [CrossRef]

74. Kendal, W.S.; Jørgensen, B. Tweedie convergence: A mathematical basis for Taylor's power law, $1/f$ noise, and multifractality. *Phys. Rev. E* **2011**, *84*, 066120. [CrossRef] [PubMed]

75. Brechtl, J.; Chen, S.Y.; Xie, X.; Ren, Y.; Qiao, J.W.; Liaw, P.K.; Zinkle, S.J. Towards a greater understanding of serrated flow in an Al-containing high-entropy-based alloy. *Int. J. Plast.* **2019**, *115*, 71–92. [CrossRef]

76. Xie, X.; Lo, Y.C.; Tong, Y.; Qiao, J.; Wang, G.; Ogata, S.; Qi, H.; Dahmen, K.A.; Gao, Y.; Liaw, P.K. Origin of serrated flow in bulk metallic glasses. *J. Mech. Phys. Solids* **2019**, *124*, 634–642. [CrossRef]

77. Brechtl, J.; Chen, B.; Xie, X.; Ren, Y.; Venable, J.D.; Liaw, P.K.; Zinkle, S.J. Entropy modeling on serrated flows in carburized steels. *Mater. Sci. Eng. A* **2019**, *753*, 135–145. [CrossRef]

Article

Effects of Manganese and Zirconium Dispersoids on Strain Localization in Aluminum Alloys

Elena Jover Carrasco [1], Juliette Chevy [2], Belen Davo [2] and Marc Fivel [1,*]

1   SIMaP, University Grenoble Alpes, CNRS, 38000 Grenoble, France; Elena.Jovercarrasco@grenoble-inp.fr
2   C-TEC Constellium Technology Center, Metallurgy Department, 38341 Voreppe CEDEX, France; Juliette.Chevy@constellium.com (J.C.); Belen.Davo@constellium.com (B.D.)
*   Correspondence: Marc.Fivel@grenoble-inp.fr; Tel.: +33-0476826336

Abstract: Strain localization in aluminum alloys can cause early failure of the material. Manganese and zirconium dispersoids, often present in aluminum alloys to control the grain size, have been found to be able to homogenize strain. To understand the effects of dispersoids on strain localization, a study of slip bands formed during tensile tests is carried out both experimentally and through simulations using interferometry and discrete dislocations dynamics. Simulations with various dispersoid size, volume fraction, and nature were carried out. The presence of dispersoids is proven to homogenize strain both is the experimental and numerical results.

Keywords: plasticity; slip bands; discrete dislocations dynamics; Al-Cu-Li alloys; Mn dispersoids; Zr dispersoids

check for **updates**

Citation: Jover Carrasco, E.; Chevy, J.; Davo, B.; Fivel, M. Effects of Manganese and Zirconium Dispersoids on Strain Localization in Aluminum Alloys. *Metals* **2021**, *11*, 200. https://doi.org/10.3390/met11020200

Academic Editor: Marcello Cabibbo
Received: 19 November 2020
Accepted: 15 January 2021
Published: 22 January 2021

Publisher's Note: MDPI stays neutral with regard to jurisdictional claims in published maps and institutional affiliations.

1. Introduction

The aerospace industry has largely benefited from the development of optimized aluminum alloys, which could maximize the performances of structural parts for minimum weight. Recently developed Al-Cu-Li alloys optimally meet these challenging aeronautic specifications requiring both improved strength and damage tolerance [1,2].

One important factor influencing damage tolerance is strain localization. In the case of cyclic loading, it has been shown that planar slip can be advantageous to prevent fatigue crack propagation, since it leads to more reversibility in the accumulated plastic strain [3]. On the other hand, in the case of monotonic loading, strain localization should be avoided, since the stress concentration induced at the grain boundaries could lead to premature failure. Strain localization is influenced by different microstructural parameters, such as hardening precipitation, solid solutions, grain size, and orientation (texture).

In Aluminum alloys, Li addition leads to the formation of the strengthening δ'-Al_3Li precipitates, while the addition of Cu together with Li leads to the additional formation of the more stable T_1-Al_2CuLi phase. Both phases are reported to be shear-able, but the shearing mechanism differs—leading to significant differences in the strain localization process [4]. These localization mechanisms and the resulting properties regarding of fatigue life and toughness have been widely studied [5,6].

Certain addition elements like Zr, Mn, Sc, or Cr are used to control the grain morphology. They precipitate into dispersoids, which consist of fine intermetallic particles [7]. They typically form during the homogenization process, and their size varies from 10 nm to 200 nm depending on their nature and the homogenization parameters [8]. The main purpose of the dispersoids is to pin the grain boundaries to control grain size and recrystallization of the material during the various stages of the alloy fabrication and to form, but they also play a major role in toughness and fatigue resistance as they influence strain localization through their interactions with dislocations [1,2,8,9]. This paper investigates the effect of these dispersoids on strain localization. In this work, we will focus on Mn and Zr dispersoids as they are the most commonly found in commercial Al-Cu-Li alloys.

161

Manganese dispersoids are present in aluminum alloys in the form of $Al_{20}Cu_2Mn_3$ ellipsoidal particles that can be up to 500 nm long by 100 nm large. Manganese being a eutectic, manganese dispersoids are concentrated at the grain boundaries [8]. In rolled samples, manganese dispersoids are mostly aligned and can act as nucleation sites for precipitates [10]. Manganese dispersoids are reported to spread and homogenize the plastic strain in aluminum alloys, increasing the toughness of said alloy [11]. Manganese dispersoids are incoherent with the matrix, which can lead to microvoid coalescence reducing the fracture toughness [11–13], and thus, counterbalancing the positive effect of manganese on the mechanical properties of the alloys.

Zr-containing dispersoids, in the from Al_3Zr, vary in number density within a given aluminum alloy, creating precipitate rich and precipitate free zones (PFZ) around the cast grain boundaries, due to zirconium's peritectic behavior in aluminum alloys. These dispersoids are usually sphere-shaped with a radius of about 20 nm [14]. Due to their small size, when the applied strain is increased, they are first by-passed by the dislocations forming Orowan loops around them, and they are later sheared when the applied strain/stress becomes high enough [15]. Zirconium effect on strain localization is less known. Due to the small size of the zirconium dispersoids, their effect on strain homogenization is less prominent than the manganese dispersoids [16].

Strain localization is associated with the formation of slip bands inside the material. The slip bands are shown by the slip traces printed on the surface of the specimen [17,18]. They are formed by dislocation activity in close slip planes. Due to its high stacking fault energy (150 mJ·m^{-2} [19]), the thermally activated cross-slip mechanism is very frequent in Aluminum. This mechanism augments the number of slip systems available to achieve large strains, which increases the ductility. Obviously, the cross-slip probability strongly affects the mechanisms of slip band multiplication. As an example, double cross-slip will give the possibility of dislocations to by-pass particles. It will also grow the slip bands by forming Koehler sources.

As said before, the formation of slip bands is controlled by many parameters related to the alloy composition. Texture and loading direction are other influential parameters. For aluminum alloys, texture components Brass, Copper, R, and S form more slip bands than Goss, Q, or P [20]. It has also been shown that the orientation of the tensile axis has an effect on the evolution of the slip bands [21]. AFM observation of Al single crystal has demonstrated that [001] uniaxial loading leads to a saturation in the slip band activity after a few percent of plastic strain. Inversely, [112] loading direction leads to continuously increasing localization in the slip bands.

This paper aims to elucidate the individual contribution of Mn and Zr dispersoids on the strain localization in Al-Cu-Li alloys using a combination of experiments and simulations. The experiments consist of tensile tests performed on Al-Cu-Li alloys specially designed to contain the investigated Zr and Mn dispersoids. The simulations are performed using 3D Discrete Dislocations Dynamics (DDD). DDD is indeed the ideal tool to address the complex dislocation/particle interaction mechanism [22], since it intrinsically contains the physics of the cross-slip mechanism as opposed to Finite Element of full field Fast Fourier Transform modeling, which could only take care of cross-slip via ad-hoc equations in the constitutive model. As an example, DDD simulations have successfully shown the role of particles in strain localization in fatigue [23].

The effect of each type of dispersoids, their size, and volume fraction on the strain localization will be investigated and compared to experiments. Both for the simulations and the experiments, the slip localization is measured from line profiles.

2. Materials and Methods

2.1. Experimental Procedure

Four alloys of the Al-Cu-Li alloy family in T8 condition were used in this study. The strain localization of "high Li" extruded products and "Low Li" rolled products was examined containing either only Zr dispersoids, or both Zr and Mn dispersoids, as indicated in Table 1. The microstructure was mainly unrecrystallized for all of them, and they presented typical deformation texture. Four samples were tested as described in the following table.

Table 1. Composition of the four samples investigated in the experiments.

Name	Zr	Mn	wt%			
	Dispersoids	Dispersoids	Li	Cu	Mn	Zr
Low Li, Zr-Mn	Yes	Yes	0.9	4.3	0.33	0.13
Low Li, Zr	Yes	No	0.9	4.3	-	0.14
High Li, Zr-Mn	Yes	Yes	1.6	3.0	0.31	0.14
High Li, Zr	Yes	No	1.6	3.0	-	0.14

In order to observe slip bands, sample surfaces were mirror-polished prior to their deformation to eliminate the sample's preexisting roughness and to ensure that the slip bands measured are independent of the surface state. First, 160 mm long aluminum samples were machined in a TOP 11 shape with a gauge length of 54 mm and a 12 mm × 2 mm section. The samples, coming from rolled products, were machined with the observation plane corresponding to the L-TL plane. This surface orientation was chosen due to the larger grain size in that particular plane, which visualizes longer slip bands.

To study the effect of dispersoids, other factors impacting the slip band formation need to be controlled. Textures found in rolled samples are limited, making it easier to find grains with similar orientations on different samples.

A tensile strain is then applied along the L direction of the samples at a displacement rate of 20 mm/min up to a total tensile strain of 4.3%. This strain was chosen to be high enough to induce visible slip bands at the free surface but low enough for the sample to not be overrun by slip bands.

Once the slip bands are visible, the samples are observed with differential interference contrast microscopy to qualitatively study the slip band organization and to select zones of interest to be further studied. As shown in Figure 1, slip bands are easily observed, and their direction and intensity depend on the grain orientation. These selected zones were first scanned in EBSD to determine the grain orientations. Once the orientation was known, those zones were mapped using an interferometer to quantify the roughness induced by the slip bands at the surface. Ten profiles perpendicular to the slip bands are then taken and analyzed in each grain.

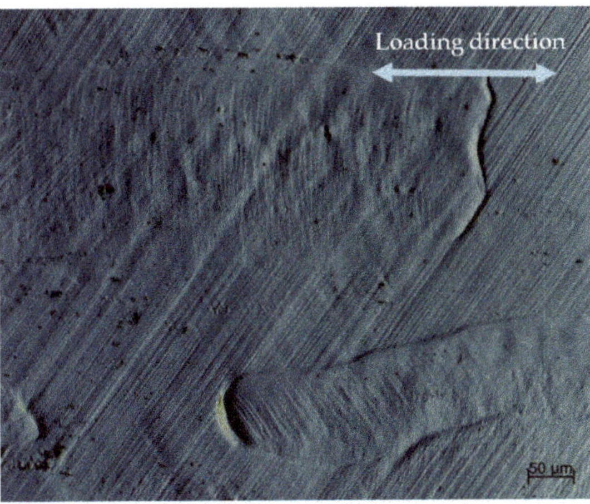

Figure 1. Typical slip bands formed in Aluminum alloy after 4.3% straining observed with differential interference contrast microscopy.

It is worth mentioning that the signal treatment of the interferometry results was kept to a minimum to ensure slip bands remained visible. Finally, the mean height of profile elements (P_c) is calculated together with the spacing between the bands P_{sm} (See Figure 2). The magnitude of the strain localization is finally quantified by the local shear, γ_{loc}, defined as the ratio P_c/P_{sm}.

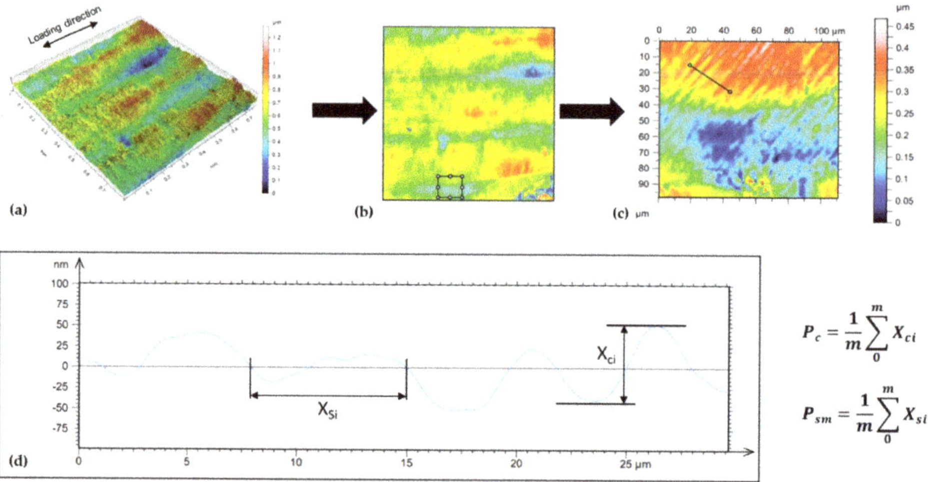

$$P_c = \frac{1}{m}\sum_{0}^{m} X_{ci}$$

$$P_{sm} = \frac{1}{m}\sum_{0}^{m} X_{si}$$

Figure 2. Interferometric mapping of a High-Li aluminum alloy. (**a**) 0.8 mm × 0.8 mm 3D surface mapping. (**b**) In plane projection of the surface heights. (**c**) zoom on a 100 µm × 100 µm area. (**d**) Typical line profile is taken perpendicular to the slip bands from which the quantities P_c and P_{sm} are evaluated.

2.2. Dislocation Dynamics Simulations

DDD simulations are performed using the edge-screw model TRIDIS [24] with the material parameters corresponding to Aluminum recalled in Table 2. Note that this code is precisely the one that was previously used to address dislocation-precipitate interactions in Reference [22], and applied to the case of fatigue of Waspaloys [23] and creep of Ni Superalloys [25].

Table 2. Discrete dislocations dynamics (DDD) simulation parameters.

Burgers Vector	Shear Modulus	Time Step	Cross-Slip Parameters		
b [Ang.]	G [GPa]	δt [s]	τ_{III} [MPa]	V [b³]	T [K]
2.86	32	10^{-10}	32	350	300

The DDD principle is detailed in Reference [24], but the main ingredient used in TRIDIS are recalled hereafter. Dislocation lines are discretized in piece-wise edge and screw segments. At each time step, the effective resolved shear stress is evaluated at the segment centers as the superposition of the applied stress and the internal stress induced by all the segments contained in the simulated volume. In the present simulations, the applied stress tensor is taken as homogeneous and corresponds to a pure tensile load.

The enforced boundary conditions aim at representing an isolated grain embedded in a much bigger polycrystalline sample. The grain boundaries consist of sets of facets that prevent the dislocations from escaping the simulated box. Here the simulated volume consists of a spherical grain of 5 µm of diameter approximated by a set of 24 facets. The particles are also geometrically defined by sets of triangular facets acting as impenetrable obstacles. The simulated volume contains distributed spherical particles representing zirconium dispersoids, and ellipsoidal particles representing manganese dispersoids, as shown in Figure 3.

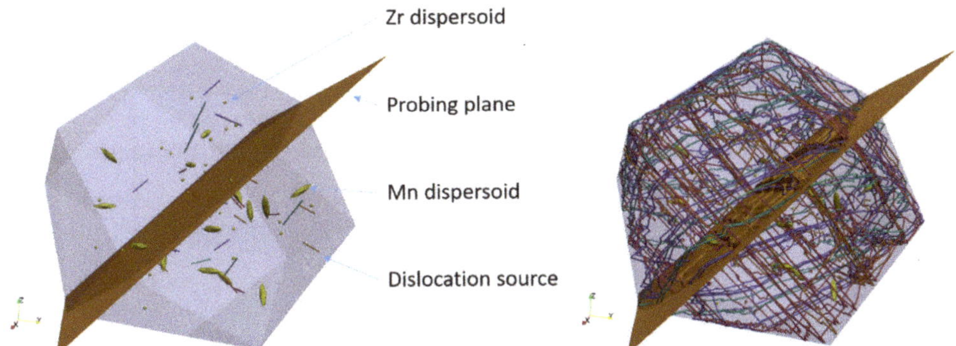

Figure 3. Initial and final configuration of a typical DDD simulation of a spherical volume of 5 µm diameter containing an assortment of small spherical zirconium dispersoids, elongated manganese dispersoids, and Frank Read dislocation sources. Strain localization will be estimated from the plastic steps printed in the probing plane inserted in the middle of the simulated volume.

The load is applied as a homogeneous tensile stress tensor applied along the $[-101]$ direction. The loading sequence is defined as a plastic strain rate monitoring mode at a rate $\dot{\varepsilon}_p = 80\ \mathrm{s}^{-1}$ along the tensile direction. Such a strain rate is low enough to allow the dislocations to reach an equilibrium position during the entire loading stage.

For each particle type, the dimension and volume fraction are varied. Configurations containing one type, both types, or no particles are systematically studied to understand the effect of each particle individually and together. To determine the effect of each particle, simulations are always carried out with the exact same configurations, i.e., with the same positions of the dispersoids and dislocation sources. Simulations with the same parameters but different configurations are used to determine the reproducibility of the results. The ranges of the investigated parameter values are presented in Table 3, together with the associated figures where the simulation results are presented.

Table 3. Range of the particle radii and volume fractions investigated in the DDD simulation campaign. The last column gives the figure where the results will be presented.

Dispersoid	Volumic Fraction	Size (µm)	Results
Zr	0.00003	0.02	Figure 6
Mn	0.00045	0.5 × 0.2	
Zr	0	0	
Zr	0.00003	0.025	Figure 7
Zr	0.00003	0.035	
Zr	0	0	
Zr	0.00002	0.025	Figure 8
Zr	0.00003	0.025	
Zr	0.00004	0.025	
Mn	0	0	
Mn	0.00045	0.5 × 0.2	Figure 9
Mn	0.00045	0.075 × 0.3	
Mn	0	0	
Mn	0.0003	0.5 × 0.2	
Mn	0.00045	0.5 × 0.2	Figure 10
Mn	0.0006	0.5 × 0.2	

Particles are placed randomly in the volume, while avoiding overlapping each other. As for the manganese dispersoids, their orientation is randomly selected. While the random orientation does not represent what is usually found in aluminum alloys, it allows the results to be independent of the particle orientation.

Each simulation is run for about 15,000 steps to ensure that the plastic strain cumulated along the [−101] loading direction is larger than 10^{-4}. Analyses are all performed for a cumulated tensile strain of 0.01%. Since the amount of cumulated plastic strain simulated by DDD is much smaller than in the experiments (a few percent), the absolute values of the local shear cannot be compared. However, we think that the slip localizations found by DDD are the premises of the slip bands that will further develop for higher strains. Thus, DDD can help in the understanding of the strain localization mechanism.

The applied stress direction is [−101], which simultaneously activates the dislocations from the slip systems C1, C5, A2 and A6 (See Table 4 for the slip system nomenclature). Those four slip systems are gliding on two different slip planes (A and C), which is evidenced at the probing plane by the two families of slip traces (see Figure 4).

Table 4. Slip system nomenclature following Schmid and Boas notation [26]. The letter indicates the slip plane normal, and the number corresponds to the Burgers vector direction.

Vector	C1	C5	A2	A6
normal	(−1−11)	(11−1)	(−111)	(−111)
Burgers	[011]	[1−10]	[0−11]	[110]

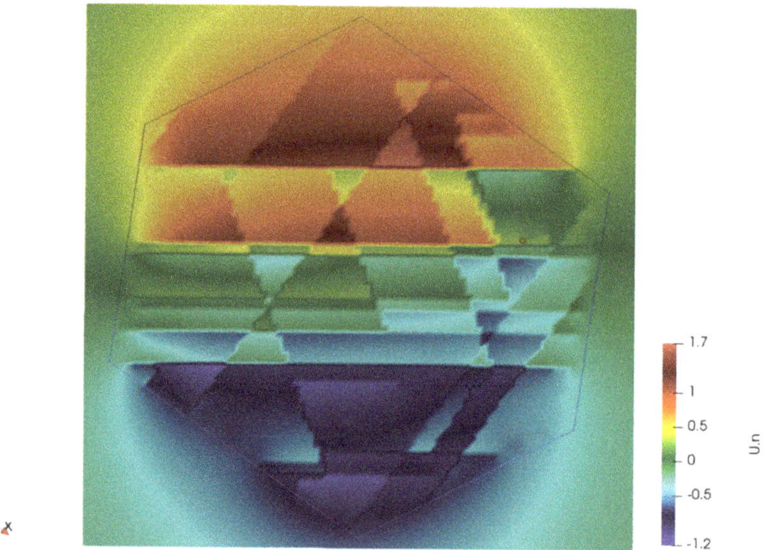

Figure 4. Typical plastic steps printed on the 5 μm × 5 μm probing plane defined in Figure 3 after 10^{-4} cumulated plastic tensile strain obtained by a DDD simulation. Scale is in nanometers. Colors correspond to the projection of the dislocation displacement field along the plane normal [0−11].

Initially, all simulations contain the same number and type of dislocations, but those are placed in the volume randomly according to the particle position to avoid overlap. There are 24 dislocations initially introduced in the volume, six dislocations per each of the four active slip systems C1, C5, A6, and A2. These dislocations are 260 nm long, and their initial direction and type are randomly selected. A typical DDD simulation lasts 2.5 h on a single Intel Xeon E5405 processor with eight threads. Once the simulations are completed, a 5 μm squared (0–11) calculation plane is inserted in the volume (see Figure 3) to visualize the spreading of the plastic strain inside the grain. The calculation plane is discretized into a grid of 100×100 points.

As shown in Figure 4, the plastic steps printed by the dislocations crossing the probing plane are then calculated using the optimized formula of the dislocation's displacement field from Fivel and Déprés [27]. Finally, five line profiles crossing the slip traces are extracted from the surface relief to quantify the magnitude of the strain localization. The local shear is then estimated at each grid point as the derivative of the profile height with the coordinate along with the line profile.

3. Results and Discussion

3.1. Experimental Results

Interferometric mapping results were post-treated, and ten surface profiles were extracted for each studied grain. For each profile, the mean height of the profile elements, Pc, is calculated. Pc represents the height of the slip band extrusions. The higher P_c is, the more the strain is concentrated on said bands. For each sample, the measurements were performed on a Brass oriented grain to avoid introducing an additional effect, due to the activation of different slip systems.

Results are pictured in Figure 5. For both alloys, samples containing both Zr and Mn dispersoids have lower P_c values, showing that the simultaneous presence of both dispersoids tends to homogenize the plastic strain. With the addition of Mn dispersoids, the extrusion height is reduced by 14.7% for Li-Zr alloys and 18.8% for Li-Zr alloys showing that Mn is beneficial to reduce slip activity in slip bands.

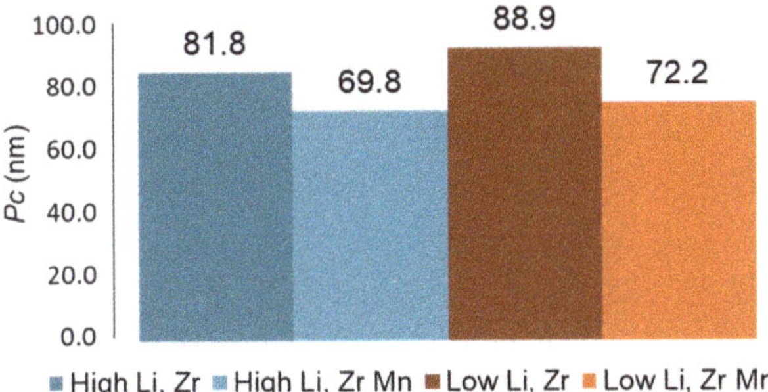

Figure 5. Average slip band extrusion height on slip bands measured for two aluminum alloys with two configurations of dispersoids on brass grains using interferometry.

Unexpectedly, it is also observed that the high lithium alloys have a slightly lower strain localization magnitude than the low lithium alloys (8% reduction for Li-Zr and 3.7% for Li-Zr-Mn). As discussed before, δ′ (Al_3Li) precipitates are supposed to promote strain localization so that the measured effect might come from either a precision error, or from the fact that we study a single specific orientation, for which T_1 (Al_2CuLi) precipitation might be significantly sheared. Other material parameters, including the effect of Cu and Li content in solid solution, are not taken into account in this study and could also impact strain localization. It can be mentioned that other grain orientations can lead to the activation of different slip systems, and thus, different slip patterns.

Table 5 gives the measurements of the local shear (defined as Pc/Psm) in the case of High Lithium alloys. Here, again, it is found that the presence of Mn dispersoids reduces the strain localization by 28%.

Table 5. Local shear is estimated from the experimental measurements.

Measured Quantities	No Mn	With Mn
Psm (μm)	5.3	6.3
Standard Deviation	**1.6**	**3.3**
Pc (μm)	81.8	69.8
Standard Deviation	**18.6**	**26.3**
Local shear = *Pc/Psm*	15.4	11.1

3.2. Simulation Results

As for the simulations, three simulations are first conducted to verify if the experimental observations could be reproduced by DDD. A first simulation containing both Zr and Mn dispersoids is performed. Then a second simulation is run with only the Zr dispersoids and finally, a third one is run with only Mn dispersoids. Size and volume fractions are given in Table 3.

Five line profiles taken along direction [0−11] are then extracted from the probing plane. The profile direction is chosen so that it is parallel to the slip traces of systems C1 and C5 so that it will give an estimate of the dislocation activity on systems A2 and A6, similarly to the experimental procedure where the line profile was perpendicular to a given slip band orientation.

The histograms of the local shear are plotted in Figure 6 together with the cumulative percentages, which better indicate the average values. The simulation with only Zr dispersoids admits an average local shear of γ_{loc} = 0.001. The local shear measured from the simulation with only Mn dispersoids is decreased by 30%: 0.0007, and when both particles are present, the local shear reduces to γ_{loc} = 0.0002 which is 5 times less than the case of solely Zr dispersoids. It can be concluded that the combined presence of Mn and Zr dispersoids is the most efficient manner to reduce the strain localization, which is consistent with the experiments presented in the previous section.

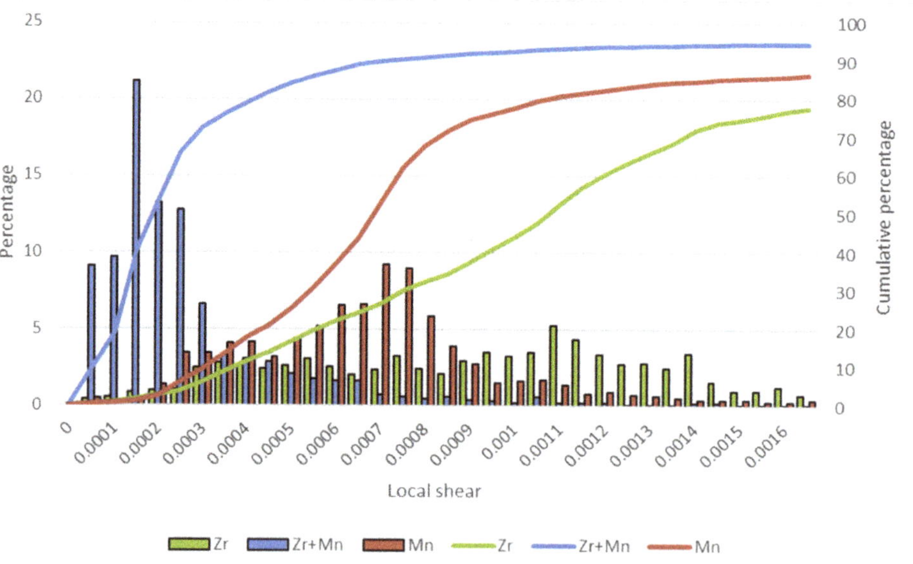

Figure 6. Local shear distribution from DDD simulations at ε = 0.01% for three comparable simulations to identify the effect of each dispersoid type. For each of the three simulations, the position of the dispersoids and Frank–Read sources are kept the same.

In order to identify the effect of the volume fractions and the size of each dispersoid type, a DDD simulation campaign is now conducted. To access more data, the post-treatment of the surface relief is slightly modified: The direction of the line profiles is chosen so that it can intercept both slip traces. Figure 7 shows the effect of the size of the Zr dispersoids for two values of the sphere particle radius: 25 nm and 35 nm and for the given volume fraction fv = 0.00003. The case of a simulated volume without any dispersoids is also plotted for reference (labeled as 0 In Figure 7). It is shown that the smaller particle size (R = 25 nm) has nearly no effect on the local shear distribution. On the other hand, the bigger particles induce a decrease of the local shear by a factor 2. It can be concluded that there is a minimum particle size that can reduce strain localization. This could be explained by the role of the cross slip, which can be triggered on the Orowan dislocation loops deposited around the particles. When the particles are bigger, the curvature of the dislocation loops is bigger, and the cross-slip probability is increased. The dislocations emitted from the Frank–Read sources can then populate new slip planes that homogenize the plastic strain.

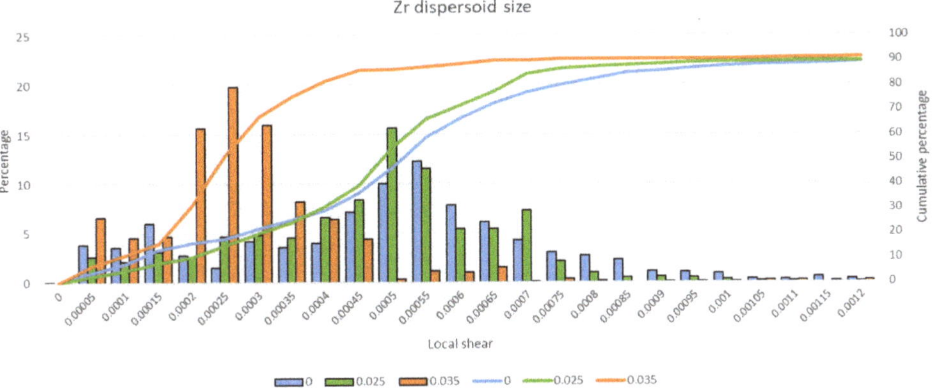

Figure 7. Local shear distribution from DDD simulations showing the isolated effect of Zr dispersoid size (fv = 0.003%). Radii are in μm. Label 0 refers to a simulation without any particle.

In Figure 8, the size of the Zr dispersoid is fixed to 25 nm, and the volume fraction is now increased to verify if there is a possibility to modify the local shear when the number of small Zr particles becomes high enough. It can be seen that for this small size of the Zr particle, changing the volume fraction from 0.002% to 0.004% has no effect on the slip dispersion. It is concluded again that the particle radius is too small to have an impact on the dislocation activity.

The same investigations are now conducted in the case of the Mn dispersoids, without the presence of Zr. Figure 9 shows the effect of the particle sizes on the local shear distribution for a fixed volume fraction fv = 0.045%. It is shown that increasing the particle size from 200 nm × 50 nm to 300 nm × 75 nm does not affect the strain localization. Looking at the simulation details, we could observe that the Orowan loops deposited around the particles are difficult to unpin because of the large shape factor of the Mn particles. When cross slip happens, the cross-slip segments are immediately in contact with the dispersoid and cannot escape. This locking phenomenon is enhanced when the size of the dispersoids is increased. In other words, the Mn dispersoids play the role of forest obstacles for the dislocation motion so that the strain is spread in the volume.

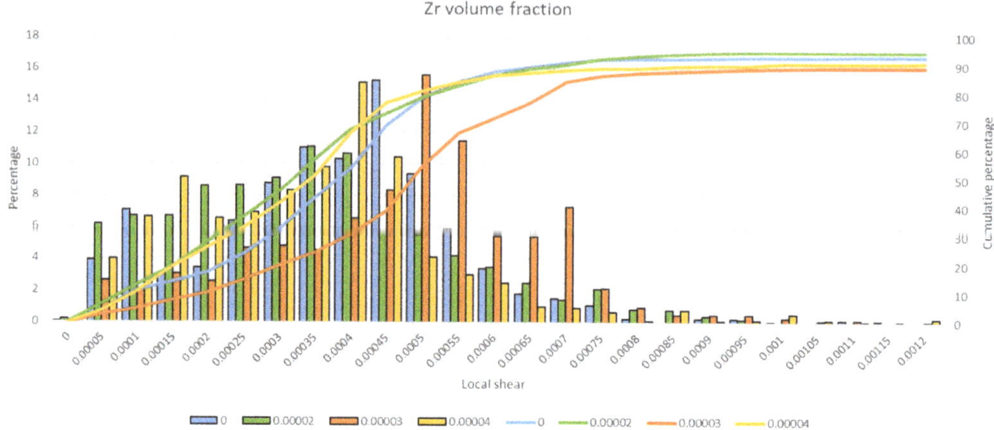

Figure 8. Local shear distributions from DDD simulations, showing the effect of Zr dispersoid volume fraction for particle size R = 25 nm. Label 0 refers to a simulation without any particle.

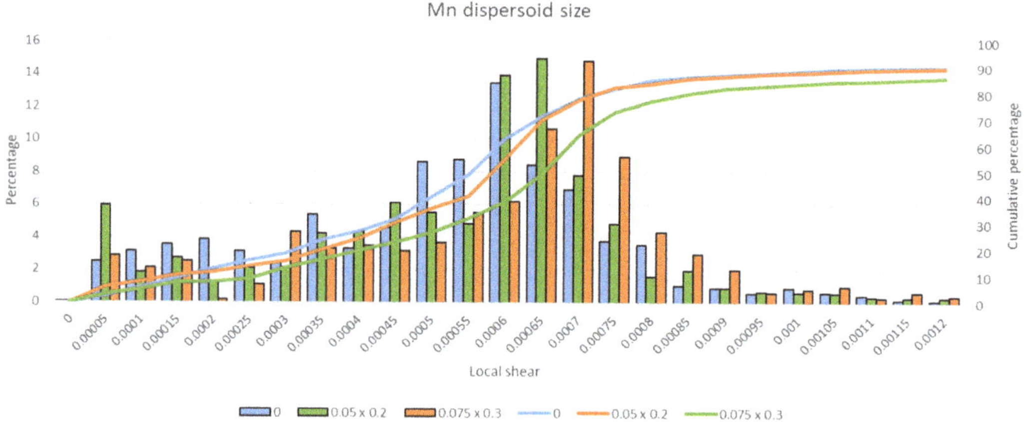

Figure 9. DDD simulation results from Mn dispersoid size on the local for fv = 0.0045%. Sizes are in μm. Label 0 refers to a simulation without any particle.

Finally, the volume fraction of Mn dispersoids is increased with a particle size fixed to 200 nm × 50 nm. Results are given in Figure 10 where it is shown that all the investigated volume fractions have a similar positive effect on the strain localization. One possible explanation is that when particles are more numerous, the decrease of strain localization induced by cross-slip and by-passing the particles is counterbalanced by the fact that the possible paths for the dislocations to move between the particles is limited in number, thus leading to strain localization within these paths.

Although the geometry of the Zr and Mn dispersoids is largely different, their impact on slip band behavior is similar. As mentioned beforehand, both dispersoids have a similar effect on the dislocation activity: The bigger they are, the more they can spread the plastic strain. The elongated shape of the Mn dispersoids is locking the by-passing mechanism by cross-slip. For zirconium spherical dispersoids, they can easily be looped by dislocations and promote cross-slip for the bigger sizes, which also leads to strain dispersion.

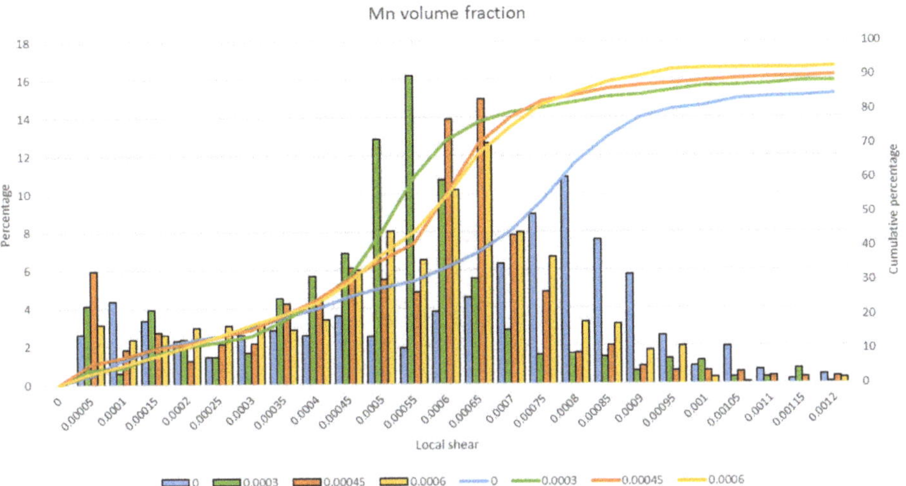

Figure 10. DDD simulation results from the volume fraction of Mn dispersoids on the local shear for a size 200 nm × 50 nm for the ellipsoid particles. Label 0 refers to a simulation without any particle.

4. Conclusions

In aluminum alloys, dispersoids affect the strain localization; the combined presence of Mn and Zr dispersoids in Al-Cu-Li alloys reduces the plastic strain amplitude in the slip bands by 28%. Discrete dislocations dynamics simulations show the same trend as in the experiments. It is also found that there exists a critical size for the Zr dispersoids that can lead to a decrease in the strain localization. Because of their aspect ratio, Mn dispersoids always tend to homogenize the plastic strain, thus reducing the strain localization. The best results should be achieved when both zirconium and manganese dispersoids are present in the alloy. The effect of the Mn volume fraction is negligible, but large Zr particles (R > 35 nm) are recommended.

Author Contributions: Conceptualization, E.J.C., M.F., J.C. and B.D.; methodology, J.C. and B.D.; software, E.J.C. and M.F.; validation, J.C.; formal analysis, E.J.C., M.F., J.C. and B.D.; investigation, E.J.C.; resources, J.C. and B.D.; data curation, E.J.C.; writing—original draft preparation, E.J.C.; writing—review and editing, M.F.; visualization, E.J.C.; supervision, M.F.; project administration, M.F., J.C. and B.D.; funding acquisition, J.C. and B.D. All authors have read and agreed to the published version of the manuscript.

Funding: This research received no external funding.

Institutional Review Board Statement: Not applicable.

Informed Consent Statement: Informed consent was obtained from all subjects involved in the study.

Data Availability Statement: The data presented in this study are available on request from the corresponding author. The data are not publicly available due to confidential issue from the industrial partner.

Conflicts of Interest: The authors declare no conflict of interest.

References

1. Dorin, T.; Vahid, A.; Lamb, J. *Fundamentals of Aluminum Metallurgy*, 1st ed.; Lumley, R.N., Ed.; Woodhead Publishing series in Metals and Surface Engineering: Cambridge, UK, 2018; Chapter 11 Aluminum Lithium Alloys; pp. 387–438, ISBN 978-1-84569-654-2.
2. Prasad, E.E.; Gokhale, A.; Wanhill, R.J.H. *Aluminum-Lithium Alloys: Processing, Properties, and Applications*; Eswara, N., Gokhale, A.A., Wanhill, R.J.H., Eds.; Elsevier BH: Oxford, UK, 2014; ISBN 978-0-12-401698-9.
3. Srivatsan, T.; Coynejr, E. Cyclic Stress Response and Deformation Behaviour of Precipitation-Hardened Aluminium-Lithium Alloys. *Int. J. Fatigue* **1986**, *8*, 201–208. [CrossRef]
4. Deschamps, A.; Decreus, B.; De Geuser, F.; Dorin, T.; Weyland, M. The Influence of Precipitation on Plastic Deformation of Al-Cu-Li Alloys. *Acta Mater.* **2013**, *61*, 4010–4021. [CrossRef]
5. Starke, E.A.; Sanders, T.H.; Palmer, I.G. New Approaches to Alloy Development in the Al-Li System. *JOM* **1981**, *33*, 24–33. [CrossRef]
6. Lavernia, E.J.; Grant, N.J. Aluminium-Lithium Alloys. *J. Mater. Sci.* **1987**, *22*, 1521–1529. [CrossRef]
7. Sainfort, P.; Dubost, B. Coprecipitation Hardening in Al-Li-Cu-Mg Alloys. *Le J. Phys. Colloq.* **1987**, *48*, C3-407–C3-413. [CrossRef]
8. Starke, E.; Staley, J. Application of Modern Aluminum Alloys to Aircraft. *Prog. Aerosp. Sci.* **1996**, *32*, 131–172. [CrossRef]
9. Robson, J.D.; Prangnell, P. Dispersoid Precipitation and Process Modelling in Zirconium Containing Commercial Aluminium Alloys. *Acta Mater.* **2001**, *49*, 599–613. [CrossRef]
10. Dumont, D.; Deschamps, A.; Brechet, Y. On the Relationship between Microstructure, Strength and Toughness in AA7050 Aluminum Alloy. *Mater. Sci. Eng. A* **2003**, *356*, 326–336. [CrossRef]
11. Uguz, A.; Martin, J. The Effect of Different Dispersoids in the Ductile Fracture Toughness of an Al-Zn6Mg Alloy. In Proceedings of the 7th International Conference on Fracture, Houston, TX, USA, 20–24 March 1989.
12. Tocci, M.; Donnini, R.; Angella, G.; Pola, A. Effect of Cr and Mn Addition and Heat Treatment on AlSi3Mg Casting Alloy. *Mater. Charact.* **2017**, *123*, 75–82. [CrossRef]
13. Walsh, J.A.; Jata, K.V.; Starke, E.A. The Influence of Mndispersoid Content and Stress state on Ductile Fracture of 2134 Type Al Alloys. *Acta Metall.* **1989**, *37*, 2861–2871. [CrossRef]
14. Weidner, A.; Beyer, R.; Blochwitz, C.; Holste, C.; Schwab, A.; Tirschler, W. Slip Activity of Persistent Slip Bands in Polycrystalline Nickel. *Mater. Sci. Eng. A* **2006**, 540–546. [CrossRef]
15. Rao, K.T.V.; Ritchie, R.O. Fatigue of Aluminium—Lithium Alloys. *Int. Mater. Rev.* **1992**, *37*, 153–186. [CrossRef]
16. Nakai, M.; Eto, T. New Aspect of Development of High Strength Aluminum Alloys for Aerospace Applications. *Mater. Sci. Eng. A* **2000**, *285*, 62–68. [CrossRef]
17. Friedel, J. *Dislocations*; Pergamon Press LTD.: London, UK, 1964; ISBN 978-0-08-013523-6.
18. Lee, D.H.; Kim, K.C.; Park, D.S.; Nam, S.W. Enhancement of the Fatigue Properties in a Weldable High-Strength Al-Zn-Mg-Mn Alloy by Means of Mn-dispersoids. *Int. J. Fatigue* **1999**, *21*, 383–391. [CrossRef]
19. Wu, X.; Wang, R.; Wang, S.-F.; Wei, Q.-Y. Ab Initio Calculations of Generalized-Stacking-Fault Energy Surfaces and Surface Energies for FCC Metals. *Appl. Surf. Sci.* **2010**, *256*, 6345–6349. [CrossRef]
20. Li, F.; Liu, Z.; Wu, W.; Zhao, Q.; Zhou, Y.; Bai, S.; Wang, X.; Fan, G. Slip Band Formation in Plastic Deformation Zone at Crack Tip in Fatigue Stage II of 2xxx Aluminum Alloys. *Int. J. Fatigue* **2016**, *91*, 68–78. [CrossRef]
21. Kramer, D.; Savage, M.; Levine, L.E. AFM Observations of Slip Band Development in Al Single Crystals. *Acta Mater.* **2005**, *53*, 4655–4664. [CrossRef]
22. Shin, C.; Fivel, M.; Kim, W.W. Three-Dimensional Computation of the Interaction between a Straight Dislocation Line and a Particle. *Model. Simul. Mater. Sci. Eng.* **2005**, *13*, 1163–1173. [CrossRef]
23. Shin, C.S.; Robertson, C.F.; Fivel, M. Fatigue in Precipitation Hardened Materials: A Three-Dimensional Discrete Dislocation Dynamics Modelling of the Early Cycles. *Philos. Mag.* **2007**, *87*, 3657–3669. [CrossRef]
24. Verdier, M.; Fivel, M.; Groma, I. Mesoscopic Scale Simulation of Dislocation Dynamics in Fcc Metals: Principles and Applications. *Model. Simul. Mater. Sci. Eng.* **1998**, *6*, 755–770. [CrossRef]
25. Chang, H.-J.; Fivel, M.C.; Strudel, J.-L. Micromechanics of Primary Creep in Ni Base Superalloys. *Int. J. Plast.* **2018**, *108*, 21–39. [CrossRef]
26. Schmid, E.; Boas, W. *Kristallplastizität*; Springer: Berlin, Germany, 1936; ISBN 978-3662342619.
27. Fivel, M.; Déprés, C. An Easy Implementation of Displacement Calculations in 3D Discrete Dislocation Dynamics Codes. *Philos. Mag.* **2014**, *94*, 3206–3214. [CrossRef]

Article

Understanding the Interdependence of Penetration Depth and Deformation on Nanoindentation of Nanoporous Silver

Yannick Champion [1,*], Mathilde Laurent-Brocq [2], Pierre Lhuissier [1], Frédéric Charlot [3], Alberto Moreira Jorge Junior [1,4,5] and Daria Barsuk [1]

[1] Univ. Grenoble Alpes, CNRS, *SIMaP*, 38000 Grenoble, France; pierre.lhuissier@simap.grenoble-inp.fr (P.L.); moreira@ufscar.br (A.M.J.J.); dashaadak91@gmail.com (D.B.)

[2] Institut de Chimie et des Matériaux Paris-Est, CNRS-UPEC, 2 rue Henri Dunant 94320 Thiais, CEDEX, France; laurent-brocq@icmpe.cnrs.fr

[3] Univ. Grenoble Alpes, *CMTC*, 38000 Grenoble, France; frederic.charlot@grenoble-inp.fr

[4] Univ. Grenoble Alpes, CNRS, *LEPMI*, 38000 Grenoble, France

[5] Federal University of São Carlos, DEMa, São Paulo 13565-905, Brazil

* Correspondence: yannick.champion@simap.grenoble-inp.fr; Tel.: +33-476-826-749

Received: 31 October 2019; Accepted: 11 December 2019; Published: 14 December 2019

Abstract: A silver-based nanoporous material was produced by dealloying (selective chemical etching) of an $Ag_{38.75}Cu_{38.75}Si_{22.5}$ crystalline alloy. Composed of connected ligaments, this material was imaged using a scanning electron microscope (SEM) and focused ion-beam (FIB) scanning electron microscope tomography. Its mechanical behavior was evaluated using nanoindentation and found to be heterogeneous, with density variation over a length scale of a few tens of nanometers, similar to the indent size. This technique proved relevant to the investigation of a material's mechanical strength, as well as to how its behavior related to the material's microstructure. The hardness is recorded as a function of the indent depth and a phenomenological description based on strain gradient and densification kinetic was proposed to describe the resultant depth dependence.

Keywords: nanoporous metallic alloy; nanoindentation; hardness; strain gradient

1. Introduction

Metallic nanoporous (NP) materials present interesting perspectives owing to their nanometric-size microstructure and large specific surface (extremely high open-porosity level). From the first feature, mechanical strength, unique plasticity, and multiphysical properties are expected [1–5]. The second is related to surface interactions and chemical reactivity, with perspectives in domains such as electrochemical energy storage and conversion, catalysts [6,7], biomedical implants [8], and filtering and purification [9]. Based on its design and fabrication, this novel material family has been the subject of intensive fundamental research [10–12], as well as characterization at various levels of microstructure, topology, and morphology [13–16].

In this article, we report on the mechanical behavior of silver nanoporous materials via the analysis of instrumented nanoindentation measurements. Materials were produced by dealloying (selective chemical etching) of the $Ag_{38.75}Cu_{38.75}Si_{22.5}$ crystalline alloy, initially studied for its electrocatalytic properties [17]. Usually, indentation testing provides a preliminary insight into the mechanical properties of materials (strength, ductile/fragile character, toughness) [18,19]. For crystalline solids, the strength σ_s is simply derived from the Tabor rule for metallic alloys: $H \approx 3\sigma_s$ where H is the Vickers hardness.

The specific feature of indentation investigated in this paper was the dependence between the hardness measured and the indent size [20]. Such dependence can be explained using the Ashby rule: a non-symmetric plastic deformation (as in an indentation where deformation is confined) generates a strain gradient in the deformation zone. For crystalline solids, the strain gradient leads to extra hardening, which Ashby described in terms of geometrically necessary dislocations (GND) [21]. Using this approach, Nix and Gao showed that the depth-dependence hardness for crystalline and ductile solids follows the rule: $H/H_0 = \sqrt{1 + h^*/h}$ where h^* is a material constant and H_0 the hardness at a hypothetical infinite indent size [22].

For non-crystalline solids, the rule is obviously different since dislocations are nonexistent. In metallic glasses, deformation is localized in thin shear bands leading to the absence of macroscopic plasticity when they are tested using tension or compression. However, plasticity can be observed in the indentation owing to the confined character of the testing. Similarly with crystalline solids, a depth dependence of the hardness was observed but with a slightly different rule: $H/H_0 = 1 + \sqrt{h^*/h}$ [23].

Whatever the material and defects involved in the plastic deformation, it was observed that the Ashby rule on strain gradient is generalized. As far as a strain gradient is defined, size dependence can be analytically expressed as this was demonstrated for metallic glasses [24]. Here we report on depth dependence in the nanohardness testing of nanoporous silver, with the rule that follows: $H/H_0 = 1 + h^*/h$. A phenomenological description of the nanohardness variation with regard to indent size was proposed, fitting the experimental observations. The modeling was based on the strain gradient induced by local densification, which was related to ligament rearrangement. This analysis provided a proper evaluation of the mechanical strength of nanoporous metal, as well as giving an insight into material dynamical evolution during deformation.

2. Experiments

2.1. Formation of Nanoporous Materials

Ag-based polycrystalline alloy with the composition $Ag_{38.75}Cu_{38.75}Si_{22.5}$ was prepared by arc melting pure Ag, Cu, and Si (99.99% purity, Alfa Aesar, Kandel Germany) in a helium atmosphere. Five successive melting steps were employed to ensure alloy homogeneity, and then rapid solidification casting of the alloy was performed on a rotating copper wheel. Foils with thicknesses ranging from 20 to 60 μm were produced by varying the rotating speed of the copper wheel. As reported in [17], the cast ribbon analyzed by X-ray diffraction was constituted by fcc Ag(Si) solid solution and hexagonal Cu_3Si. Dealloying of the as-cast ribbons was performed at ambient temperature using two chemical etchants [25]: 13.4 wt% 2.12 M of HNO3, and 0.67 M of HNO3 and 0.64 M of HF in deionized water. The second etchant was used to accelerate the dissolution of Si atoms, resulting in a more regular microstructure and composition of the matrix, minimizing the content of residual Si and Cu elements. Full details of the materials' synthesis and characterization are reported in [17].

The materials were characterized using SEM (Zeiss Ultra 55, Oberkochen, Germany) complemented by tomographic imaging, using focused ion-beam slice cutting and image reconstruction (Carl ZEISS Cross Beam NVision FIB, Ga ion with an acceleration voltage of 30 kV and current intensity of 300 pA). After each etching, imaging of the freshly exposed material cross-section was recorded using the SEM. The stack of images was processed to generate the dissected volume of interest, producing a three-dimensional representation of the alloy's microstructure. To capture the nanostructure with an average length scale of about 50 nm, slices of 5 nm were cut (voxel of $5 \times 5 \times 5$ nm^3) and the thickness analyzed was of the order of 250 nm.

Reconstruction of the 3D images were performed using ImageJ/Fiji (NIH, Bethesda, MD, USA) and Avizo software (Thermo Scientific, Waltham, MA, USA) as well as a homemade plugin [26]. Nanoindentations were performed using a HysitronTI950 (Bruker, Eden Prairie, MN, USA) with a Berkovitch diamond tip. The surface topography and indents were imaged using the scanning-probe microscopy mode of the nanoindenter, which consisted of scanning the sample surface with the nanoindenter tip. Focused ion-beam (FIB) cutting of an indented zone of the material was also performed for detailed analysis and modeling.

First, quasistatic measurements, with a maximum load of 900 and 10,000 μN, a loading rate of 100 μN/s, and a holding time of 2 s, were performed. Second, dynamic mechanical analysis (DMA), which is also known as the continuous stiffness measurement (CSM), was used to continuously measure hardness and the Young modulus as a function of depth. A constant strain-rate loading of 0.15 s^{-1} was applied—up to a maximum load of 10,000 μN—and was maintained for 2 s, after which unloading at a constant rate was then performed. The oscillating frequency was set to 220 Hz.

In this study, two nanoporous materials NP1 and NP2 were compared. They were dealloyed for 45 and 180 min, respectively, and showed different structures in terms of density, ligament size, and connectivity. The average size and diameter of ligaments was estimated from a series of measurements using the Straight Line function of ImageJ and statistical analyses. The apparent porosity level was estimated from the surface of the SEM image after thresholding and evaluation of a fraction of the white pixels present. With these materials, multiple nanoindentation experiments at various positions on the samples' surface were carried out and averaged. Finally, the hardness and Young's modulus were determined using the Oliver–Pharr method [27].

2.2. Nanoporous Structure

Comparison between nanoporous metals NP1 and NP2 (Figure 1) was made in an attempt to reveal structural characteristics involved in nanoindentation size effect. This approach was consistent, as the materials showed different sizes in their microstructure but similar microstructural characteristics (morphology and topology). From a nanoporous formation, where coarsening is controlled by the surface diffusion, self-similarity (scale independent microstructure) is usually expected, although the existence of phenomenon is still subject to debate. Here the self-similarity is the ability of the material to coarsen during dealloying while maintaining its structural characteristics. This is seldom observed, however, because the time scale of the experiments usually places them in a transient domain of the process [16].

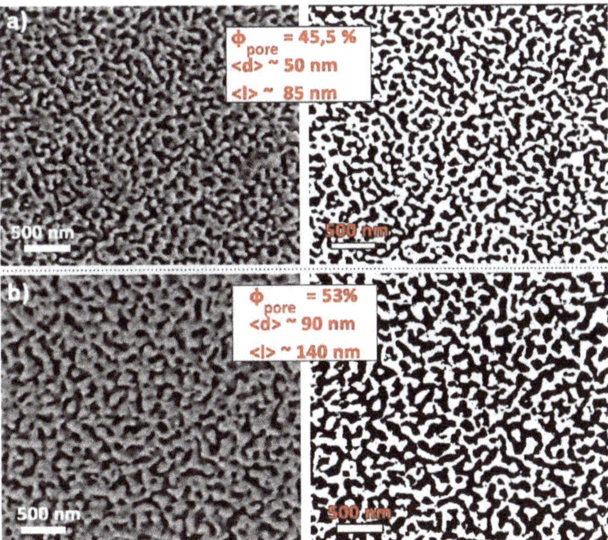

Figure 1. SEM (scanning electron microscope) secondary electron images of a nanoporous (NP) material and corresponding threshold patterns as a binary generated image: (**a**) NP1, dealloying time of 45 min and (**b**) NP2, dealloying time of 3 h.

A FIB tomographic image of a nanoporous silver similar to NP1 and NP2 showed the microstructural complexity in terms of morphological and topological characteristics. This type of analysis provided us with the local surface shape (Figure 2a) and curvedness level (Figure 2b), and from them a surface shape–curvedness statistical distribution (Figure 2c). This was a useful structural function of the material for properties analysis (shape index and curvedness are calculated from the principal curvatures κ_1 and κ_2)

$$S = \frac{2}{\pi}\arctan\frac{\kappa_1 + \kappa_2}{\kappa_1 - \kappa_2} \text{ and } C = \sqrt{\frac{\kappa_1^2 + \kappa_2^2}{2}}$$

These parameters were of particular interest in the understanding and modeling of an NP material's properties. For example, its curvature ended up affecting its mechanical properties through the Laplace law and chemical properties through the Gibbs-Thomson law, with a dominant effect at the nanoscale level. This complexity makes it rather difficult to prove a real similarity between different specimens.

However, to help in our work, simple visual comparison was made between NP1 and a rescaled NP2, though differences in porosity did not allow for full adaptation. The scaling factor used was the ratio between different porosities, and the superimposition in Figure 3 makes a convincing case for near-similarity, at least on the 2D SEM images.

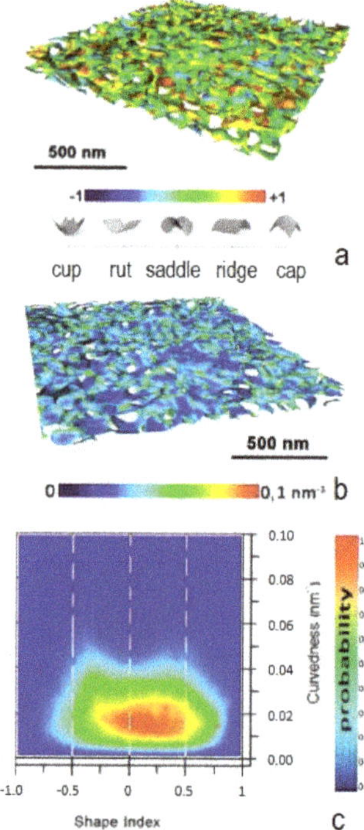

Figure 2. Tomographic perspective views of the nanoporous silver via focused ion-beam (FIB) scanning electron microscope slice cutting and computer image reconstruction. Color mapping surface according to its classification of shape index (**a**); and curvedness (**b**); (**c**) 2D view of the surface shape and curvedness distributions.

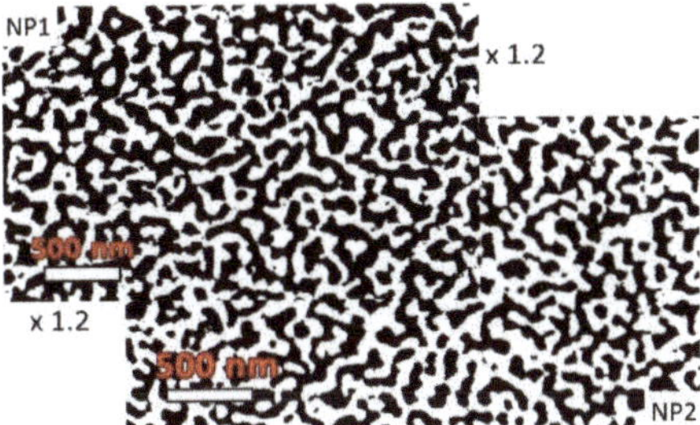

Figure 3. Visual evaluation of self-similarity between NP1 and NP2 by superimposition after rescaling NP1 by a factor of 1.2.

2.3. Nanoindentation

With a probe (indentation tip) size in the order of the length scale of the cellular microstructure, interpretations and quantitative analyses from the nanoindentation of nanoporous materials were challenging and subject to debate. Preliminary studies were undertaken to evaluate these length–scale issues. A series of more than 50 nanoindentations in quasistatic condition were performed on NP1 and NP2. A selection of 4 different indentations taken from the NP2 series and compared with a bulk silver sample were as seen in Figure 4. Every loading curve (see the inset of Figure 4) exhibited 2 successive extended bumpy regions (diffuse serrations).

The first region had a penetration depth of ~30 nm, which was probably not relevant and likely due to the Berkovitch tip edge imperfections (roughness, irregularities). The second one extended from about 50 to 150 nm, which was of the order of the NP microstructure length scale and over which the tip was alternately probing pores and ligaments. Despite fluctuations in the loading curves, it is interesting to observe from Figure 5, that an indent was already clearly formed at an approximate depth of 70 nm in the bumpy region, and likely useable throughout our statistical series analyses.

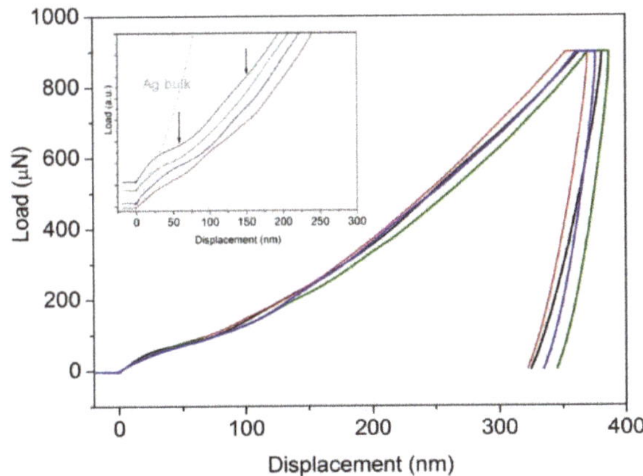

Figure 4. Selection of indentation curves on various positions of the sample of NP2. The inset shows the details of the diffuse serrations at an early stage of deformation and its comparison with bulk silver.

As usual, an apparent (size-effect affected) hardness was drawn from the maximum load $F_{\max}$ and depth h using $H = F_{\max}/24.5 \times h^2$, averaged over the indentation series for NP1 and NP2, respectively, and thus giving $H_{NP1} = 250 \pm 2$ MPa and $H_{NP2} = 240 \pm 10$ MPa. From the unloading part of the curves, $E_r = \frac{dF}{dh} \times \frac{1}{C\sqrt{A_{\max}}}$, which was a good estimation of the elastic modulus of the material. The modulus values were $E_{NP1} = 20.7 \pm 3.2$ GPa and $E_{NP2} = 12.7 \pm 5.5$ GPa for NP1 and NP2, respectively.

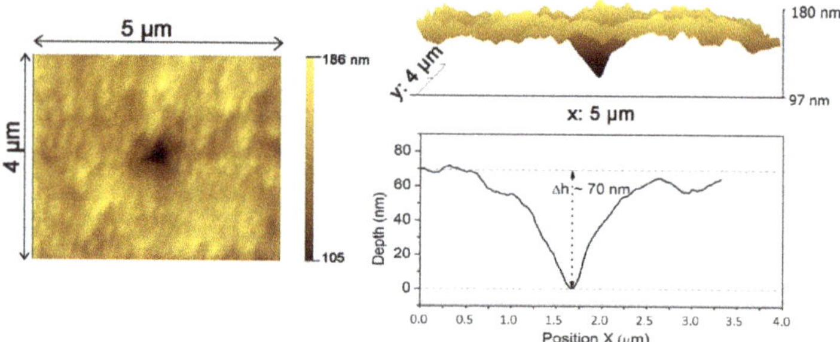

Figure 5. Scanning probe microscopy of the surface, and profile image of the slow-rate quasistatic indent, which reached a final depth of 70 nm after elastic relaxation.

The depth dependence was investigated using the DMA mode of the indenter, allowing us to keep an ongoing measurement of the hardness as a function of depth. In Figure 6a,b, the overall plot of H as a function of $1/h$ showed a maximum value at about 30 nm, which was of the order of the lower limit for a relevant measure due to tip imperfections (Figure 6a,b). H as a function of $1/h$ was suggested as the best linear plot for the test series seen in Figure 6c,d. For h larger than 40–50 nm, which corresponded to the upper part of the bumpy region, the data are plotted Figure 6c,d for NP1 and NP2, respectively. From the linear fit, hardness at a hypothetical infinite indent size ($h \rightarrow \infty$) when none were affected by size effect was obtained respectively for NP1 and NP2, $H_{01} = 58 \pm 10$ MPa and $H_{02} = 163 \pm 18$ MPa. The slope (α) related to the size-effect sensitivity informed other deformation dynamics of the NP metal: $\alpha_1 = 67 \pm 8$ GPa·nm and $\alpha_2 = 47 \pm 15$ GPa·nm for NP1 and NP2, respectively.

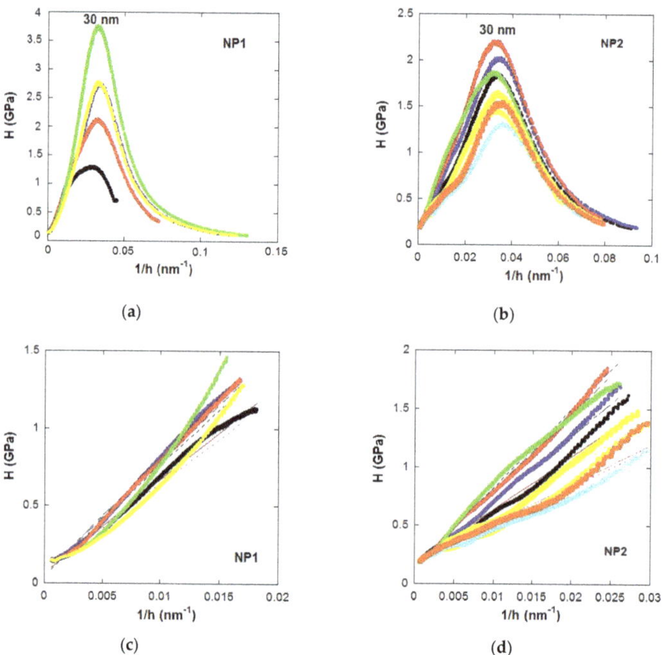

Figure 6. Plots of series of hardness as a function of 1/h from dynamic mechanical analysis (DMA) analyses for NP1 (**a,c**) and NP2. (**b,d**). (**c,d**) are zoomed in at a low 1/h of (**a,b**). Linear fittings of the curves were plotted as straight or dashed lines on (**c,d**).

3. Depth Dependence of Hardness

An analytical description of the size effect in a phenomenological approach was proposed to interpret the hardness value in the experimental data. The hardness H was the ratio of the applied load F to the area $A(h)$ produced by the penetration of the indent in the materials. F was arbitrarily fixed and the resulting area $A(h)$ was such that H became a function of h. This could be written as the sum of a constant hardness and an additional variable hardness dependent on h:

$$H = \frac{F}{A(h)} = \frac{F_0(h)}{A(h)} + \frac{f(h)}{A(h)}$$
$$\text{with, and } \forall h,\ \frac{F_0(h)}{A(h)} = H_0, H \to H_0 \text{ or } \frac{f(h)}{A(h)} \to 0 \text{ for } h \to \infty \tag{1}$$

The projected area was $A(h) = Ah^2$ with $A = 24.56$ for Berkovitch indent geometry (Figure 7). Then, the local force on an infinitesimal surface $dA(h)$ at height h was derived as:

$$df(h) = \sigma(h) \times 2Adh \tag{2}$$

where $\sigma(h)$ was the local back stress induced by the confinement of the deformation. The Ashby rule regarding strain gradient [21] predicted the stress $\sigma(h)$ at height h, following: $(\sigma(h)/\sigma_0)^2 = l\gamma/h$, where σ_0 and l were the stress and length scale parameters, respectively.

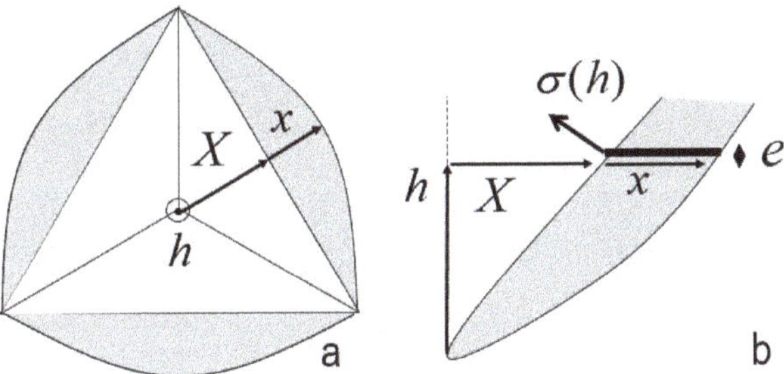

Figure 7. Scheme of a Berkovitch indent with the strained zone denoted in grey. (a) Top view, (b) cross-section view of a third of the indent. The parameters used in this modeling are as shown: indent shear X applied at height h, and the resulting shear displacement x of an NP infinitesimal thickness e. $\sigma(h)$ was the resulting back stress produced at height h from the strain gradient.

For highly porous materials made of plastic ligaments, densification was the main deformation mechanism when compressed. In this way, the strain gradient could be readily attributed to the variation in densification along the indent surface. This assumption was verified qualitatively by cutting an indent using the FIB technique and analyzing ligament distributions at various heights. This was imaged by plotting the greyscale distribution, along with a line scan for two different zones of the indent (Figure 8). We observed that the frequency of the greyscale variation was larger at more profound locations than when compared to the zone just below the sample surface.

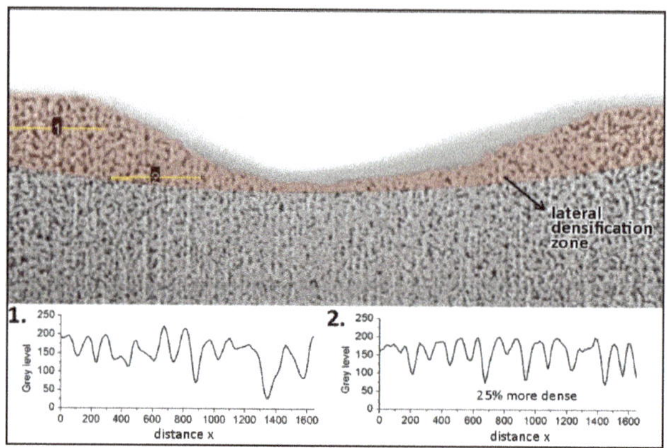

Figure 8. SEM image displaying the FIB (focused ion-beam) cross-section of an indent made on an NP1, up to 1.4 µm in depth. The insets show grey-level variation along lines 1 and 2 that correspond to the less and more densified regions of the deformed volume under the indent.

Our modeling consisted of defining the pertinent expression for the Ashby strain gradient to the insert in Equation (1) of the hardness. At the height h, strain was defined as $\gamma = 1 - \frac{x}{X}$ (as seen in Figure 7a,b). X was the displacement produced by the indent, and x the resulting displacement in the material. The difference between X and x was due to densification which varied as a function of h as shown on Figure 8. In a thin slice e at height h, densification was evaluated via volume variation. Considering the relative density $D = \rho/\rho_0$, with ρ and ρ_0 representing, respectively, the volume weight of the nanoporous and bulk materials, it can be shown that (The width of the impacted zone was proportional to h, with p a proportionality factor depending on the indent shape, v the volume of the densified region (Figure 7), and with $v = phex$, at h, the volume variation in the thin slice e was $dv = phedx$. Dividing by the weight of the slice and introducing the volume weight ρ it becomes, and eventually $\frac{dD}{D} = d\gamma$)

$$\frac{dD}{D} = d\gamma \tag{3}$$

The densification, which was a positive variation of D in the porous media, could be attained by the densification dynamics of powders proposed by Shapiro and Kolthoff et al. [28]. It has been conjectured that the porosity $(1 - D)$ variation as a function of the applied pressure follows a 1st-order kinetics law, which at a constant stress rate can be written:

$$-\frac{d(1 - D)}{d\sigma} = k(1 - D) \tag{4}$$

and with (3): $\frac{d\gamma}{d\sigma} = k(\frac{1}{D} - 1)$.

Deriving the Ashby rule, $2\sigma(h)d\sigma(h)/\sigma_0^2 = ld\gamma/h$ gave us an equation for the back stress generated by the densification gradient: $\sigma(h) = \frac{\sigma_0^2 l}{2h} \times \frac{d\gamma}{d\sigma(h)}$. Combined with Equation (4) gave us:

$$\sigma(h) = \frac{\sigma_0^2 lk}{2} \times (\frac{1}{D} - 1) \times \frac{1}{h} \tag{5}$$

It was easily shown (see [24]) that $\frac{f(h)}{A(h)} = \sigma(h)$, and with Equation (1) the hardness dependence in relation to the indent depth was eventually derived:

$$H = H_0 + \frac{\sigma_0^2 lk}{2} \times (\frac{1}{D} - 1) \times \frac{1}{h} \tag{6}$$

As observed experimentally, this relation shows the dependance the material hardness on the inverse of the depth considered.

4. Discussion

The mechanical properties of foam, cellular, or porous materials as a function of their porosity were systematically examined with respect to the Gibson–Ashby scaling law on hardness and elastic modulus [29]. In our analyses, the hardness of NP1 and NP2 could not be evaluated because the hardness of the material (Ag) forming ligament was unknown. In contrast to this, elastic modulus was could be well evaluated from quasistatic nanoindentation experiments.

From the Gibson–Ashby law $E_{NP} = E \times D^2$ and with an elastic modulus of bulk Ag measured under the same conditions $E \approx 76 \pm 2.5$ GPa, the moduli of NP1 and NP2 were calculated respectively as $E_{NP1}^{calc} \approx 22.6$ GPa and $E_{NP2}^{calc} \approx 16.8$ GPa, and compared to measured values $E_{NP1} = 20.7 \pm 3.2$ GPa and $E_{NP2} = 12.7 \pm 5.5$ GPa.

Our work mainly focused on the size dependence of hardness in nanoindentation of NP metals where their analytical description brought some interesting insights into NP deformation as controlled by densification. The parameters in Equation (6) describing H as a function of $1/h$ are hereafter discussed. First and most satisfactorily of all, was our finding that parameter $(\frac{1}{D} - 1)$ implied that the material was porous and had densification ability. A drastic limitation on this finding, however, was that to uphold linear dependence this parameter must be also nearly constant, which meant that the variation of the relative density produced during deformation was negligible before D.

From the Ashby rule on strain gradient, σ_0 was the stress scaling parameter and was related to the elastic properties of the ligament material. l was a scaling length parameter and comes from the rule that for a given strain gradient, the back stress is as large as l is large. This indicated that l was related to the densification ability of the NP material. Similarly, the parameter k from the Shapiro–Kolthoff relation also characterizes the ability of the porous material to increase in density under a given incremental stress. We showed that k correlated to the yield strength of the ligament material in its bulk form, following $k \propto 1/\sigma_y$ where σ_y was the yield strength [30] (Note: In the original work, this was the powder particle's material strength in its bulk form).

Comparing NP1 and NP2, the measured slopes ($\alpha_1 = 67 \pm 8$ GPa·nm and $\alpha_2 = 47 \pm 15$ GPa·nm.) show that NP1 had a larger propensity for densification than NP2. This was in contrast to the respective porosity levels of the NPs. The porosity of NP2 was greater, $((\frac{1}{D} - 1) = 0.83$ and 1.13 for NP1 and NP2, respectively) and the resultant larger space should therefore allow for a greater degree of densification. Assuming σ_0 was nearly the same for both NPs, the dominant effect was related to the NPs ability for densification (characterized by $k \times l$) which was shown as larger for NP1. From a structural point of view, it was obvious that the presence of smaller ligaments also meant a larger ligament density and therefore a greater possibility for ligament rearrangement and densification.

For the indent-size effect on hardness, a greater ability for densification meant that a larger, more pronounced gradient density had developed in the deformation zone. We emphasize here that use of a bad practice during nanoindentation can lead to an incorrect evaluation of the materials' properties. Of course, first, the depth dependence may lead to overestimation of the materials' hardness (this is true for all materials). Second, and probably much more detrimental, is misleading comparisons. In our research, NP1 softer than NP2 shows a larger slope in the plot of H as a function of $1/h$ (Figure 6c,d). This leads to an inversion in apparent hardness; NP1 has a higher hardness when compared to NP2 below an average critical depth of around 150 nm. A similar trend was observed for metallic glasses [24].

5. Conclusions

The depth dependence of an NP's hardness, as measured using nanoindention in DMA mode, was studied for nanoporous silver prepared by dealloying an $Ag_{38.75}Cu_{38.75}Si_{22.5}$ precursor. Two near-similar NPs were analyzed and compared. These materials were composed of connected ligaments with various levels of surface shape and curvedness, which was revealed in our experiment by tomographic imaging. For the two NPs, the hardness-dependence indent depth followed a rule of the form $H/H_0 = 1 + h^*/h$, first identified empirically and then compared to a description based on the strain-gradient approach proposed by Ashby and the densification rule for granular materials proposed by Shapiro–Kolthoff.

From these analyses, two main conclusions must be emphasized. First, heterogeneity of NPs at the length scale of the indent probe produced significant scattering in the results but systematic linear dependence of H as a function of $1/h$. This enabled us to evaluate the relevant hardness with far less scattering and by extrapolating at an infinite indent size $(1/h \rightarrow 0)$. Second, the proposed analytical description confirmed that the porosity level (as in the Gibson–Ashby scaling law) was not the only essential parameter controlling densification during the NPs' plastic deformation process.

The length scale was derived from the Ashby strain-gradient approach and the kinetic constant from the 1st order law of Shapiro–Kolthoff, which became dominant at the nanometer scale. These two parameters were both related to the local structure of ligaments and their ease of rearrangement, which also included their size and overall topological features. Nanoindentation was proven to be a useful technique for investigating the structural properties of NPs but further analysis should be performed in connection with 3D tomography characterization so as to relate NP structural characteristics with their mechanical behavior.

Author Contributions: D.B. prepared nanoporous metals and participated to all characterization and data analyses; A.M.J.J. and Y.C. supervised the work and results interpretation; F.C. carried out F.I.B. experiments. M.L.-B. carried out nano-indentation experiments; P.L. performed and analysed 3D imaging. Y.C. developped the model; All authors participated in writing and reading the paper.

Funding: This work was supported by the European ITN Network project "VitriMetTech" (grant No. FP7-PEOPLE-2013-ITN-607080) in the framework of the Marie Sklodowska-Curie actions program. The program and works on nanoporous metals were initiated by Alain Reza Yavary (SIMaP CNRS G-INP). YC, AMJJ, and DB dedicate this article to his memory.

Conflicts of Interest: The authors declare no conflict of interest.

References

1. Weissmuller, J.; Newman, R.C.; Jin, H.J.; Hodge, A.M.; Kysar, J.W. Nanoporous Metals by Alloy Corrosion: Formation and Mechanical Properties. *MRS Bull.* **2009**, *34*, 577–586. [CrossRef]
2. Mameka, N.; Wang, K.; Markmann, J.; Lilleodden, E.T.; Weissmuller, J. Nanoporous Gold-Testing Macro-scale Samples to Probe Small-scale Mechanical Behavior. *Mater. Res. Lett.* **2016**, *4*, 27–36. [CrossRef]
3. Badwe, N.; Chen, X.Y.; Sieradzki, K. Mechanical properties of nanoporous gold in tension. *Acta Mater.* **2017**, *129*, 251–258. [CrossRef]
4. Jin, H.J.; Weissmuller, J. A Material with Electrically Tunable Strength and Flow Stress. *Science* **2011**, *332*, 1179–1182. [CrossRef] [PubMed]
5. Jin, H.J.; Weissmuller, J.; Farkas, D. Mechanical response of nanoporous metals: A story of size, surface stress, and severed struts. *MRS Bull.* **2018**, *43*, 35–42. [CrossRef]
6. Fujita, T.; Tokunaga, T.; Zhang, L.; Li, D.W.; Chen, L.Y.; Arai, S.; Yamamoto, Y.; Hirata, A.; Tanaka, N.; Ding, Y.; et al. Atomic Observation of Catalysis-Induced Nanopore Coarsening of Nanoporous Gold. *Nano Lett.* **2014**, *14*, 1172–1177. [CrossRef]
7. Chen, Q.; Ding, Y.; Chen, M.W. Nanoporous metal by dealloying for electrochemical energy conversion and storage. *MRS Bull.* **2018**, *43*, 43–48. [CrossRef]
8. Seker, E.; Berdichevsky, Y.; Staley, K.J.; Yarmush, M.L. Microfabrication-Compatible Nanoporous Gold Foams as Biomaterials for Drug Delivery. *Adv. Healthc. Mater.* **2012**, *1*, 172–176. [CrossRef]
9. El-Safty, S.A.; Hoa, N.D.; Shenashen, M.A. Topical Developments of Nanoporous Membrane Filters for Ultrafine Noble Metal Nanoparticles. *Eur. J. Inorg. Chem.* **2012**, *2012*, 5439–5450. [CrossRef]
10. Juarez, T.; Biener, J.; Weissmuller, J.; Hodge, A.M. Nanoporous Metals with Structural Hierarchy: A Review. *Adv. Eng. Mater.* **2017**, *19*, 1700389. [CrossRef]
11. Seker, E.; Reed, M.L.; Begley, M.R. Nanoporous Gold: Fabrication, Characterization, and Applications. *Materials* **2009**, *2*, 2188–2215. [CrossRef]

12. Vargas-Martinez, J.; Estela-Garcia, J.E.; Suarez, O.M.; Vega, C.A. Fabrication of a Porous Metal via Selective Phase Dissolution in Al-Cu Alloys. *Metals* **2018**, *8*, 378. [CrossRef]

13. Ziehmer, M.; Hu, K.X.; Wang, K.; Lilleodden, E.T. A principle curvatures analysis of the isothermal evolution of nanoporous gold: Quantifying the characteristic length-scales. *Acta Mater.* **2016**, *120*, 24–31. [CrossRef]

14. Jiao, J.; Huber, N. Deformation mechanisms in nanoporous metals: Effect of ligament shape and disorder. *Comput. Mater. Sci.* **2017**, *127*, 194–203. [CrossRef]

15. Storm, J.; Abendroth, M.; Emmel, M.; Liedke, T.; Ballaschk, U.; Voigt, C.; Sieber, T.; Kuna, M. Geometrical modelling of foam structures using implicit functions. *Int. J. Solids Struct.* **2013**, *50*, 548–555. [CrossRef]

16. Lilleodden, E.T.; Voorhees, P.W. On the topological, morphological, and microstructural characterization of nanoporous metals. *MRS Bull.* **2018**, *43*, 20–26. [CrossRef]

17. Barsuk, D.; Zadick, A.; Chatenet, M.; Georgarakis, K.; Panagiotopoulos, N.T.; Champion, Y.; Jorge, A.M. Nanoporous silver for electrocatalysis application in alkaline fuel cells. *Mater. Des.* **2016**, *111*, 528–536. [CrossRef]

18. Lawn, B.; Wilshaw, R. Indentation fracture principles and applications. *J. Mater. Sci.* **1975**, *10*, 1049–1081. [CrossRef]

19. Hutchings, I.M. The contributions of David Tabor to the science of indentation hardness. *J. Mater. Res.* **2009**, *24*, 581–589. [CrossRef]

20. Yuan, Z.W.; Li, F.G.; Chen, B.; Xue, F.M. The correlation between indentation hardness and material properties with considering size effect. *J. Mater. Res.* **2014**, *29*, 1317–1325. [CrossRef]

21. Ashby, M.F. The deformation of plastically non-homogeneous materials. *Philos. Mag.* **1970**, *21*, 399–424. [CrossRef]

22. Nix, W.D.; Gao, H.J. Indentation size effects in crystalline materials: A law for strain gradient plasticity. *J. Mech. Phys. Solids* **1998**, *46*, 411–425. [CrossRef]

23. Huang, Y.J.; Shen, J.; Sun, Y.; Sun, J.F. Indentation size effect of hardness of metallic glasses. *Mater. Des.* **2010**, *31*, 1563–1566. [CrossRef]

24. Champion, Y.; Perriere, L. Strain Gradient in Micro-Hardness Testing and Structural Relaxation in Metallic Glasses. *Adv. Eng. Mater.* **2015**, *17*, 885–892. [CrossRef]

25. Zhang, M.; Jorge Junior, A.M.; Pang, S.J.; Zhang, T.; Yavari, A.R. Fabrication of nanoporous silver with open pores. *Scr. Mater.* **2015**, *100*, 21–23. [CrossRef]

26. Boulos, V.; Salvo, L.; Fristot, V.; Lhuissier, P.; Houzet, D. Investigating performance variations of an optimized GPU-ported granulometry algorithm. In Proceedings of the 18th International European Conference on Parallel and Distributed Computing, Rhodes Island, Greece, 27–31 August 2012.

27. Oliver, W.C.; Pharr, G.M. An improved technique for determining hardness and elastic-modulus using load and displacement sensing indentation experiments. *J. Mater. Res.* **1992**, *7*, 1564–1583. [CrossRef]

28. Shapiro, I.; Kolthoff, I.M. Studies on the Aging of Precipitates and Coprecipitation. XXXVIII. The Compressibility of Silver Bromide Powders. *J. Phys. Chem.* **1947**, *51*, 483–493.

29. Ashby, M.F.; Evans, A.G.; Fleck, N.A.; Gibson, L.J.; Hutchinson, J.W.; Wadley, H.N.G. *Metal Foams: A Design Guide*; Butterworth Heinemann: Boston, MA, USA, 2000.

30. James, P.J. Particle deformation during cold isostatic pressing of metal powders. *Powder Metall.* **1977**, *20*, 199–204. [CrossRef]

Article

Shear Banding in a Contact Problem between Metallic Glasses

Anne Tanguy [1,2,*], **Peifang Chen** [1], **Thibaut Chaise** [1] and **Daniel Nélias** [1]

[1] LaMCoS, INSA-Lyon, CNRS UMR5259, University Lyon, F-69621 Villeurbanne CEDEX, France; chenpeifang8@outlook.com (P.C.); Thibaut.Chaise@insa-lyon.fr (T.C.); Daniel.Nelias@insa-lyon.fr (D.N.)

[2] ONERA, Université Paris-Saclay, Chemin de la Hunière, BP 80100, 92123 Palaiseau, France

* Correspondence: Anne.Tanguy@insa-lyon.fr

Abstract: The case of a frictionless contact between a spherical body and a flat metallic glass is studied using a mesoscopic description of plasticity combined with a semi-analytical description of the elastic deformation in a contact geometry (code ISAAC). Plasticity is described by irreversible strain rearrangements in the maximum deviatoric strain direction, above some random strain threshold. In the absence of adhesion or friction, the plastic deformation is initiated below the surface. To represent the singularities due to adhesion, initial rearrangements are forced at the boundary of the contact. Then, the structural disorder is introduced in two different levels: either in the local strain thresholds for plasticity or in the residual plastic strains. It is shown that the spatial organization of plastic rearrangements is not universal, but it is very dependent on the choice of disorder and external loading conditions. Spatial curved shear bands may appear below the contact but only for a very specific set of parameters, especially those characterizing the random thresholds compared to externally induced strain gradients.

Keywords: metallic glasses; plasticity; contact mechanics; shear banding

Citation: Tanguy, A.; Chen, P.; Chaise, T.; Nélias, D. Shear Banding in a Contact Problem between Metallic Glasses. *Metals* **2021**, *11* , 257 . https://doi.org/10.3390/met11020257

Academic Editors: Mikhaïl A. Lebyodkin and Vincent Taupin

Received: 31 December 2020

Accepted: 29 January 2021

Published: 3 February 2021

Publisher's Note: MDPI stays neutral with regard to jurisdictional claims in published maps and institutional affiliations.

1. Introduction

Bulk metallic glasses are made of a disordered assembly of metallic atoms. Thus, they exhibit an atomic structure that is very close to the liquid one, but below the glass transition temperature, they are trapped into a metastable mechanical equilibrium identified by the absence of significant viscous flow at rest (viscosity larger than 10^{13} poise). This similarity to the liquid structure and the absence of both long-range order and characteristic length, despite the interatomic distance, is characteristic of amorphous materials. The original microstructure of amorphous materials bestows them very specific properties, such as a very smooth and moldable surface, low rigidity, but very high hardness, and a strong capacity to withstand very high loads before failure [1]. Indeed, the atomistic disorder precludes the formation of extended linear defects such as dislocations and thus forces to maintain irreversible deformation restricted to point-like defects. This makes them highly interesting for their ability to withstand heavy loads while maintaining their elasticity [2]. However, these properties are very dependent on the temperature: they can change drastically at the glass transition temperature, which is between 100 and 1000 K depending on the composition [3]. At finite temperatures [4] or under high-frequency acoustic waves [5], some metallic glasses can even exhibit superplasticity, with up to 160% irreversible plastic flow strain without breaking, at centimeter scales. Moreover, at low temperature or strain rate, irreversible (plastic) deformation is observed at small scale, and it can organize along macroscopic shear bands, which break the apparent homogeneity of the deformation and are the precursors of catastrophic failures [6–8]. Thus, it is important to understand the origin of the plastic deformation (irreversible non-viscous deformation) and its spatial organization.

The microscopic origin of plastic deformation in metallic glasses was initially related to the emergence and diffusion of local voids described successfully within the framework

185

of the free volume theory [9,10] and then to isochoric shear transformation zones analogous to dislocation loops [11]. Recently, it was proved that the constitutive laws of plastic flow results from the accumulation of plastic deformations described as Eshelby inclusions [12]. These Eshelby inclusions result from a loss of material stability [13–15]. Thus, they unfold in the direction of instability. The latter depends on the glass composition. In metallic glasses, it is mainly due to the local shear. It is generally given by the direction of the maximum deviatoric strain [16].

The occurrence of shear bands may have two origins: either the unfolding of the initial instability along a plane (elementary shear bands [17–19]), or the progressive accumulation of damage encouraged by the self-sustainability of the shear bands [20–22]. The accumulation of plasticity without self-sustainability can also give rise to the sparse distribution of shear bands, allowing the occurrence of a yielding transition appearing as a universal critical phenomenon [23,24] but without permanent shear banding. Finally, the nucleation of the shear bands may be intrinsic (with a volumetric origin) or extrinsic (from surface defects at the free surfaces, for example) [20], the later favoring permanent shear banding. Thus, under the same loading conditions, the presence of interfaces and specific boundary conditions may induce shear bands that would be absent elsewhere [25,26].

In this paper, we focus on the nucleation of shear bands under a nanoindenter. It was shown that under such a loading condition, glasses with low free volume, such as metallic glasses, and contrary to pure silica for example, have a tendency to develop shear bands [7,8]. In Zr-based metallic glasses, Su and Anand [27] have shown a nice and regular pattern of curved shear bands that was interpreted with the help of a specific constitutive equation. This continuum constitutive law was supposed to relate the local shearing rate to the local shear stresses, weighted by the local pressure, and also to give rise to a kinetic increase of the free volume and consequently to post yield–strain softening [27]. The crossing between the shear bands was characterized by an angle of 84°. Budrikis et al. [28] have observed similar features in a mesoscopic modeling of plasticity under a nanoindenter. They related them to a universal critical phenomenon for shear bands nucleation at the yield strain for macroscopic plastic flow. This critical phenomenon would be similar to what happens under more simple loadings such as pure shear or hydrostatic compression [23,24]. However, Shi and Falk [29] have shown using molecular dynamics simulations that this pattern depends indeed on the processing history of the glass (cooling rate) and also on the strain rate. Moreover, Amon and Crassous [30] have observed in granular materials a progressive transition from initially curved shear band patterns to straight bands across the sample. Thus, it is not clear whether the shape of such a shear band is universal or not, and how it is nucleated. To understand the necessary ingredients for curved shear band nucleation, we have run a mesoscopic model of local plasticity based on microscopic known processes and a realistic description of long-range elastic couplings.

In Section 2, we will describe the model used and the parameters involved. In Section 3, we will discuss the results, more specifically the role of disorder in the plastic transformation compared to the role of disorder in the local plastic threshold. Section 4 will discuss the conditions for getting shear bands in a contact problem. Section 5 summarizes the results.

2. Materials and Methods

The material studied is a typical bulk metallic glass. The material is not studied at the atomic scale, nor at a continuous scale, but the numerical model (combining the software ISAAC with local fortran programs, following the steps described in Figure 1 (right) used in this article is inspired from a set of mesoscopic models proposed in the last twenty years to model plasticity in disordered materials at intermediate length-scales (nm–μm) [31]. These models keep the necessary ingredients for local plasticity in solids, inspired from continuum mechanics with local criteria for the nucleation of plastic rearrangements [32], or more recently fed by molecular dynamics data concerning the details involved in the local criteria when they exist [33]. These models can indeed be shown as a good description of the atomistic modeling, but at a slightly larger coarse-graining length scale [34]. In these models, the effect of disorder is mainly located near the plastic rearrangements, and the elastic couplings are described within the framework of continuous linear isotropic and homogeneous elasticity. We will describe in this part the local plastic criteria and plastic events, then the numerical tools used for long-range elasticity in a contact mechanics problem, and finally the details of the contact conditions.

2.1. Local Plastic Criteria and Plastic Transformations

Plasticity is clearly related to an instability process [35], that is, a loss of mechanical stability due to the crossing of an energy barrier, and revealed by the loss of positive definiteness of the dynamical matrix of the full system (3N × 3N matrix of the second-order derivatives of the potential energy with N the number of atoms). In amorphous materials, such a loss of stability manifests itself by the nucleation of a high local strain concentration [13] that becomes unstable, and whose unfolding gives rise to a single Eshelby inclusion, or to a bidimensional alignment of Eshelby inclusions along elementary (embryonic) shear bands [12]. Although the instability process is a collective one (occurrence of a negative eigenvalue in the dynamical matrix of the full system), in the presence of disorder, the unstable mode can be described as a set of Eshelby inclusions, only one of them being the initiator of the mechanical instability. Thus, the mesoscopic models [31,32,36,37] are based on the nucleation of local Eshelby inclusions, interacting through long-range isotropic elasticity characteristic of the mechanical response of amorphous solids at large scale [38]. The nucleation criteria are still debated [33], since the instability process is indeed not local but global. However, a pragmatic analysis of the numerical data at the atomic scale shows that the nucleation of the local Eshelby inclusion may be related to large local strains, or low local elastic modulus. In the case of bulk metallic glasses, the plastic rearrangement has mainly a shear component. Thus, the nucleation criterion is related to the maximum local deviatoric strain, and naturally, the direction of the instability unfolding is the eigendirection related to this maximum deviatoric strain in the deviatoric part of the local strain tensor.

The implementation of the model is the following. The first two materials of Hertzian geometry are put in contact with a vertical load F_z giving rise to a spherical contact area with a radius a (see Section 2.3). Starting from this initial configuration, a threshold ε^d_{thresh} is chosen as detailed later. If the maximum local deviatoric strain becomes larger than ε^d_{thresh}, then an Eshelby inclusion is nucleated at this place. ε^d_{thresh} is random with

$$\varepsilon^d_{thresh}(i) = \varepsilon^0_{thresh} + Q.r(i) \tag{1}$$

where ε^0_{thresh} is a homogeneous strain threshold, and $r(i)$ is a Gaussian random number with zero average and variance unity, which is characteristic of the site i before rearrangement occurs. Q is a parameter: it is the amplitude of the random contribution to the local strain threshold. In the paper, we varied Q between 0 and 10^{-1} (the maximum value is one order of magnitude larger than the maximum deviatoric strain induced by the Hertzian contact for the chosen load F_z). ε^0_{thresh} is chosen in order to allow for a progressive nucleation of plastic events, with only a limited number of sites. It is adjusted to the external load in order to get a given number of sites (5, 10, or 100) with a deviatoric strain larger than ε^d_{thresh}, that is to nucleate a fixed number of plastic events (5, 10, or 100 respectively) at each step. This allows imposing an average constant plastic strain rate. In practice, the sample is discretized into sites with lateral size r_0 corresponding to the size of the local rearrangements. r_0 is chosen to be equal to 10^{-3} times the radius of curvature R at the contact (that is 1 nm for $R = 1$ µm). At each deformation step, the deviatoric strain will be computed on every site in the sample. The random value $(-Q.r(i))$ will be added at each site i to quantify the departure from the local plastic threshold. A fixed number of sites (5, 10, or 100) with the largest value of $(\varepsilon^d_i - Q.r(i))$ is identified, and these sites undergo a plastic rearrangement. After a plastic rearrangement occurred, a new value of $r(i)$ is chosen at this site.

The unfolding of the plastic rearrangements thus nucleated gives rise to a spherical Eshelby inclusion with radius r_0. After the nucleation sites have been identified, an eigenstrain $\varepsilon*(i)$ is attributed to each site i. The strain fields are computed again, taking into account the presence of the (5, 10, or 100) newly nucleated Eshelby inclusions, and the procedure is repeated with the calculation of the largest deviatoric strains—$Q.r\,(i)$ in the new configuration. The method is summarized in Figure 1 (right).

The residual plastic strain inside the inclusion is the eigenstrain ε^*. In the absence of disorder, the residual plastic strain is proportional to $(\varepsilon_i^d - \varepsilon_{thresh}^d)$ (see Figure 1 (left)) with ε_i^d the deviatoric strain measured at site i. In the presence of disorder, the eigenstrain ε^* is chosen as

$$\varepsilon * (i) = S.\max(\varepsilon_i^d - \varepsilon_{thresh}^d, 0) + Q * .r * (i) \qquad (2)$$

with $r*(i)$ another gaussian random number with zero average and variance unity, and S and Q^* are two parameters of the model. S quantifies the exceeding of the local threshold. It is chosen in the absence of disorder in order to get similar values in the maximum deviatoric strain after the first nucleations of plastic rearrangements (too small values will stop the plastic deformation in the absence of disorder, and too large values will drive too much plastic rearrangement in the next step). For example, in the case of the contact problem described in Section 2.3, the values of the maximum deviatoric strain obtained after a given number n_{sites} of plastic events have been nucleated simultaneously in the first step, as summarized in Table 1.

Table 1. Values of the maximum local deviatoric strain $\varepsilon^d{}_1$ in a contact problem without disorder after the first step, compared to the initial value of the maximum deviatoric strain $\varepsilon^d{}_0$. The contact problem is described in Section 2.3, here for different values of the parameter S and different numbers n_{sites} of simultaneously nucleated sites.

n_{sites}	S = 0.01	S = 0.10	S = 1.00
5	~0.015 $\varepsilon^d{}_0$	~0.99 $\varepsilon^d{}_0$	~1.5 $\varepsilon^d{}_0$
10	~0.015 $\varepsilon^d{}_0$	~0.99 $\varepsilon^d{}_0$	~1.5 $\varepsilon^d{}_0$
100	~0.015 $\varepsilon^d{}_0$	~0.98 $\varepsilon^d{}_0$	~1.5 $\varepsilon^d{}_0$

It appears in Table 1 that the value $S = 0.10$ allows staying at the same values of deviatoric strains as in usual mesoscopic models [28] and will thus be kept in the following. Thus, Q and Q* are the only remaining parameters of the model. They will allow discussing the role of disorder in the organization of plastic events in our system.

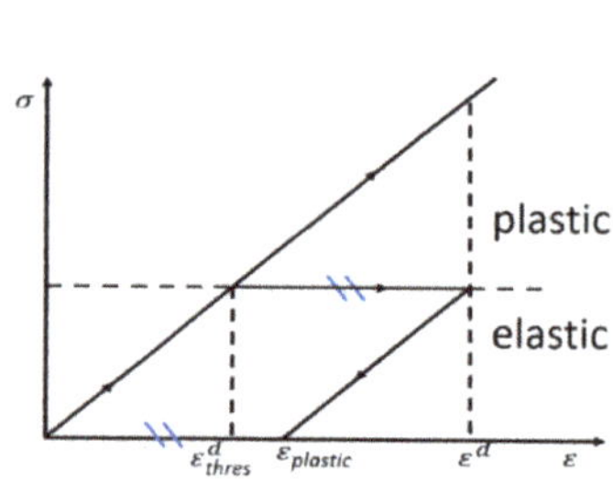

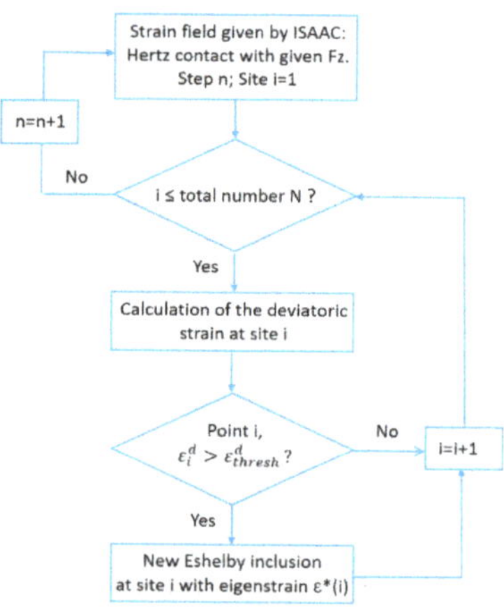

Figure 1. (**left**) Schematic description of residual plastic strain in a linear elastic–perfectly plastic model. (**right**) Schematic diagram of the mesoscopic model.

2.2. A Semi-Analytical Method for Long-Range Elasticity

We face to solve here the calculation of elastic strain in a contact problem and in the presence of Eshelby inclusions. The contact problem is first considered as a purely elastic contact between two linear elastic bodies with neither friction nor adhesion. One surface is flat, while the other has a radius of curvature R: it is a Hertz contact [39]. The calculation of the strain field beneath the contact can be performed with the help of superposition theorem and green functions for any given or computed pressure applied at the surface (similar to a Hertzian pressure) [40]. The Eshelby inclusions [41,42] can be computed thanks to the use of images symmetric to the interface, in order to still fulfill the limit boundary conditions [43].

The method was implemented in the software ISAAC (Version 2020, LaMCoS, Villeurbanne, France) [44] developed by D. Nélias et al. and adapted by T. Chaise, among others, for the calculation of residual plastic strain [45]. The semi-analytical method proposed to solve the contact problem is based on Eshelby's formalism but uses 3D Fast Fourier Transforms to speed up the computation. Thus, the time and memory necessary are greatly reduced in comparison with the classical finite element method. It allows a realistic computation of the stress and strain fields, taking into account the exact boundary conditions. Here, only an elastic contact is studied, but the same method can be extended to frictional contacts [46] and to adhesive contacts [47]. The only approximation used is that the only interface that matters is the contact surface: the bodies are considered as semi-infinite and treated within the framework of linear elasticity. The Eshelby inclusions and the corresponding heterogeneous solution are described using enrichment fields that are superimposed to the homogeneous problem [43]. To perform the simulations, a standard Young modulus $E = 115$ GPa and Poisson's ratio $v = 0.3$ is used for both materials. These values correspond to the elastic moduli of Ti-based and Fe-based metallic glasses, the range of Young moduli spanning generally from 40 (La-based glasses) to 150 GPa approximately [1], and the Poisson's ratio being between 0.3 and 0.4 in bulk metallic glasses. The discretization of space is made through $N = 65{,}559$ nodes, which are separated by a distance $dx = dy = 2r_0$ horizontally and $dz = r_0$ vertically. Note that our goal here is not to describe precisely a given material but to highlight the disorder-induced mechanisms of localization of plastic deformation in a contact geometry.

2.3. Contact Problem

The initial contact problem is represented in Figure 2. It is made of a flat surface with Young modulus E and Poisson ration v in contact with a spherical surface having a radius of curvature R and the same elastic properties. The bodies are pressed together using a normal force F_z. Only the strain field inside the flat surface is studied here. The scaling relations in the Hertz contact are the following [39]:

The radius of the spherical contact surface is

$$a = \sqrt[3]{\frac{3RF_z(1 - v^2)}{2E}}.$$

(3)

The vertical displacement in the center is

$$U_z = \frac{a^2}{R}.$$

(4)

The maximum shear stress appears below the surface, at a distance $z_{max} \approx 0.48\,a$ for $v = 0.3$, and its value is

$$\tau_{max} \approx 0.31\frac{aE}{\pi R(1 - v^2)}.$$

(5)

corresponding to a maximum deviatoric strain

$$\varepsilon^d_{max} \approx 0.31\frac{a}{\pi R(1 - v)}.$$

(6)

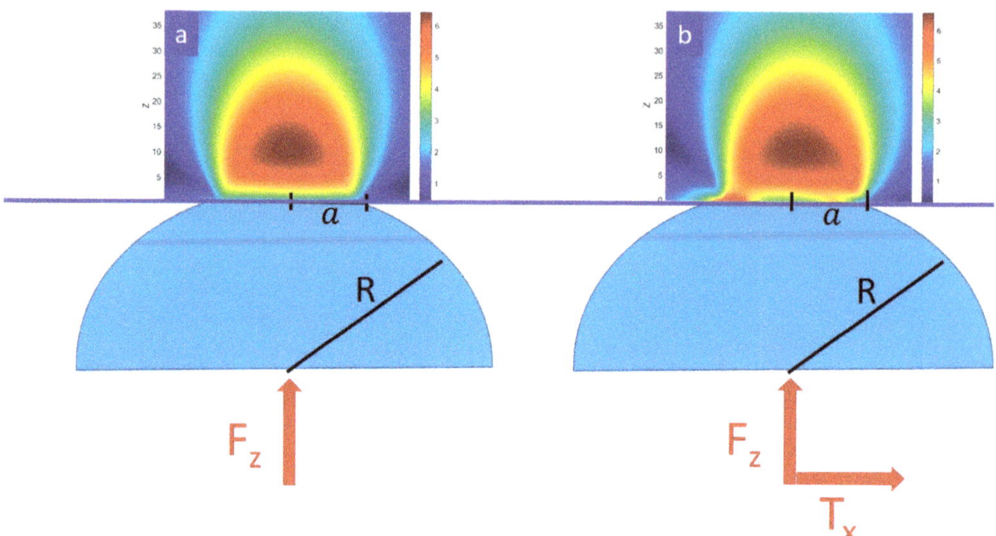

Figure 2. (**a**): Hertzian contact and resulting deviatoric strain field inside the flat body compared to the same configuration (**b**): but with a friction coefficient $\mu = 0.3$ at the interface between the two bodies and with a tangential load $T_x = 0.24\,F_z$.

In our case, the initial force is imposed, thus fixing the initial contact radius a, and then the deformation is induced by the succession of plastic rearrangements.

In the case of a purely elastic contact, the initiation of plasticity cannot appear at the surface, because the maximum shear stress τ_{max} is below the surface. In the absence of structural heterogeneities, the nucleation of shear bands is necessarily homogeneous in this case and located close to the depth z_{max}. However, as soon as some adhesion or a sufficiently high static friction coefficient ($\mu \geq 0.3$) is present, then the shear stress becomes maximum at the surface [39,48]. Thus, the nucleation of plasticity and of shear banding is heterogeneous (i.e., from the surface) in these cases.

In order to allow heterogeneous nucleation or to modelize the effect of structural disorder, we have compared two cases: the case of isotropic and homogeneous linear elastic solids, and the case where two Eshelby inclusions are already located, close to the boundary of the contact, that is at a distance a from the center of the contact zone. In this latter case, in order to get a real competition between the nucleation in the center and nucleation at the surface, the Eshelby inclusions have a shear eigenstrain slightly larger than ε^d_{max}.

3. Results

We will now compare the shape in the organization of the successive plastic events taking place in the flat solid for the different parameters studied.

3.1. Homogeneous vs. Heterogeneous Nucleation

The deviatoric stress induced by the presence of pre-existing inclusions competes with the deviatoric stress due to the Hertz contact. Figure 3 gives an idea of the spatial distribution of deviatoric strain (that is, the proximity to the local plastic yield strain assumed constant here as a first step) for different strengths of the two pre-existing defects. From left to right, the role of the two defects on the strain field increases. In the last case, where the deviatoric strain field at the defects is 10 times stronger than ε^d_{max}, the plastic deformation should localize on the defects, but in the two first cases with lower strain field, a connection is expected between the pre-existing defects at the surface and the subsurface zone of high shear strain due to the strain gradients induced by the Hertz contact.

Figure 4 shows the position of the first plastic events in three cases: without defects, with defects comparable to the Hertz fields, with defects far stronger (10.1 times) than the Hertz fields. In the first two cases, the initial plastic events tend to localize all near $z_{max} \approx 0.48\ a$, and there is no reason for the occurrence of shear bands. However, in the case of very strong defects, the simultaneous accumulation of plastic events near the defects and at z_{max} suggests a future connection between them, for example along plastic bands. Clearly, from Figure 4, it appears that when the defects are not sufficiently strong, the plastic activity remains grouped below the surface at z_{max}, while the presence of strong defects favors shear banding between z_{max} and the surface. Thus, in the following, we will always keep the pre-existing defects at their highest value (with $\varepsilon_{xy} = 10.1\ \varepsilon^d_{max}$).

Here, we have not taken into account the effect of disorder neither in the local yield strain nor on the eigenstrain characterizing the core of the unfolded inclusion. We will now discuss the effect of disorder on the strain fields and consequently on the nucleation of plastic events in the metallic glasses.

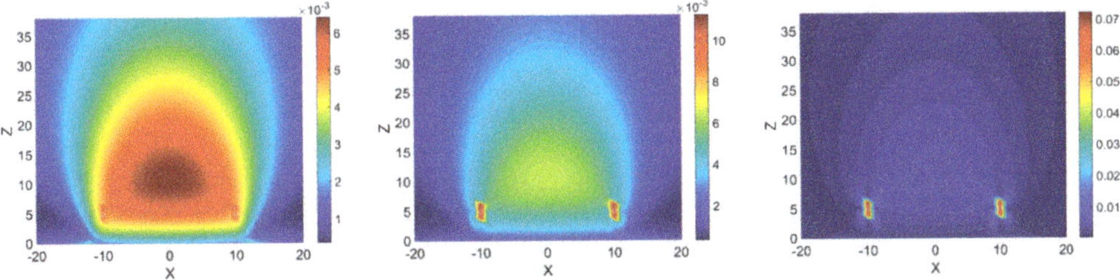

Figure 3. Deviatoric strain in the flat solid with two inclusions located very close to the surface. The shear eigenstrain supported by the inclusions is (from left to right): $\varepsilon_{xy} = 0.101\ \varepsilon^d_{max}$, $\varepsilon_{xy} = 1.01\ \varepsilon^d_{max}$ and $\varepsilon_{xy} = 10.1\ \varepsilon^d_{max}$.

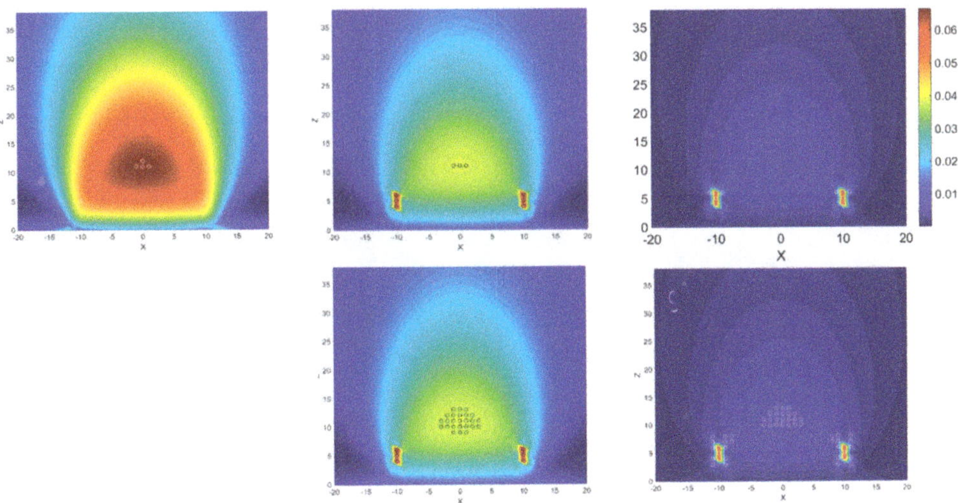

Figure 4. Deviatoric strain and position of the inclusions (white or black circles). Top: at the first deformation step, for $n_{sites} = 10$ new inclusions per step, without pre-existing defects at the surface (**left**), with pre-existing defects at the surface shown as Eshelby inclusions with the shear eigenstrain $\varepsilon_{xy} = 1.01\ \varepsilon^d_{max}$ (**middle**), and with pre-existing defects at the surface shown as Eshelby inclusions with the shear eigenstrain $\varepsilon_{xy} = 10.1\ \varepsilon^d_{max}$ (**right**). Bottom: at step 10 of the deformation ($n_{sites} = 10$) with pre-existing defects with the shear eigenstrain $\varepsilon_{xy} = 1.01\ \varepsilon^d_{max}$ (**left**) and $\varepsilon_{xy} = 10.1\ \varepsilon^d_{max}$ (**right**).

3.2. Role of Disorder in the Residual Plastic Eigenstrain

Figure 5 shows the effect of the two parameters S and Q^* on the position of plastic events after 250 plastic rearrangements took place (25 steps with 10 rearrangements at each step). In these figures, there is no disorder in the yield strain ($Q = 0$), and the pre-existing defects are 10 times stronger than the Hertz field, as detailed in the previous part. It appears clearly from this figure that the parameter S, which is used to counterbalance the lack of precision on the determination of the local yield strain and allow for a better definition of plastic flow, induces spatial correlations in the location of plastic events: the deviatoric strain is higher close to the Eshelby inclusions, thus giving rise to higher eigenstrains that will contribute to self-sustain the process. The location of the plastic events shows a better alignment close to the pre-existing defects, and it looks more sparse close to z_{max}. As discussed in Section 2.1 for the disorder-free case, the value of $S = 0.1$ will be kept in the following.

Let's look now at the effect of the disorder through the parameter Q^*. In Figure 6, the distribution of deviatoric strains in the sample is shown for different values of Q^*. Increasing Q^* not only contributes to enlarging the distribution of deviatoric strains as compared to the disorder-free Hertz case, but it can even give rise to very large cumulated deviatoric strains, up to 10% strain for $Q^* = 0.01$ at the 25th step (Figure 6 bottom). Let us remind that $Q^* = 0.01$ corresponds to a Gaussian distribution of eigenstrains with variance $Q^* = 1\%$, owing in our case at the 25th step extreme eigenstrains of about 3.5% only (Figure 6 top). In Figure 5, increasing Q^* up to the 1% strain clearly contributes to counterbalance the role of the Hertz fields and of their spatial gradients. The plastic rearrangements localize close to the pre-existing defects. The resulting shear aligns along large shear bands, propagating straightly from the surface defects. While the role of defects is clearly enhanced in this case, it does not gives rise to curved shear bands, but only to straight bands initiated from the surface, even after a large number of events.

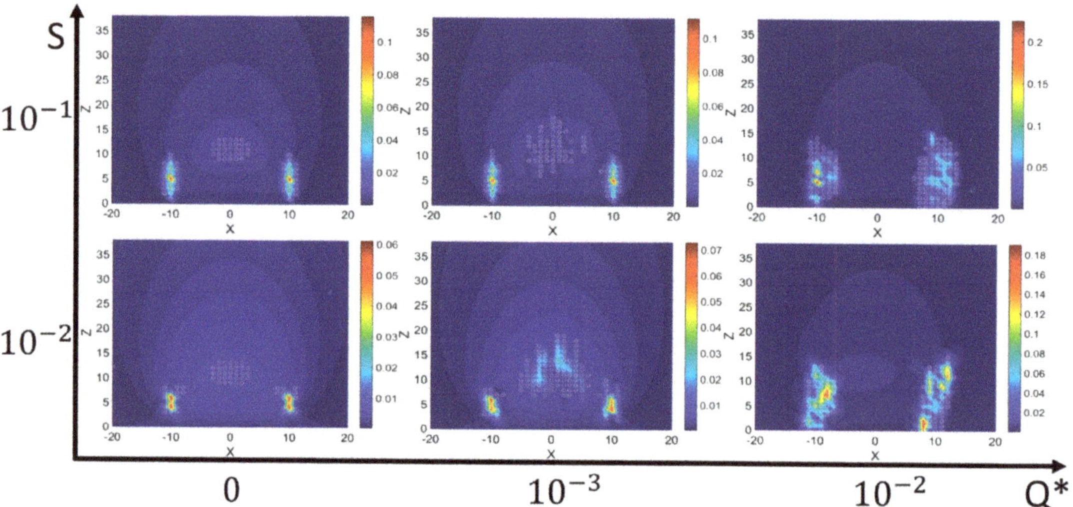

Figure 5. Effect of S and Q^*, parameters characterizing the eigenstrain ε^* of the nucleated events Equation (2), on the localization of plastic rearrangements at the 25th step of deformation with 10 events at each step, for $Q = 0$ (no disorder in the local thresholds) and initial defects $\varepsilon_{xy} = 10.1\ \varepsilon^d_{max}$. With our choice of loading F_z, the maximum deviatoric strain ε^d_{max} due to Hertz contact in the absence of disorder is equal to 0.65×10^{-2} here.

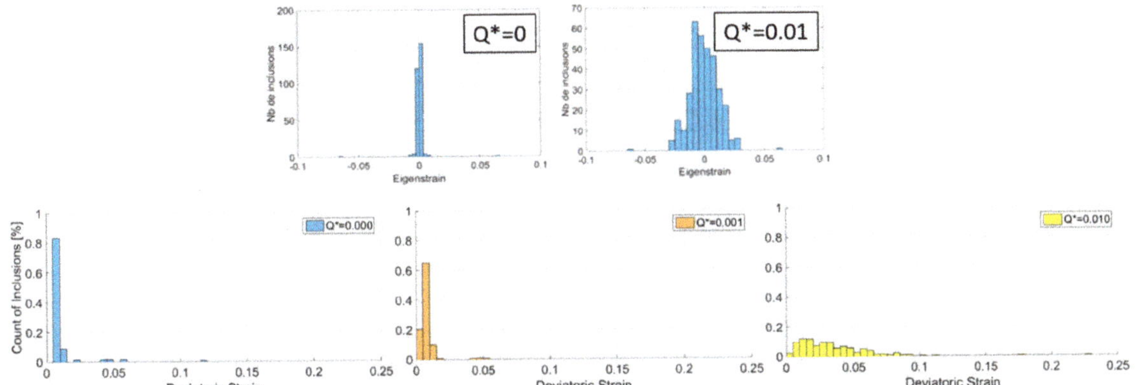

Figure 6. Top: Distribution of eigenstrains ε^* in the sample for different values of Q^*, at steps 25 with 10 events at each steps, with $S = 0.10$ and $Q = 0$. Bottom: Global distribution of deviatoric strains ε^d in the sample, for the same conditions. With our choice of loading F_z, the maximum deviatoric strain ε^d_{max} due to Hertz contact in the absence of disorder was equal to 0.65×10^{-2} here.

3.3. Role of Disorder in the Local Plastic Thresholds

Let us now look at the effect of disorder in the yield strains. Figure 7 summarizes the effect of the disorder induced by the increase in the parameter Q in the random part of the local nucleation threshold of Eshelby inclusions. The colored map shows the local deviatoric strain —$Q.r\ (i)$. Thus, the plastic events always take place at the brightest colors (maximum values) in this map. While for low values of Q, the gradients of the Hertz field are visible, they progressively disappear for large disorder values. The previous values of $S = 0.10$ and $Q^* = 0.01$ are kept in this case.

For low disorder $Q = 10^{-3}$, large shear bands are initiated from the surface defects, as discussed before in Section 3.2. When Q increases, the nucleation of plastic events is less focused on the surface defects, and some of them become sensitive to the Hertz fields and nucleate preferentially close to z_{max}. When the disorder is very strong and far larger than the maximum deviatoric strain ε^d_{max} induced by the Hertz contact, the events are sparse and can occur everywhere. However, the Hertz fields will introduce a bias that will favor a nucleation close to z_{max}. In this latter case, the plastic events do not organize clearly along shear bands, and more importantly, they do not show any sensitivity to the surface that appears completely ignored in this organization.

Interestingly, between these two extreme cases, for a contribution of the disorder to the local yield strain that is comparable to ε^d_{max}, a connection is made between events nucleated at the surface, and events deeper below the surface. For $Q \approx 2\varepsilon^d_{max} \approx 10^{-2}$, the characteristic shape of curved shear bands initiated from the strongest defects at the surface is recognizable.

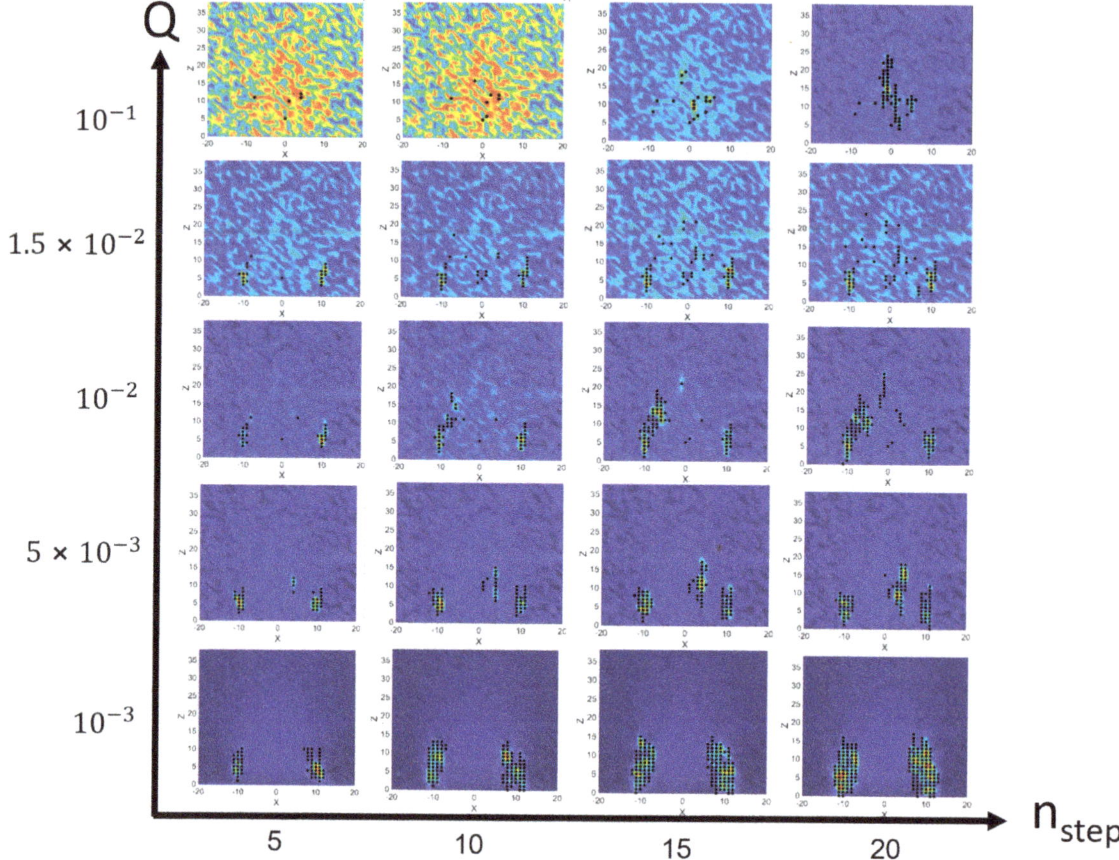

Figure 7. Effect of Q characterizing the disorder in the yield strain from Equation (1), on the localization of plastic rearrangements, for different number of steps, with $Q^* = 0.01$ and $S = 0.100$ and initial defects with $\varepsilon_{xy} = 10.1\,\varepsilon^d_{max}$. With our choice of loading F_z, the maximum deviatoric strain ε^d_{max} due to Hertz contact in the absence of disorder is equal to 0.65×10^{-2} here.

4. Discussion

We have studied here with the help of a mesoscopic model the spatial organization of plastic events inside a disordered material in a contact geometry. The results obtained in this context underline the fact that the sliding conditions at the surface play an important role on the plasticity pattern when the structural disorder is sufficiently low. The role of adhesion or friction at the interface is modeled with the help of pre-existing inclusions located near the free surface of the solid and at the boundary of the contact zone. Indeed, adhesion or friction with small tangential load induces very strong deviatoric strains at the boundary of the contact [39]. It has been shown (Figure 4) that the presence of initial defects—or equivalently non sliding conditions—is crucial to force the emergence of shear bands at the surface, as observed experimentally in many cases [27–29,49]. These pre-existing defects are also representative of the possible role played by a very inhomogeneous structural disorder or by surface defects. This is in agreement with the possibility to generate extrinsic nucleation of shear bands [20], and this enhanced surface sensitivity can be used for specific materials design [26], for example to prevent subsurface damage. We have discussed in the present paper the role of the eigenstrains included in these pre-existing defects, and we focused mainly on the case of defects whose eigenstrain is about 10 times larger than the maximum deviatoric strain induced by the Hertz contact beneath the surface. Indeed, this high value appears necessary to avoid a concentration of plasticity only below the surface. Moreover, we have shown (Figure 5) that when the same amount of disorder appears as well in the local strain rearrangements (value of Q^* in the eigenstrains), then plasticity localizes along straight lines initiated from the initial defects. However, this type of disorder is not sufficient to induce curved shear bands, as sometimes observed [27–29,49,50].

The occurrence of curved shear bands is shown here to result only from a specific set of parameters in the random local strain thresholds for plasticity. To get well marked permanent shear bands (Figure 7), it is necessary to have an amount of disorder in the local strain threshold that is comparable to the maximum deviatoric strain due to the Hertz contact between curved surfaces. However, this one depends not only on the contact geometry (radius of curvature R) but also on the external load (Equation (6)). Thus, the occurrence of curved shear bands is not universal as initially affirmed [28] but depends on the comparison between intrinsic heterogeneities (local strain thresholds for plasticity) and extrinsic parameters (geometry, external load). It is very important here to note that the structural disorder in itself is not sufficient to predict the shape of shear bands but must be put in perspective of the loading conditions: size of the contact area, radius of curvature, and external load, as detailed in Equation (6). The respective role of the disorder in the eigenstrains of the unfolded defects and in the strain thresholds may also explain why the curved shape of initial shear bands is not maintained for higher loads in easily rearranging materials [30]. The disorder in the eigenstrain or some kind of self-sustainability [20–22] may finally dominate and move away from the initial range of parameters. This may also explain the variety of the plasticity patterns observed in metallic glasses under a nanoindentor [29].

These effects were not discussed in [28], aiming at proving the universal character of plastic yield in amorphous solids for any kind of load. However, indeed, many assumptions were made in [28] concerning the relative values of disorder. First, the yield strains and the eigenstrains are not chosen independently, but the eigenstrain is assumed to be proportional to the yield strain. Moreover, non-sliding contact, meaning very strong solid friction, has been assumed that favors nucleation at the surface and finite size effects, as discussed in our article.

The disorder in the strain thresholds is usually related to the structural disorder characteristic of amorphous materials [33,51,52]. Experimentally, the amount of disorder depends on the preparation protocol, such as the quenching rate or the annealing for example [53,54], the free volume [9,55], or even defects induced by neutron irradiation [56]. The amount of disorder is known to affect the elastic and the plastic properties of the glass [57]. In all cases, increasing the effective temperature of the glass, or the amount of soft—equivalently less relaxed—zones [58] will make the glass more ductile, less fragile, and contribute to homogenize the deformation. In case of very strong disorder, the role of surface defects becomes negligible (Figure 7). Thus, the extreme values of strains are sensitive again to the Hertzian field. Consequently, the deformation, while irregular, concentrates below the surface, at the maximum deviatoric strain due to the Hertz contact. Such a kind of deformation without shear bands was already observed experimentally with in situ X-ray diffraction in Zr-based metallic glass samples [59]. However, the disorder was unfortunately not characterized in these samples. Thus, the absence of shear bands may also be simply due to perfectly sliding conditions at the surface.

In general, very different kinds of plasticity patterns have been observed below an indentor. Perepezko et al. [60] have identified experimentally on the same sample at least two kinds of events: at the surface and below the surface. Luo et al. [6] have observed different plasticity patterns depending on the amount of free volume induced by chemical strengthening in silicate glasses. The decrease in the free volume always favors localization along shear banding; this may result from a lower quenching rate [29] or the increase in the relaxation induced by longtime annealing, for example [51,53]. This effect also appears in our analysis of the role of the amplitude of disorder in the local strain thresholds: a strong disorder will make every site equivalent, thus masking the role of surface or isolated defects and preventing the nucleation of shear bands. Then, the average mechanical response becomes that of a homogeneous sample.

Note that the size of the samples is shown in experiments to play a role on the ductile behavior of bulk metallic glass samples [61,62]. In general, diminishing the size of the sample, as in films or in micropillars, will induce a more homogeneous repartition of plastic deformation. In this case, the increased role of surface defects (due to the large surface over volume ratio of the samples) may be counterbalanced by the lack of disorder or low effective temperature of the sample, due to its small size. However, on the contrary, the enhanced ductile behavior could also be due to an increase in the effective temperature due to the pressure increase and the confinement. As suggested in this paper, contact force experiments with different friction coefficients, or with different radius of curvature, could help disentangle these explanations by looking at the shear band patterns.

In our mesoscopic model, the formation of shear bands results only from the succession of discrete Eshelby events. Instantaneous shear bands (also called elementary shear bands [17] or embryonic shear bands [20]) are not properly described here, as it is the case in all mesoscopic models [31]. Elementary shear bands result indeed from a specific out-of-equilibrium unfolding process with a collective kinetic alignment of events. However, mesoscopic models are able to reproduce some characteristics of these bands: the resulting spatial organization of plastic deformation is finally not very far from quasi-static atomistic simulations [51]. This may be due to the fact that the driving force in the unfolding of the instability is the elastic force, as in the mesoscopic models. However, kinetic effects may play a role. For example, inertial effects could affect the results, as already discussed in [58]. The goal of the present paper is to focus on the primary role of disorder. To compare different shear rates, we have imposed a different number of simultaneous plastic events at each step in the model before redistributing the elastic energy. This does not induce any significant difference in the results, at least in the low disorder case. Thus, we have chosen to keep, in the majority of the simulations presented here, an intermediate number of simultaneous plastic events corresponding to a low shear rate. However, the role of the shear rate should be deepened, since it is known to induce different spatial organizations of events [63]. The role of inertia and of shear rates should deserve further studies if needed.

5. Conclusions

The choice of the approximate mesoscopic model used in this study allows keeping the main ingredients for the nucleation and succession of plastic events in an amorphous material: amorphous materials are linear elastic at large scale, with homogeneous and isotropic elasticity [38], and plastic rearrangements are described as Eshelby inclusions in agreement with recent molecular dynamics simulations [12].

In this work, two important effects were evidenced that have never been discussed in previous studies: the first one is the role of initial defects, and the second one is the relative role of disorder vs. Hertz mechanical inhomogeneities. Concerning the role of initial defects, it is clear that if these defects are absent, then the plastic deformation will be self-sustained below the contact at the maximum deviatoric strain location due to the Hertzian contact, as in homogeneous solids. When the deviatoric strain supported by the pre-existing defects becomes comparable or larger to the maximum Hertzian deviatoric strain, then the extrinsic nucleation of shear bands occurs, and new processes take place. We saw in Section 3.2 that the role of initial defects is reinforced when the residual plastic strain (eigenstrain of the nucleated Eshelby inclusions) has a random component. However, this type of disorder is not sufficient to induce well-defined curved shear bands inside the sample. It only contributes to enlarging the plastic zone around an initial one. We saw also in the simple model used here that the sensitivity of the residual plastic strain on the amount of deviatoric strain at the nucleation of the plastic event induces a small memory effect, tending to localize the plastic activity around its initial place. Concerning the role of disorder, we compared different sources of disorder in the amorphous sample.

Interestingly, we saw that the spatial organization of the plastic activity and the connection between different places inside the sample (that would correspond to extended shear bands) depend mainly on the amount of disorder in the local yield strain only.

This can be understood by the fact that the high strain disorder due to structural origins should counterbalance the strain inhomogeneities induced by the external load. For very high disorder, the differences in extreme values responsible for the nucleation of plastic events will be reinforced by Hertzian inhomogeneities and become certainly insensitive to surface effects. Two simple cases have been easily identified: the case of low disorder where pre-existing defects will control the nucleation of shear bands (initiated at the surface when these defects represent adhesive or frictional contact). In this case, the shear band extends straight inside the sample. The second limit case is the case of very high disorder where plasticity is sparse and insensitive to surface defects. At intermediate values of disorder, more precisely when the contribution of disorder in the yield strain becomes comparable to the maximum Hertzian deviatoric strain, well-defined curved shear bands connecting the surface and the subsurface are shown. This latter case occurs for different amounts of disorder, depending on the geometry and on the external load: as shown in Equation (3), increasing the load or decreasing the radius of curvature at the contact increase the deviatoric strain to which the amount of disorder in the strain threshold must be compared. However, the disorder that is needed to nucleate shear bands and inhomogeneous deformation has less influence on strain when the contact radius of curvature is smaller. Note that in the absence of surface defects, the plastic activity also remains located at the place of maximum Hertzian deviatoric strain without extended shear banding.

These different cases were all observed in experiments: less relaxed, fast quenched, or irradiated samples giving rise to more homogeneous plastic deformation, as observed here in the strong disorder case. However, it is difficult to have a more quantitative estimation of the degree of structural disorder needed. Indeed, in bulk metallic glasses, species are mixed, and it is quite difficult in general to characterize the disorder [53], unless in very specific cases [64]. In sodo-silicate glasses, at least the nanoporous silicon skeleton and channels of mobile ions give some keys of spatial characterization of the mattering structural disorder [65–67].

To conclude, boundary conditions, and also the relative amount of intrinsic strain disorder (to some extend in the residual plastic strain, but more importantly in the local yield strain), compared to the maximum Hertzian deviatoric subsurface strain, have a significant impact on the spatial organization of plasticity in amorphous materials in a contact geometry. The boundary conditions as well as the Hertzian strain depend on the external load, making the mechanical behavior of glasses non-universal in a contact geometry.

Author Contributions: Conceptualization: A.T.; methodology: A.T. and D.N.; ISAAC software: D.N. and T.C.; adaptation of the ISAAC software to local plasticity with Eshelby inclusions: T.C. and P.C.; simulations: P.C.; validation: P.C. and A.T.; writing—review and editing: A.T.; figures: P.C. and A.T. All authors have read and agreed to the published version of the manuscript.

Funding: This research was funded by the laboratory LaMCoS project ECO-VERRES. P.C. did this study during his last year internship at INSA Lyon.

Institutional Review Board Statement: Not applicable.

Informed Consent Statement: Not applicable.

Data Availability Statement: Data are stored in the laboratory LaMCoS, Villeurbanne, France.

Acknowledgments: A.T. thanks D. and C. Aumaître for interesting discussions.

Conflicts of Interest: The authors declare no conflict of interest.

References

1. Ashby, M.; Greer, A. Metallic glasses as structural materials. *Scr. Mater.* **2006**, *54*, 321–326. [CrossRef]
2. Ma, M.Z.; Liu, R.P.; Xiao, Y.; Lou, D.C.; Liu, L.; Wang, Q.; Wang, W.K. Wear resistance of Zr-based bulk metallic glass applied in bearing rollers. *Mater. Sci. Eng. A* **2004**, *386*, 326–330. [CrossRef]
3. Wang, L.; Bian, X.; Hu, L.; Wu, Y.; Guo, J.; Zhang, J. Glass transition temperature of bulk metallic glasses: A linear connection with the mixing enthalpy. *J. Appl. Phys.* **2007**, *101*, 103540. [CrossRef]
4. Liu, Y.H.; Wang, G.; Wang, R.J.; Zhao, D.; Pan, M.X.; Wang, W.H. Super Plastic Bulk Metallic Glasses at Room Temperature. *Science* **2007**, *315*, 1385–1388. [CrossRef]
5. Li, X.; Wei, D.; Zhang, J.; Liu, X.; Li, Z.; Wang, T.; He, Q.; Wang, Y.-J.; Ma, J.; Wang, W.; et al. Ultrasonic plasticity of metallic glass near room temperature. *Appl. Mater. Today* **2020**, *21*, 100866. [CrossRef]
6. Gross, T.; Tomozawa, M.; Koike, A. A glass with high crack initiation load: Role of fictive temperature-independent mechanical properties. *J. Non-Cryst. Solids* **2009**, *355*, 563–568. [CrossRef]
7. Luo, J.; Lezzi, P.J.; Vargheese, K.D.; Tandia, A.; Harris, J.T.; Gross, T.M.; Mauro, J.C. Competing Indentation Deformation Mechanisms in Glass Using Different Strengthening Methods. *Front. Mater.* **2016**, *3*, 52. [CrossRef]
8. Schuh, C.A.; Hufnagel, T.C.; Ramamurty, U. Mechanical behavior of amorphous alloys. *Acta Mater.* **2007**, *55*, 4067–4109. [CrossRef]
9. Spaepen, F. A microscopic mechanism for steady state inhomogeneous flow in metallic glasses. *Acta Met.* **1977**, *25*, 407–415. [CrossRef]
10. Cohen, M.H.; Turnbull, D. Molecular Transport in Liquids and Glasses. *J. Chem. Phys.* **1959**, *31*, 1164–1169. [CrossRef]
11. Argon, A. Plastic deformation in metallic glasses. *Acta Met.* **1979**, *27*, 47–58. [CrossRef]
12. Albaret, T.; Tanguy, A.; Boioli, F.; Rodney, D. Mapping between atomistic simulations and Eshelby inclusions in the shear de-formation of an amorphous silicon model. *Phys. Rev. E* **2016**, *93*, 053002. [CrossRef]
13. Maloney, C.; Lemaître, A. Universal Breakdown of Elasticity at the Onset of Material Failure. *Phys. Rev. Lett.* **2004**, *93*, 195501. [CrossRef]
14. Gartner, L.; Lerner, E. Nonlinear modes disentangle glassy and Goldstone modes in structural glasses. *Sci. Post Phys.* **2016**, *1*, 16. [CrossRef]
15. Boioli, F.; Albaret, T.; Rodney, D. Shear transformation distribution and activation in glasses at the atomic scale. *Phys. Rev. E* **2017**, *95*, 033005. [CrossRef]
16. Rottler, J.; Robbins, M.O. Yield conditions for deformation of amorphous polymer glasses. *Phys. Rev. E* **2001**, *64*, 051801. [CrossRef]
17. Tanguy, A.; Leonforte, F.; Barrat, J.L. Plastic response of a 2d lennard-jones amorphous solid: Detailed analyses of the local rearrangements at very slow strain rate. *Eur. Phys. J. E* **2006**, *20*, 355–364. [CrossRef]
18. Tanguy, A.; Mantisi, B.; Tsamados, M. Vibrational modes as a predictor for plasticity in a model glass. *EPL Eur. Phys. Lett.* **2010**, *90*, 16004. [CrossRef]
19. Rodney, D.; Tanguy, A.; Vandembroucq, D. Modeling the mechanics of amorphous solids at different length scale and time scale. *Model. Simul. Mater. Sci. Eng.* **2011**, *19*, 083001. [CrossRef]
20. Albe, K.; Ritter, Y.; Şopu, D. Enhancing the plasticity of metallic glasses: Shear band formation, nanocomposites and nanoglasses investigated by molecular dynamics simulations. *Mech. Mater.* **2013**, *67*, 94–103. [CrossRef]
21. Albano, F.; Falk, M.L. Shear softening and structure in a simulated three-dimensional binary glass. *J. Chem. Phys.* **2005**, *122*, 154508. [CrossRef] [PubMed]
22. Li, L.; Homer, E.R.; Schuh, C.A. Shear transformation zone dynamics model for metallic glasses incorporating free volume as a state variable. *Acta Mater.* **2013**, *61*, 3347–3359. [CrossRef]
23. Ghosh, A.; Budrikis, Z.; Chikkadi, V.; Sellerio, A.L.; Zapperi, S.; Schall, P. Direct Observation of Percolation in the Yielding Transition of Colloidal Glasses. *Phys. Rev. Lett.* **2017**, *118*, 148001. [CrossRef] [PubMed]
24. Shrivastav, G.P.; Chaudhuri, P.; Horbach, J. Yielding of glass under shear: A directed percolation transition precedes shear-band formation. *Phys. Rev. E* **2016**, *94*, 042605. [CrossRef]
25. Varnik, F.; Bocquet, L.; Barrat, J.-L.; Berthier, L. Shear Localization in a Model Glass. *Phys. Rev. Lett.* **2003**, *90*, 095702. [CrossRef]
26. Qu, R.; Zhao, J.; Stoica, M.; Eckert, J.; Zhang, Z. Macroscopic tensile plasticity of bulk metallic glass through designed artificial defects. *Mater. Sci. Eng. A* **2012**, *534*, 365–373. [CrossRef]
27. Su, C.; Anand, L. Plane strain indentation of a Zr-based metallic glass: Experiments and numerical simulation. *Acta Mater.* **2006**, *54*, 179–189. [CrossRef]
28. Budrikis, Z.; Castellanos, D.F.; Sandfeld, S.; Zaiser, M.; Zapperi, S. Universal features of amorphous plasticity. *Nat. Commun.* **2017**, *8*, 15928. [CrossRef]
29. Shi, Y.; Falk, M.L. Stress-induced structural transformation and shear banding during simulated nanoindentation of a metallic glass. *Acta Mater.* **2007**, *55*, 4317–4324. [CrossRef]
30. Le Bouil, A.; Amon, A.; McNamara, S.; Crassous, J. Emergence of Cooperativity in Plasticity of Soft Glassy Materials. *Phys. Rev. Lett.* **2014**, *112*, 246001. [CrossRef]
31. Nicolas, A.; Ferrero, E.E.; Martens, K.; Barrat, J.-L. Deformation and flow of amorphous solids: A review of mesoscale elas-toplastic models. *Rev. Mod. Phys.* **2018**, *90*, 045006. [CrossRef]

32. Baret, J.-C.; Vandembroucq, D.; Roux, S. Extremal Model for Amorphous Media Plasticity. *Phys. Rev. Lett.* **2002**, *89*, 195506. [CrossRef] [PubMed]
33. Patinet, S.; Vandembroucq, D.; Falk, M.L. Connecting Local Yield Stresses with Plastic Activity in Amorphous Solids. *Phys. Rev. Lett.* **2016**, *117*, 045501. [CrossRef] [PubMed]
34. Castellanos, D.F.; Roux, S.; Patinet, S. Insights from the quantitative calibration of an elasto-plastic model from a Lennard-Jones atomic glass. *Comptes Rendus Phys. l'Académie Sci.* **2021**. to appear.
35. Gottschalk, H.P.; Eisner, E.; Hosalkar, H.S. Medial epicondyle fractures in the pediatric population. *J. Am. Acad. Orthop. Surg.* **2012**, *20*, 223–232. [CrossRef]
36. Martens, K.; Bocquet, L.; Barrat, J.L. Spontaneous formation of permanent shear bands in a mesoscopic model of flowing disor-dered matter. *Soft Matter* **2012**, *8*, 4197–4205. [CrossRef]
37. Dahmen, K.A.; Ben-Zion, Y.; Uhl, J.T. Micromechanical Model for Deformation in Solids with Universal Predictions for Stress-Strain Curves and Slip Avalanches. *Phys. Rev. Lett.* **2009**, *102*, 175501. [CrossRef]
38. Tsamados, M.; Tanguy, A.; Goldenberg, C.; Barrat, J.-L. Local elasticity map and plasticity in a model Lennard-Jones glass. *Phys. Rev. E* **2009**, *80*, 026112. [CrossRef]
39. Johnson, K.L.; Keer, L.M. Contact Mechanics. *J. Tribol.* **1986**, *108*, 659. [CrossRef]
40. Love, A.E.H. *A Treatise on the Mathematical Theory of Elasticity*, 4th ed.; Unabridged and Unaltered Republ. of the 4. (1927) ed; Dover Publ: New York, NY, USA, 1990.
41. Eshelby, J.D. The Determination of the Elastic Field of an Ellipsoidal Inclusion, and Related Problems. *Math. Phys. Eng. Sci.* **1957**, *241*, 376–396.
42. Eshelby, J.D. The Elastic Field Outside an Ellipsoidal Inclusion. *Math. Phys. Eng. Sci.* **1959**, *252*, 561–569.
43. Koumi, K.E.; Zhao, L.; Leroux, J.; Chaise, T.; Nélias, D. Contact analysis in the presence of an ellipsoidal inhomogeneity within a half space. *Int. J. Solids Struct.* **2014**, *51*, 1390–1402. [CrossRef]
44. Jacq, C.; Nélias, D.; Lormand, G.; Girodin, D. Development of a Three-Dimensional Semi-Analytical Elastic-Plastic Contact Code. *J. Tribol.* **2002**, *124*, 653–667. [CrossRef]
45. Chaise, T.; Nélias, D. Contact Pressure and Residual Strain in 3D Elasto-Plastic Rolling Contact for a Circular or Elliptical Point Contact. *J. Tribol.* **2011**, *133*, 041402. [CrossRef]
46. Boucly, V.; Nelias, D.; Liu, S.; Wang, Q.; Keer, L. Contact analyses for bodies with frictional heating and plastic behaviour. *ASME J. Tribol.* **2005**, *127*, 355–364. [CrossRef]
47. Medina, S.; Dini, D. A numerical model for the deterministic analysis of adhesive rough contacts down to the nano-scale. *Int. J. Solids Struct.* **2014**, *51*, 2620–2632. [CrossRef]
48. Ghaednia, H.; Wang, X.; Saha, S.; Xu, Y.; Sharma, A.; Jackson, R.L. A Review of Elastic–Plastic Contact Mechanics. *Appl. Mech. Rev.* **2017**, *69*, 060804. [CrossRef]
49. Seleznev, M.; Vinogradov, A. Shear Bands Topology in the Deformed Bulk Metallic Glasses. *Metal* **2020**, *10*, 374. [CrossRef]
50. Keryvin, V.; Crosnier, R.; Laniel, R.; Hoang, V.H.; Sanglebœuf, J.-C. Indentation and scratching mechanisms of a ZrCuAlNi bulk metallic glass. *J. Phys. D: Appl. Phys.* **2008**, *41*, 1–7. [CrossRef]
51. Tyukodi, B.; Vandembroucq, D.; Maloney, C.E. Avalanches, thresholds, and diffusion in mesoscale amorphous plasticity. *Phys. Rev. E* **2019**, *100*, 043003. [CrossRef]
52. Richard, D.; Ozawa, M.; Patinet, S.; Stanifer, E.; Shang, B.; Ridout, S.A.; Xu, B.; Zhang, G.; Morse, P.K.; Barrat, J.-L.; et al. Predicting plasticity in disordered solids from structural indicators. *Phys. Rev. Mater.* **2020**, *4*, 113609. [CrossRef]
53. Li, W.; Bei, H.; Tong, Y.; Dmowski, W.; Gao, Y.F. Structural heterogeneities induced plasticity in bulk metallic glasses: From well-relaxed fragile glass to metal-llike behavior. *App. Phys. Lett.* **2013**, *103*, 171910. [CrossRef]
54. Makarov, A.; Afonin, G.V.; Mitrofanov, Y.; Kobelev, N.; Khonik, V. Heat Effects Occurring in the Supercooled Liquid State and Upon Crystallization of Metallic Glasses as a Result of Thermally Activated Evolution of Their Defect Systems. *Metal* **2020**, *10*, 417. [CrossRef]
55. Liu, L.; Liu, Z.; Huan, Y.; Wu, X.; Lou, Y.; Huang, X.; He, L.; Li, P.; Zhang, L.-C. Effect of structural heterogeneity on serrated flow behavior of Zr-based metallic glass. *J. Alloy. Compd.* **2018**, *766*, 908–917. [CrossRef]
56. Brechtl, J.; Wang, H.; Kumar, N.; Yang, T.; Lin, Y.-R.; Bei, H.; Neuefeind, J.; Dmowski, W.; Zinkle, S. Investigation of the thermal and neutron irradiation response of BAM-11 bulk metallic glass. *J. Nucl. Mater.* **2019**, *526*, 151771. [CrossRef]
57. Yu, P.; Bai, H. Poisson's ratio and plasticity in CuZrAl bulk metallic glasses. *Mater. Sci. Eng. A* **2008**, *485*, 1–4. [CrossRef]
58. Salerno, K.M.; Robbins, M.O. Effect of inertia on sheared disordered solids: Critical scaling of avalanches in two and three dimensions. *Phys. Rev. E* **2013**, *88*, 062206. [CrossRef]
59. Gamcova, J.; Mohanty, G.; Michalik, S.; Wehrs, J.; Bednarcik, J.; Krywka, C.; Breguet, J.M.; Michler, J.; Franz, H. Mapping strain fields induced in Zr-based bulk metallic glasses during in-situ nanoindentation by X-ray nanodiffraction. *Appl. Phys. Lett.* **2016**, *108*, 031907. [CrossRef]
60. Perepezko, J.H.; Imhoff, S.D.; Chen, M.-W.; Wang, J.-Q.; Gonzalez, S. Nucleation of shear bands in amorphous alloys. *Proc. Nat. Acad. Sci. USA* **2014**, *111*, 3938–3942. [CrossRef] [PubMed]
61. Ghidelli, M.; Idrissi, H.; Gravier, S.; Blandin, J.-J.; Raskin, J.-P.; Schryvers, D.; Pardoen, T. Homogeneous flow and size dependent mechanical behavior in highly ductile Zr65Ni35 metallic glass films. *Acta Mater.* **2017**, *131*, 246–259. [CrossRef]

62. Jang, D.; Greer, J.R. Transition from a strong-yet-brittle to a stronger-and-ductile state by size reduction of metallic glasses. *Nat. Mater.* **2010**, *9*, 215–219. [CrossRef] [PubMed]

63. Fusco, C.; Albaret, T.; Tanguy, A. Rheological properties vs Local Dynamics in model disordered materials at Low Temperature. *Eur. Phys. J. E* **2014**, *37*, 1–9. [CrossRef] [PubMed]

64. Idrissi, H.; Ghidelli, M.; Béché, A.; Turner, S.; Gravier, S.; Blandin, J.-J.; Raskin, J.-P.; Schryvers, D.; Pardoen, T. Atomic-scale viscoplasticity mechanisms revealed in high ductility metallic glass films. *Sci. Rep.* **2019**, *9*, 1–11. [CrossRef] [PubMed]

65. Mantisi, B.; Kermouche, G.; Barthel, E.; Tanguy, A. Impact of pressure on plastic yield in amorphous solids with open structure. *Phys. Rev. E* **2016**, *93*, 033001. [CrossRef]

66. Molnár, G.; Ganster, P.; Tanguy, A. Effect of composition and pressure on the shear strength of sodium silicate glasses: An atomic scale simulation study. *Phys. Rev. E* **2017**, *95*, 043001. [CrossRef]

67. Keryvin, V.; Charleux, L.; Hin, R.; Guin, J.-P.; Sangleboeuf, J.-C. Mechanical behaviour of fully densified silica glass under Vickers indentation. *Acta Mater.* **2017**, *129*, 492–499. [CrossRef]

MDPI

St. Alban-Anlage 66

4052 Basel

Switzerland

Tel. +41 61 683 77 34

Fax +41 61 302 89 18

www.mdpi.com

Metals Editorial Office

E-mail: metals@mdpi.com

www.mdpi.com/journal/metals